The Forrest Mims Circuit Scrapbook

Volume I

by Forrest M. Mims III

LLH
Technology Publishing
An Imprint of Elsevier
Eagle Rock, VA
www.LLH-Publishing.com

ISBN-13: 978-1-878707-48-2

ISBN-10: 1-878707-48-5

Cover design: Sergio Villarreal, San Diego, CA
Production services: Kelly Johnson, San Diego, CA
Developmental editing: Dr. John Carpenter, Memphis, TN

Transferred to Digital Printing, 2010

Printed and bound in the United Kingdom

LLH
Technology Publishing
An Imprint of Elsevier
Eagle Rock, VA
www.LLH-Publishing.com

Contents

Three

LED Circuits

Four

Test and Measurement Circuits

Five

Power Sources

Six

Digital Circuits

Seven

Experimenter, Hobby, and Game Circuits

Foreword

It is impossible to think of hobby electronics without thinking of Forrest Mims. Since his first appearance in *Popular Electronics* magazine back in 1970, millions and millions of readers—and that's not hyperbole or exaggeration—have learned electronics technology thanks to the fluid, engaging writing and hand-drawn circuit diagrams produced by Forrest Mims.

Over the years, Forrest's fascination with technology has resulted in some remarkable experiences and adventures. For example, in January, 1975 Forrest received an assignment to write the owner's manual for a new device called the "Altair." The Altair was the world's first commercially available personal computer. While Forrest wrote the manual, a pair of young men named Bill Gates and Paul Allen, who later founded Microsoft, were hired to create software for the Altair. Thus, Forrest holds the honor of being the author of the first personal computer book. Later that year, *The National Enquirer* asked Forrest to eavesdrop on Howard Hughes at his home in the Bahamas by bouncing laser beams off his windows. (Forrest wisely declined, although he figured out how to do it.) Over two decades later, Forrest was named a winner of Rolex's international "Spirit of Enterprise" award for his development of simple, low cost devices to measure atmospheric ozone and the establishment an international network of volunteer observers to measure ozone. Predictably, Forrest's devices were able to obtain more accurate atmospheric ozone measurements than NASA could with their multi-billion dollar network of satellites! Today, Forrest travels to such diverse locations as the Big Island of Hawaii to lecture and do scientific research with instruments of his own design. From his humble beginnings in the original *Popular Electronics*, Forrest's work now appears in such journals as *Nature* and *Scientific American.*

The Forrest Mims Circuit Scrapbook Volume I and *The Forrest Mims Circuit Scrapbook Volume II* are compilations of Forrest's columns and articles that appeared in *Popular Electronics* and *Modern Electronics* magazines. Volume I consists of material from *Popular Electronics*, while Volume II is material from *Modern Electronics.* These chronicles of Forrest's creativity and engineering wizardry are still as valid and useful to electronics hobbyists today when they were first written. Most of the components used in these circuits are still available; in cases where certain parts are no longer available, equivalent devices (often with improved performance) are generally available. Moreover, the real importance of Forrest's written work—the teaching of basic electronics principles and a philosophy of electronics design—is timeless.

For over two decades, I have been fortunate enough to be able to count Forrest as a friend. It is a great pleasure to introduce these two compilations of his work, and I look forward to the day when I can write the introduction to two more volumes!

Harry Helms

Preface

The first electrical circuit I built was a headlight for a homemade soapbox racer. The lamp was soldered inside a tin can. The switch was made from a bent nail. The light turned on just fine, but it didn't turn off completely. When the batteries lasted only a short time, I checked the circuit and found the problem. Instead of connecting the switch between the battery and lamp, I had connected it across the lamp. So when I thought the switch was in the off position, it was really short circuiting the battery.

I was 11 years old when I built that soapbox racer. The lesson learned from that improperly wired "switch" has stayed with me ever since. Even today I never apply power to a circuit I've built until checking and rechecking every connection. All the circuits in this book went through that checking process before they were first published in *Popular Electronics* and *Modern Electronics*.

The circuits in this book are among my favorites. Many are optoelectronic in nature, which means they detect or emit light. One of my first optoelectronic circuits was a light sensor that helped control a homemade analog computer that translated 20 words of English into Russian. The computer's memory was a bank of 20 miniature trimmer resistors, each set to indicate a specific word. The word to be translated was dialed into six potentiometers on a control panel one letter at a time. A battery-powered music box modified to act as a sequential switch then began connecting the resistors in the memory bank one by one to a Wheatstone bridge. When the resistance of the word dialed into the input panel matched the resistance of a sampled trimmer, a -1 to +1 milliammeter indicated 0 current. A small rectangle of aluminum foil glued to the meter's needle then blocked the light from a small bulb from reaching a small silicon solar cell mounted below the pointer. A single-transistor switch then actuated a relay that switched off the music box. One of 20 red lamps on an output panel then glowed to indicate the translated word. I built that analog computer for my senior science fair project back in 1962. Today it would be easy to simulate that computer and all of its operations with a relatively simple program that will run on virtually any personal computer.

Yet while computers can implement many circuit applications, the circuits in this book demonstrate that there are many circuits that even computers cannot emulate. That's another lesson I've learned about electronics. While I've been using personal computers since 1975, circuits like those in this book still provide the most fun and the biggest challenges. Why just yesterday I was experimenting with a new kind of Sun photometer. I used a state-of-the-art Pentium and a spreadsheet program to calculate the values of several key gain resistors based on a series of actual Sun measurements. The computer saved a few minutes of design time. But the real challenge of this project was soldering and desoldering actual resistors until the circuit provided exactly the required gain. I can hardly wait to calibrate this new instrument on my next trip to Hawaii's Mauna Loa Observatory.

Electronics offers both career opportunities and solutions to countless practical problems. Electronics is also one of the most fascinating and creative of hobbies. I hope you enjoy building and using the circuits in this book as much as I do.

Forrest M. Mims III

The Forrest Mims Circuit Scrapbook

Volume I

One
Analog Circuits

1. Analog Computer Circuits, Part 1

If you think analog computers are a relic of the past, think again. Recent developments have made possible sophisticated analog computer ICs with unprecedented accuracy. One of these new circuits, billed as the first single-chip analog computer is the Analog Devices AD534 precision multiplier-divider. Available for about the same price as many microprocessors, the AD534 can perform all the MDSSR functions (multiplication, division, squaring and square rooting) with ease and in *real time*. There are no elaborate software requirements and the AD534 will perform any of these operations while the input information is changing at up to 1 MHz.

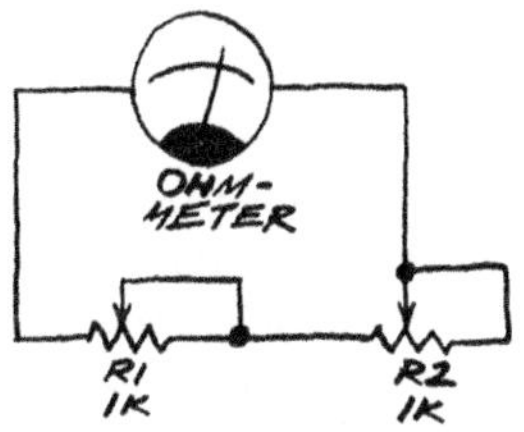

Fig. 1. Analog adder made from potentiometers,

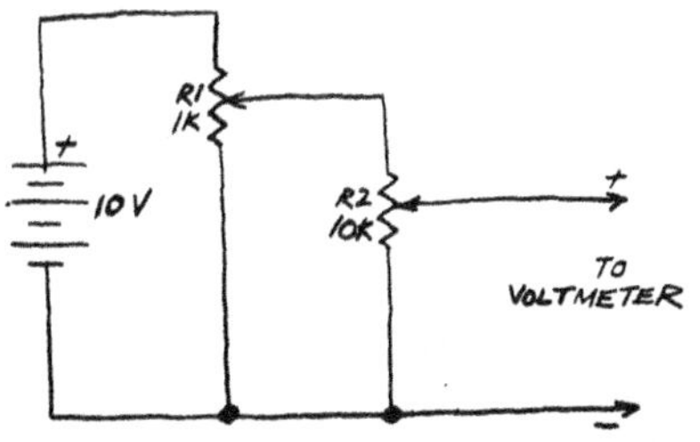

Fig. 2. Simple multiplier made from potentiometers.

Analog vs Digital. Most readers of this column already know the essential difference between analog and digital integrated circuits. Analog ICs, such as amplifiers, oscillators and timers, process or produce signals whose amplitudes are continuously variable over a given range. Digital ICs, on the other hand, process or produce signals which occupy either a low or high state.

An amazing number of applications use either digital *or* analog circuits. Lately, of course, the trend has been toward digital almost exclusively. In some ways, however, this trend (or is it a fad?) is clearly unfortunate. Consider the digital voltmeter (DVM). The unprecedented accuracy of a three- or four-digit DVM is ideal for making measurements of fixed voltages. But have you ever tried to monitor a slowly changing voltage with a DVM? If not, good luck! The stream of constantly changing numbers is almost impossible to read and makes little sense. An old-fashioned panel meter with an analog readout (pointer and scale) offers a much better solution to this common problem.

When it comes to computers, digital technology has the advantage in accuracy, programmability and computing power. Analog computers, however, are ideal for simulating a real-world system such as a structure, machine, dam, aircraft or missile. Simply rotating a few control knobs enables the computer's operator to observe the effect of changes on the system as they take place (in real time). Contrast this simple procedure with the elaborate software manipulation required to vary conditions in a digital computer's program.

Analog computers have one further advantage you should know about: low cost. You can build an analog computer capable of solving a general quadratic equation ($ax^2 + bx + c = 0$) with a dozen or so inexpensive op amps, a few dozen resistors and potentiometers and a digital voltmeter. And you can rewire (reprogram) this computer so that it can solve many other mathematical functions and equations also.

Adding with Resistors. A simple analog adding calculator can be made from a pair of linear potentiometers equipped with pointer knobs and scales and an ohmmeter. For best results, compromise with digital technology and use a digital multimeter. The circuit for the adder is shown in Fig. 1.

The total resistance of two resistors in series is the sum of their individual resistances. If the potentiometers in Fig. 1 each have a resistance of 1000 ohms, the adder will add any two numbers of up to 1000 and produce any total up to 2000. The accuracy of the adder is controlled by the accuracy of the ohmmeter, the linearity of the potentiometers and the calibration of the scales. Thanks to the high accuracy of the DMM, the error introduced by the ohmmeter can be very small. With this concession to digital technology, the potentiometers and their calibration scales become the major sources of error.

Assuming perfectly linear potentiometers are not available (a reasonable assumption), the adder can be *custom calibrated* with the help of the DMM. This is done by connecting the DMM to a pot and marking its scale with a fine index line at each 100-ohm interval. The index lines are then labeled 0, 100, 200, 300. . . .1000 and the space between index lines is further subdivided into smaller increments.

Depending upon the care exercised in calibrating the adder (a large-diameter scale helps), its accuracy can be better than 1 percent. While this is not as accurate as a pocket calculator, the analog adder operates in *real time* since it continuously adds even as the dials are being turned to new settings!

Multiplying with Resistors. Figure 2 shows a resistive circuit that multiplies. The circuit consists of two potentiometers and a DVM or conventional voltmeter. Potentiometer *R1* forms a voltage divider across the 10-volt power supply and *R2* forms a voltage divider across *R1*'s wiper and ground.

Can you figure out how the circuit multiplies? It's easy if you ignore the resistance of the two pots and consider only the positions of their wipers. Assume, for example, *R1* has a scale with eleven equally spaced lines marked 0 through 10. Also assume *R2* has a similar scale marked 0 through 1 in increments of 0.1. When *R1*'s shaft is rotated to its midpoint, its dial points to the 5 on its scale. This means the voltage at *R1*'s wiper is 5 volts. If *R2* is rotated to its midpoint or 0.5, the incoming voltage is again divided in half. The resulting voltage at *R2*'s slider is 2.5 volts, the product of the two potentiometer settings ($5 \times 0.5 = 2.5$).

Improving the Accuracy of Resistor Calculators. The adder and multiplier circuits we've been experimenting with can be made much more accurate by using ten-turn dial potentiometers. These pots are fairly expensive when new but can sometimes be purchased used or surplus from mail-order parts suppliers. They have a built-in turns indicator usually marked from 0 to 100 and can increase the accuracy of a simple resistor calculator significantly.

Op Amps Add Versatility. The simple resistor calculators we've experimented with are ideal for simple applications where input information is programmed by hand or, perhaps, by a me-

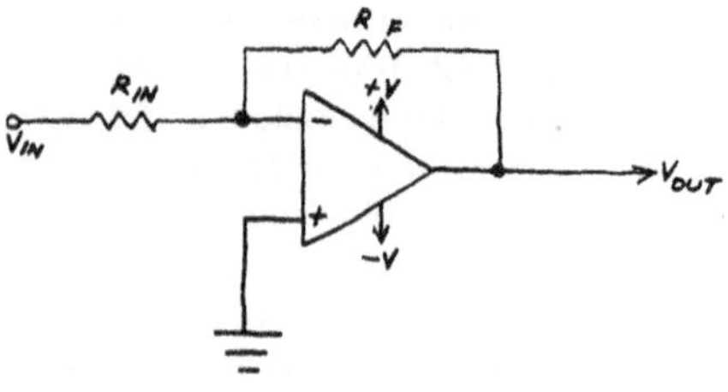

Fig. 3. Basic op amp circuit.

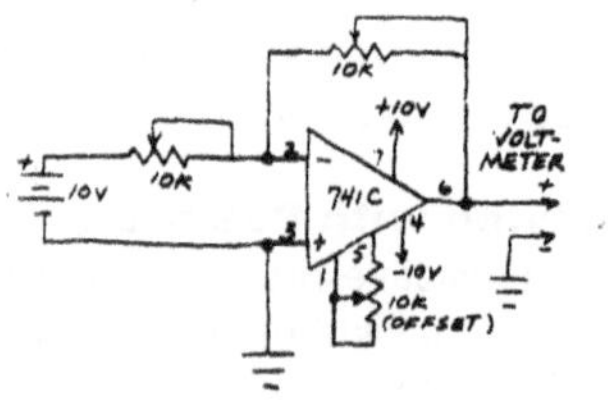

Fig. 4. Op amp multiplication.

chanical device. Many applications, however, require that the information enter the calculator in the form of a raw voltage.

Typically, a *transducer* is used to convert information such as windspeed, the rate of fluid flow through a pipe, temperature, weight or some other variable information into a representative voltage. This voltage may then be mathematically combined with the voltage from one or more other transducers to produce an output voltage.

Op Amp Fundamentals. Figure 3 illustrates the most important principle of the op amp—the output voltage (V_{OUT}) of an op amp equals the product of its feedback resistance (R_F) and the incoming voltage (V_{IN}) divided by the input resistance (R_{IN}). Algebraically, this is expressed as $V_{OUT} = -R_F V_{IN}/R_{IN}$. This simple relationship or *transfer function* means an op amp can both multiply and divide. If R_{IN} is 1 ohm, then the op amp will multiply the input voltage times the feedback resistance. Likewise, if the input voltage is a fixed 1 volt, the op amp will divide the feedback resistance by the input resistance.

Note that a negative sign appears in the transfer function. This is so because input signals are applied to the inverting input of the op amp which causes output signals to be the opposite polarity with respect to the input signals. If R_F equals R_{IN} and a +1-volt dc level is applied between R_{IN} and ground, the output of the op amp will be −1 volt—the same magnitude (absolute value), but the opposite polarity. Keep this change of sign in mind when experimenting with this and the following circuits in this two-part series, all of whose op amps are employed in the inverting mode.

You can easily demonstrate op amp multiplication and division with a common op amp such as the 741, several potentiometers and a digital voltmeter. Figure 4 shows the 741 pin connections and how to connect the pots to the 741.

The op amp's offset voltage will introduce a small error in the output voltage. Figure 4 also shows how to connect a 10,000-ohm pot to the 741 to alleviate this problem. Adjust the pot so that the 741 output is exactly zero when pin 2 is grounded.

As you've probably noticed, these methods of using an op amp to multiply and divide still require at least one potentiometer adjustment to perform a calculation. There's a clever way to use op amp to multiply and divide *without* having to make manual potentiometer adjustments, and we'll examine it in detail in the second installment of this series. Meanwhile, let's look at ways to add and subtract using op amps that *don't* require potentiometer adjustments.

Adding with an Op Amp. Figure 5 shows an op amp adder that will accept two input voltages, add them and deliver the sum as an output voltage. *Summing*

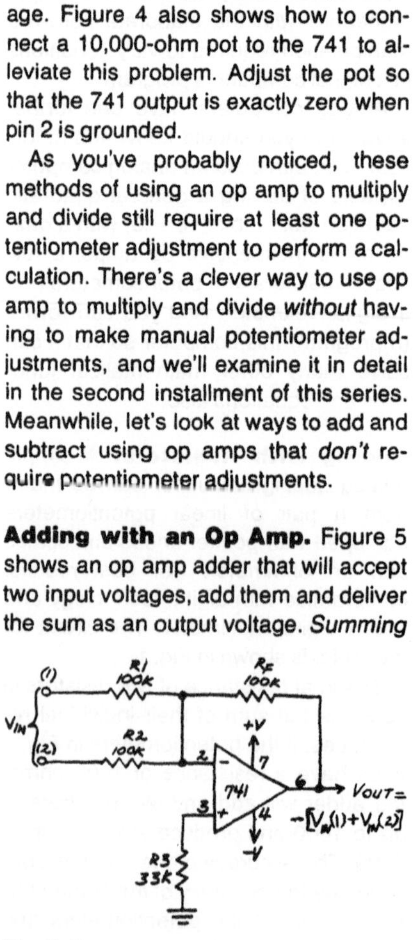

Fig. 5. Op amp summer.

amplifier is the technical name for this circuit.

Because the feedback resistor (R_F) has the same resistance as *R1* and *R2*, the voltage gain of the op amp is 1. This means the op amp has unity gain and doesn't alter the result of an addition. Changing the resistance of R_F or both *R1* and *R2* will cause the circuit to both add *and* multiply.

Resistor *R3* helps reduce errors caused by the op amp's input bias current. In non-critical applications it can be eliminated, in which case pin 2 is connected directly to ground. If you use *R3*, its value should be equal to the reciprocal of the sum of the reciprocals of *R1* and *R2*. In other words, $R3 = 1/(1/R + 1/R2) = R1R2/(R1+R2)$.

For a quick demonstration of how the adder works, connect the positive terminals of a few flashlight cells to the inputs. Connect the negative terminals of the cells to the adder's ground. If the cells each have identical voltages of 1.5 volts, the adder's output will be −3.0 volts.

For the best results, use 1-percent tolerance resistors for *R1*, *R2* and R_F. If you don't have, can't find or can't afford 1-percent resistors, use a DMM to select three resistors having values as closely matched as possible.

You can use a summing amplifier to add more than two voltages simply by connecting additional resistors to the noninverting input. The new resistors should equal the input resistors and R3 should be adjusted according to $R3 = 1/(1/R1 + 1/R2 + 1/R3 + \ldots 1/R_N)$.

Averaging with an Op Amp. A summing amplifier can also average two incoming voltages. All that's necessary is to make the ratio R_F/R_{IN} equal to the reciprocal of the number of input voltages. For example, to convert the adder circuit of Fig. 5 to an averager, leave the values of *R1* and *R2* undisturbed but change R_F to 50,000 ohms. This causes the ratio of R_F/R_{IN} (1/2) to equal the reciprocal of the number of inputs (2).

If you want to average more than two voltages, add additional input resistors and adjust the values of R_F and *R3* accordingly. The average of five incoming voltages is their sum divided by 5. Therefore, the reciprocal of R_F/R_{IN} should be 5. Because the resistance of R_{IN} is fixed at 100,000 ohms, the resistance of R_F should be 20,000 ohms.

Subtracting with an Op Amp. The final analog computer circuit we will look at this month is the subtractor or *difference amplifier* shown in Fig. 6. This circuit is based upon the ability of an op amp to amplify the difference between the two voltages applied to its two inputs. If the ratio of R_F to the circuit's input resistors is 1, the circuit has unity gain and will produce the arithmetic difference of two input voltages.

The output voltage of the difference amplifier equals $V2 - V1$. The circuit works for both positive and negative inputs and outputs. Thus, if $V1$ is +5 volts and $V2$ is +10 volts, the output will be +5 volts. Of course, the output cannot exceed the power supply voltage.

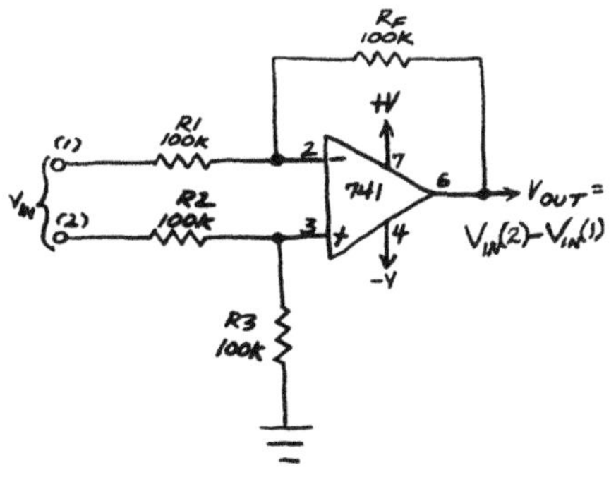

Fig. 6. Op amp subtraction.

Incidentally, an analog computer circuit that handles any of the four possible combinations of positive and negative inputs is called a *four-quadrant* device. A *two-quadrant* device responds to two of the four combinations of input polarities, and *one-quadrant* devices respond to one combination of input polarities. ■

Analog Computer Circuits, Part 2

One way to multiply or divide two voltages is to convert both to their logarithms. Multiplication is accomplished by adding the two logs with a summing amplifier. Division can be performed by subtracting the log of the divisor from the log of the dividend with a difference amplifier. The antilog of the result is the product or quotient, as the case may be.

Now that the pocket calculator has replaced the slide rule, logarithms are not used nearly as often as they once were. So let's take time out for a brief refresher course before moving on.

Logarithms. Any decimal number can be expressed as a power of ten. For example, 1,000 is 10^3 and 736 is $10^{2.8669}$. In both cases, the exponent of the base 10 is referred to as the number's logarithm. One important aspect of logarithms is revealed by the following table.

Number	Power of Ten	Logarithm
1	10^0	0
10	10^1	1
100	10^2	2
1,000	10^3	3
10,000	10^4	4
100,000	10^5	5
1,000,000	10^6	6

As you can see, a very wide range of decimal numbers occupies a very small range of logarithms. The resulting compression provides a handy shorthand method for processing very large numerical variations.

We noted earlier that two numbers can be multiplied by adding their logs or divided by subtracting their logs. That's how a slide rule works. It's also possible to add and subtract numbers using ordinary rulers. Place one ruler atop the other. Then align the 0 on the top ruler with one of the numbers being added on the bottom ruler. Next, find the second number being added on the top ruler. This number will point to the sum on the bottom ruler.

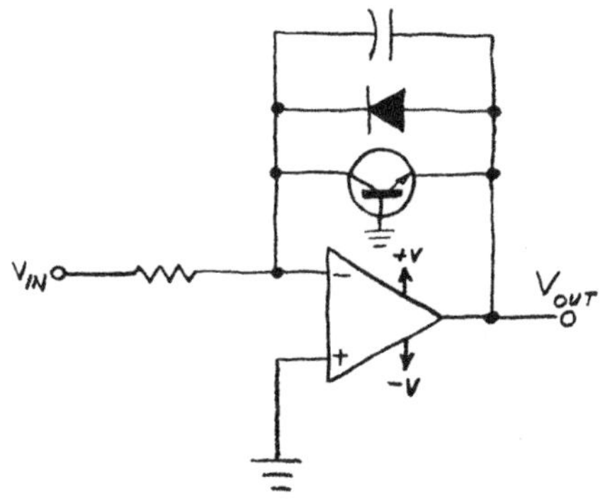

Fig. 1. Schematic of a basic logarithmic amplifier.

Rulers have a linear scale—their divisions are equally spaced. A slide rule, on the other hand, has a logarithmic or compressed scale. When you multiply two numbers with a slide rule, you are actually adding their logs.

Look back at the table and multiply 1,000 × 100 to see how this works. The log of 1,000 is 3 and the log of 100 is 2. 3 + 2 = 5 so the log of 1,000 × 100 is 5. From the table, 5 is the log of 100,000 (or 100,000 is the antilog of 5) so 1,000 × 100 = 100,000. Try dividing a few numbers in the table by subtracting the log of the divisor from the log of the dividend and taking the antilog of the remainder to obtain the quotient.

Before the advent of the pocket calculator, the use of logarithms was standard procedure when multiplying and dividing very large or very small numbers. Logs are also handy for extracting roots. The cube root of 27, for example, is

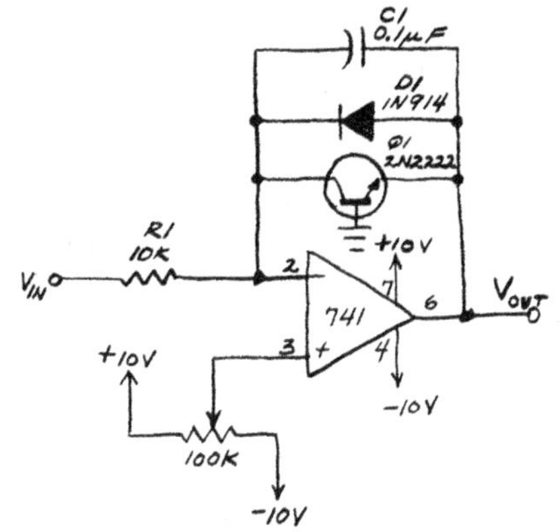

Fig. 2. A practical logarithmic amplifier circuit.

found by dividing the log of 27 by 3 and extracting the antilog of the result. (The log of 27, 1.4314, divided by 3 is 0.4771; the antilog of 0.4771 is 3, the cube root of 27.)

Incidentally, numbers in any number system can be expressed as logarithms. Can you figure out the logarithms of the binary sequence 1, 10, 100, 1000 . . . 10000000? (Hint: 1000 is 2^3.)

The Logarithmic Amplifier. The voltage drop across a diode is related logarithmically to the current flowing through it. This makes possible the conversion of a voltage into its log.

Practical log conversion is best achieved by using a transistor in a common- or grounded-base configuration instead of a diode. Figure 1 shows how the transistor is connected in place of an op-amp's feedback resistor to give what is called a *transdiode logarithmic amplifier.* Although the circuit is an amplifier, you can think of it as a log generator to avoid confusion.

Not all transistors exhibit logarithmic properties over as wide a range as might be required. Many, however, do and one readily available type is the 2N2222 (equivalent to Radio Shack type RS2009).

You can easily assemble a breadboard log amplifier with the help of a 741 or any other frequency-compensated op amp. Figure 2 shows the details of a practical version of the circuit in Figure 1. Capacitor *C1* does not assist in the log conversion process. Instead, it reduces the ac gain of the op amp and helps eliminate high-frequency oscillation which might otherwise occur. Diode *D1* protects the transistor from excessive reverse base-emitter bias from the op amp's output.

On the following page are the results I obtained from a breadboard version of the circuit in Fig. 2.

Input (mV)	**Output** (mV)
1	−322
10	−371
100	−432
1,000	−494
10,000	−557

In all cases, the output voltage was inverted (negative), but this is of no major consequence as we can either ignore the polarity or, if desired, change it with an inverting buffer.

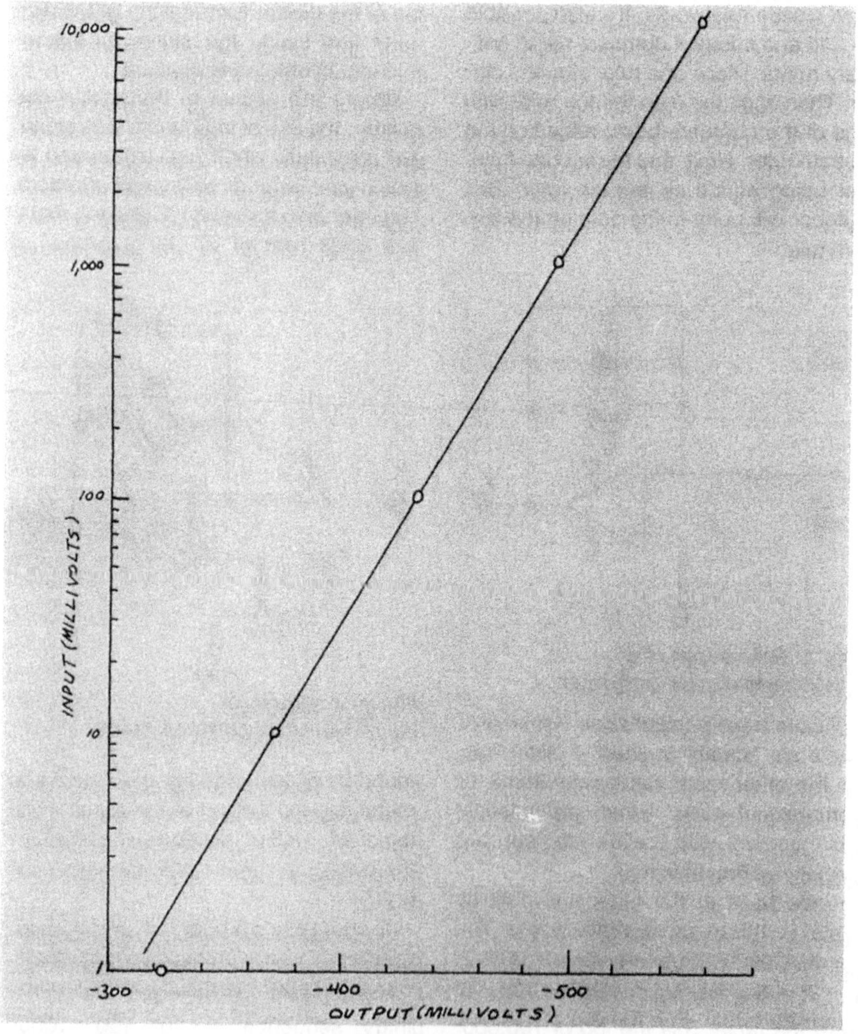

Fig. 3. Operation of a log amplifier plotted on semi-log graph paper.

Figure 3 shows the data in the table plotted on a semi-log graph. The graph is called semi-log because one axis is linear (the output voltage) and the other is logarithmic (the input voltage). A plot of the data produces a straight line on the semi-log graph, so we know the log amplifier is reasonably accurate over the given range.

Now that we've seen how a real log amplifier works, let's look at a few of its characteristics. First, notice the very small range in output voltage (a few hundred millivolts) that results from the huge swing in input voltage (10,000 millivolts). This characteristic of log amplifiers is ideal for compressing very large voltage excursions into more manageable form.

A second characteristic is that the transfer function of our log amplifier is *not* − V_{OUT} = log (V_{in}). Rather, it's approximately − $V_{out} = 0.06 \log V_{in} + K$ where *K* is a constant. For the log amplifier I built, *K* is 0.495. Your amplifier might yield a slightly different *K*. You can use a programmable calculator to compute the exact transfer function.

A third characteristic of our log amplifier is that it is temperature sensitive. That's not good because the current flowing through the 2N2222 causes heating which can alter the accuracy of the circuit. The error this introduces can be substantial, easily several percent.

Yet another characteristic of the amplifier is that the input offset voltage of the op amp can cause a substantial but predictable error when the input voltage is small. This problem can be alleviated by connecting a 10,000-ohm potentiometer to the 741 as shown in Fig. 4. Pin 2 of the 741 is then temporarily shorted to ground and the offset potentiometer is adjusted until V_{OUT} is exactly zero volt.

A more significant error is introduced by the op amp's bias current. This ranges from 80 to 500 nonoamperes for the 741. Figure 4 also shows how to compensate for this problem by temporarily replacing the components in the feedback loop with a 100,000-ohm resistor and adding a bias current potentiometer. The pot is then adjusted until $-V_{OUT}$ exactly equals V_{IN} over as wide a voltage range as you expect the amplifier to receive.

You don't have to make all these calibration adjustments when building a simple log amplifier for experimental purposes. But if you decide to build your own analog computer, for best results

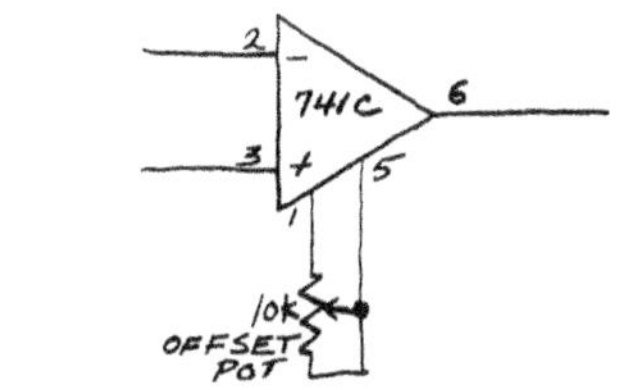

Fig. 4. Adding an offset pot to the log amplifier.

you'll need to calibrate or *trim* each op amp using the methods described.

The Antilogarithmic Amplifier. Analog computing circuits that use log amplifiers require one or more antilog amplifiers to convert results back into linear form. Antilog amplifiers can also be used to expand narrow ranging input voltages into much wider and therefore more easily resolved form.

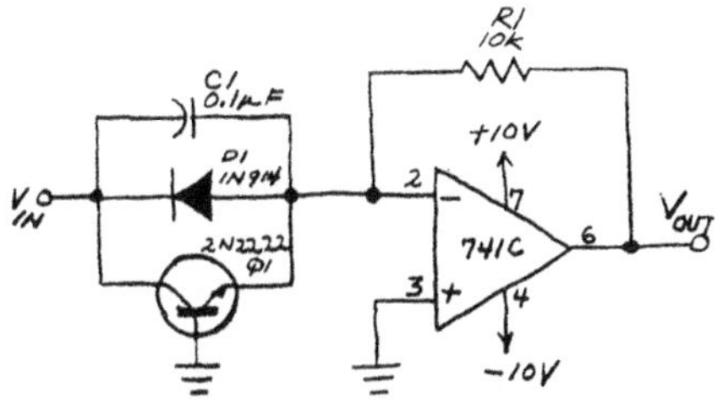

Fig. 5. Schematic of an anti-logarithmic amplifier.

If the transfer function of an ideal log amplifier is $V_{OUT} = \log (V_{IN})$ then the transfer function of an ideal antilog amplifier is $V_{OUT} = 10^{V_{IN}}$. In an actual circuit, however, the transfer function is the inverse of the log amplifier's. The differneces between ideal and actual transfer functions are therefore compensated.

Figure 5 shows the circuit for a working antilog amplifier you can make. An interesting experiment is to connect the input of the antilog amplifier to the output of the log amplifier in Figure 2. If both amplifiers are perfectly accurate, the transfer function for the combination will be $V_{OUT} = V_{IN}$.

Here are the results I obtained from a log-antilog combination with *no* calibration adjustments:

V_{IN} (mV)	V_{OUT} (mV)
1	1
10	−6
100	−111
1,000	−1,205
10,000	−11,490

As you can see, the error is fairly high. Calibrating both amplifiers using the methods previously outlined will provide much better results.

The Analog Multiplier. Now that we've built log, antilog and summing amplifiers, we can build an analog multiplier. A block diagram for the multiplier is shown in Fig. 6 and a complete circuit in Fig. 7.

The maximum error of the multiplier is easily in excess of 10 percent. Can you improve this figure over several decades of input voltage? (Hint: Use careful calibration procedures and try to keep all feedback transistors at the same temperature by, say, bonding them together with epoxy cement.)

You can convert the multiplier into an analog divider simply by changing the summing amplifier into a difference amplifier.

Single-Chip Multipliers. Contrary to my usual practice of breadboarding every circuit that appears in this column, I must confess to not having assembled the multiplier in Fig. 7. Several single IC multipliers that include all necessary amplifiers and transistors on the same silicon chip are now available, and they're much easier to use and more accurate because all circuit elements on the chip are at the same temperature. One such multiplier is Motorola's MC1595.

Single-chip multipliers like the MC1595 require many external calibration resistors, but recently Raytheon and

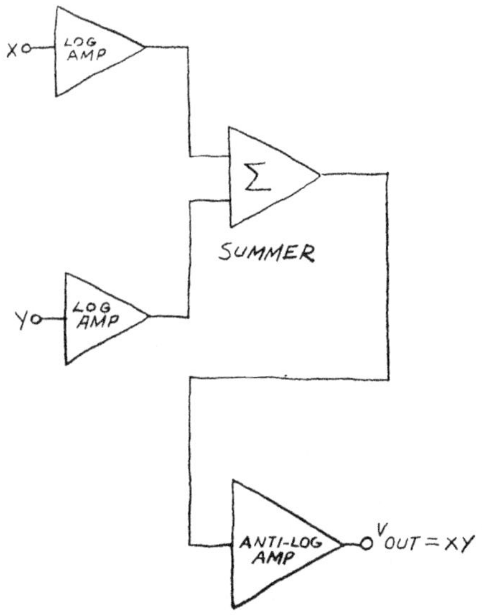

Fig. 6. Block diagram of a logarithmic multiplier.

Analog Devices introduced single-chip multipliers that include built-in error correcting features. Raytheon's chip is the 4200 and Analog Devices' is the AD534.

The 4200 is much less expensive than the AD534, but the latter is far superior to any previous single-chip multiplier because it includes 12 calibration resistors that have been factory-trimmed to a high degree of accuracy by a pulsed laser. The laser zaps away bits of thin-film calibration resistors that have been previously deposited directly on the silicon chip until a specified accuracy is reached.

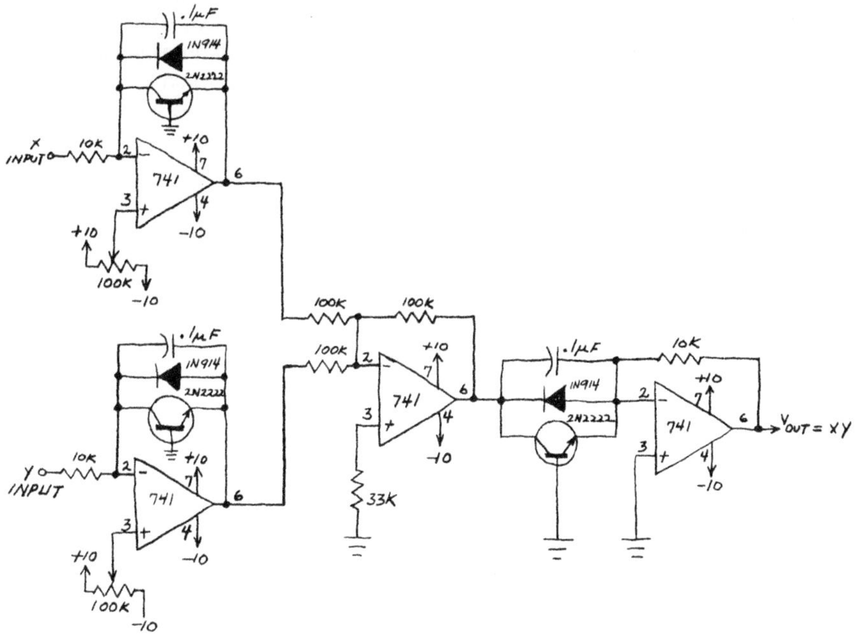

Fig. 7. Analog logarithmic multiplier circuit.

The AD534 is being billed as the first single-chip analog computer. Having experimented with both overly demanding, temperature-sensitive log amps and now the AD534, I'm more than willing to accept this enthusiastic claim. Figure 8 shows why. All the circuits shown are complete—*no* calibration resistors are required.

Here's an example of the results I obtained for an AD534 connected as a multiplier and a square rooter:

X	X^2	AD534	$\sqrt{10X}$	AD534
1	1	.95	3.16	3.09
2	4	4.08	4.47	4.51
3	9	9.20	5.48	5.52
4	16	16.24	6.32	6.36
5	25	24.40	7.07	7.11
6	36	35.20	7.75	7.72
7	49	48.20	8.37	8.42
8	64	63.20	8.94	8.90
9	81	79.80	9.49	9.50
10	100	98.70	10.00	10.05

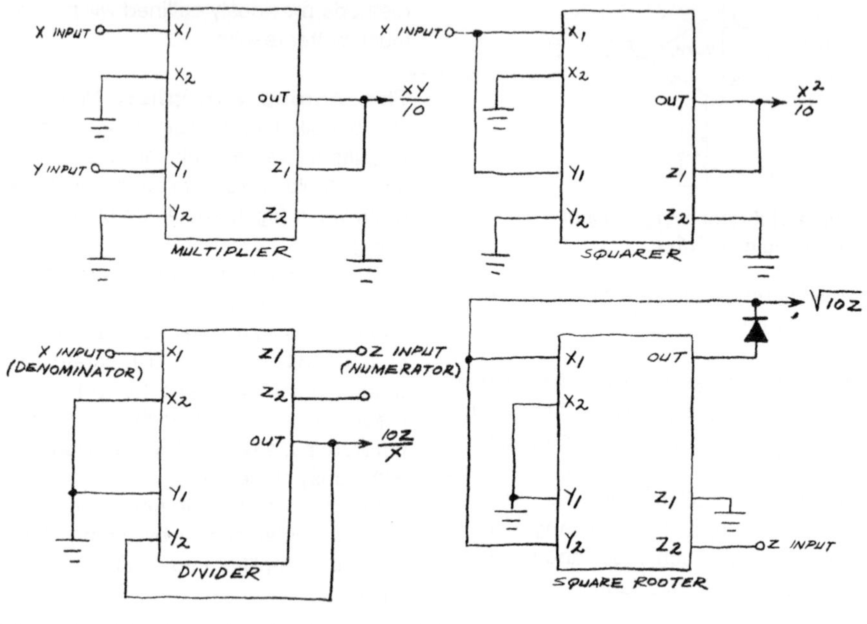

Fig. 8. Applications for the AD534 multiplier.

As you can readily see, the AD534 is exceptionally accurate. If you want to experiment with the AD534, you'll have to order one from an Analog Devices representative. Write the company for a list of reps and a specification sheet (Route One, Industrial Park, Box 280, Norwood, MA 02062). The single-quantity price is $26.00 for the lowest accuracy version (AD534J; ±1% total error). If the price seems high, look at it again after you've spent a frustrating evening trying to calibrate a homebrew multiplier. If you prefer digital circuits, consider the cost of the hardware and the time to develop software for a microprocessor that will perform the same functions. ■

2. The Analog Sample/Hold Circuit

MICROPROCESSOR enthusiasts are constantly seeking simple ways to interface small controllers and computers with the outside world of analog signals. One well-known analog circuit with many interfacing applications is the op-amp sample/hold (or sample and hold) circuit.

A sample/hold circuit stores in a capacitor the instantaneous voltage present at its input. The stored voltage, which can represent anything from the intensity of light illuminating a photocell to an audio signal, can be converted into digital form by an analog-to-digital converter for processing by a microprocessor. What's more, a sample/hold stage can be used in many different analog applications.

Simple Sample/Hold Circuit. Figure 1 schematically shows a very simple but functioning sample/hold circuit. The key component of the circuit is capacitator *C1*. When switch *S1* is momentarily toggled to its SAMPLE position, the capacitor charges to the voltage present at the input. The charge on the capacitor is monitored by *IC1*, an op amp with a very high input impedance, which should be an NE536 or similar amplifier with a FET input stage. When switch *S1* is released, it returns to its center (off) HOLD position and disconnects *C1* from the input of the circuit. Because the op amp's input impedance is very high, the charge stored in the capacitor is effectively trapped and the voltage across *C1* remains almost constant for an appreciable amount of time.

The op amp is connected as a voltage follower with unity gain. This arrangement permits the charge on the capacitor to be measured by a standard multimeter without significantly altering the amount of stored charge. Connecting a conventional, low-input-impedance voltmeter across the capacitor would, of course, quickly drain the capacitor of its charge. After the magnitude of the voltage sample has been determined, the switch can be momentarily placed in its RESET position to remove the charge from the capacitor and prepare the circuit for a new sample.

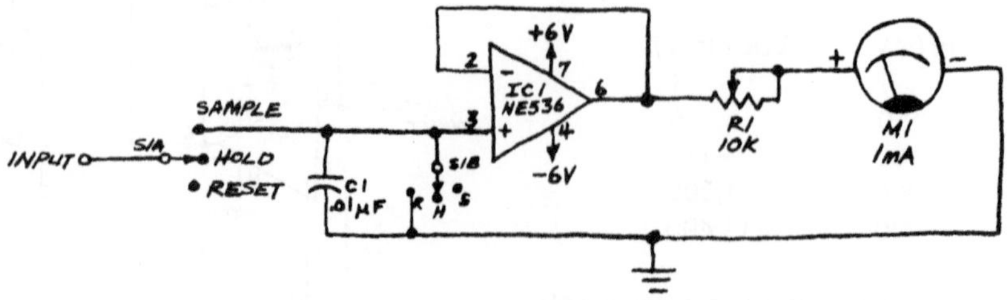

Fig. 1. Schematic of a demonstration sample/hold circuit.

It's easy to operate the circuit in Figure 1. Use a 1.5-volt cell to supply the input voltage. With the sample switch closed, adjust calibration potentiometer *R1* until the meter reads 0.15 mA, which corresponds to a voltage of 1.5 volts.

You can omit the milliammeter and potentiometer if you prefer to connect a voltmeter directly to the output of the op amp. Because the op amp is connected as a unity-gain voltage follower, the voltage at its output will be identical to that at its input. This does not mean that the op amp is superfluous. To the contrary, it provides the very high input impedance that permits the voltage stored across *C1* to be monitored without significant loss.

After you have sampled the input voltage, allow the switch to return to its HOLD position and monitor the meter reading. If *C1* is a high-quality, low-loss polystrene or Mylar unit, the sampled voltage will remain constant for a substantial period of time. Lower-quality capacitors including some ceramic discs will lose their stored charge at a much faster rate, as will be evidenced by a noticeable downward movement of the meter needle.

To increase the circuit's useful storage time, a capacitor larger than the one specified in Figure 1 can be used. But it will require a longer sample interval to charge up to the full input voltage, especially if the sample generator has a significant internal impedance.

Incidentally, if you don't have an NE536 or similar FET-input op amp on hand, you can use a standard 741 for demonstration purposes. You'll have to increase the capacitance of *C1* to 1μF or more, because the stored voltage will be lost much more rapidly than if a FET-input op amp is used, due to the 741's much lower input impedance.

Adding Digital Control. The circuit we have just described is fine for demonstration purposes, but the circuit shown in Fig. 2 is more practical because the sample/hold process is controlled by logic levels instead of mechanical switches. As you can see by comparing the two circuits, the ganged sample/hold/reset switches have been replaced by two of the analog switches in a CD4066 quad analog switch. Furthermore, the NE536 has been replaced by a CA3130 MOSFET-input op amp, but the circuit will work with the 536 as a pin-for-pin replacement without *C2*.

The analog switch is a newcomer to this column. Like the three-state gate described in the March 1978 issue, the analog switch has an ENABLE input and ports that allow a signal to enter and leave. The analog switch, however, can transmit or block both digital logic levels *and* analog (variable) voltages. Like a conventional mechanical switch, an analog switch can pass a signal in either direction. Figure 3 shows the equivalent circuit of the analog switch.

The CA3130 op amp is also a newcomer to this column. I'll have more to say about both it and the CD4066 in future columns. Meanwhile, suffice it to say that the CD4066 is one of a family of CMOS analog switches having many fascinating applications. The switches in the CD4066 are *off* when their ENABLE inputs are *low* and *on* when their ENABLE inputs are *high*. The "off" resistance is around 10^{11} ohms, and the "on" resistance is typically 80 ohms.

To sample a voltage with the circuit shown in Fig. 2, the SAMPLE ENABLE input, which is normally kept low, is allowed to go high. The sampled signal level is then stored by *C1* until the RESET ENABLE input, which also is normally kept low, goes high. This allows *C1* to discharge to ground through *IC1B*.

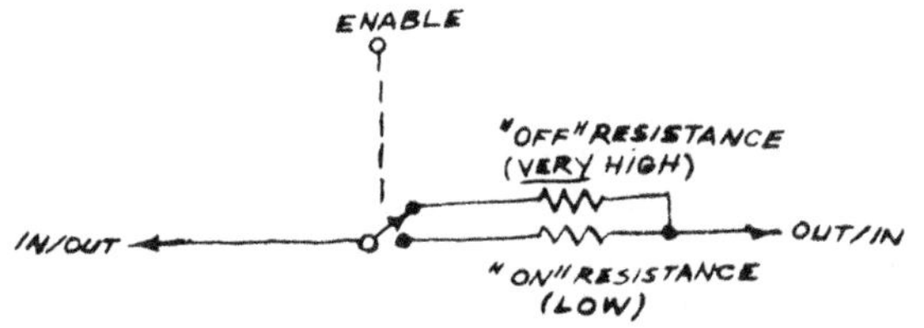

Fig. 3. Equivalent circuit of a basic analog switch.

If the RESET ENABLE input is again made low, *C1* will immediately store the voltage present at the input if the SAMPLE ENABLE input is still high. Of course, if the SAMPLE ENABLE input is low, *C1* will not receive the new sample until the SAMPLE ENABLE input is high. (Ordinarily, both ENABLE lines should *not* be allowed to go high simultaneously. Otherwise, the signal source output will be connected to ground via a low-impedance path.)

As you can see, there are several operating possibilities for the circuit, each of which can be readily selected by a two-bit logic signal. The factors governing the calibration of the output meter and the selection of *C1* are identical to those that apply to the previous circuit.

Applications. The most straightforward application for the sample/hold circuit of Fig. 2 is an analog memory circuit capable of storing a transduced temperature, light intensity, or pressure, or any other analog signal for later processing.

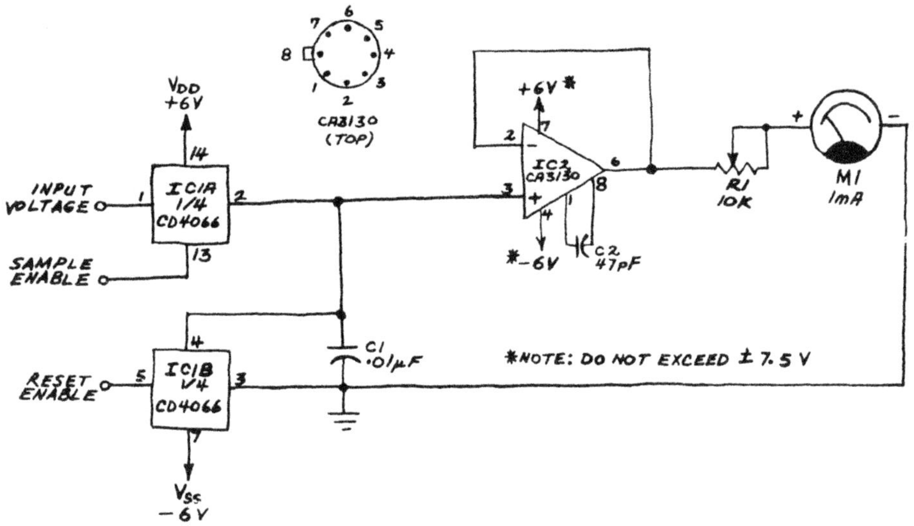

Fig. 2. Digitally controlled sample/hold circuit using CMOS chips. To sample: make sample enable high. To reset: make reset enable high.

The circuit can also be used as a timer. Replace meter *M1* and *R1* with a LED and 330-ohm series resistor. The LED will glow until the voltage across *C1* drops below the LED's turn-on threshold. Increase the capacitance of *C1* for longer time delays. Increasing the magnitude of the sampled voltage up to a maximum of V_{DD} will also give longer delays.

You can create unusual sound effects by connecting the output of the circuit to a voltage-controlled oscillator such as the 566 function generator or unijunction transistor relaxation oscillator. For a siren effect, connect a high resistance (e.g. 1 megohm) between pin 4 of *IC1B* and *C1*. When the RESET ENABLE input is activated, the output voltage will slowly decrease, causing the vco to generate a siren-like sound. The upward wail of the siren is obtained by connecting a second high-value resistor between pin 1 of *IC1A* and the INPUT VOLTAGE source. ■

3. The Analog Comparator

THE ANALOG comparator is a circuit that compares an input voltage to a reference voltage and changes the state of its output when the input exceeds the reference. This decision-making ability has many important applications, several of which we will examine here.

A simple analog comparator can be made by using an operational amplifier without a feedback resistor. The role that a feedback resistor usually plays is to pass some of the amplified signal back to the inverting input of the op amp, thus reducing the amplifier's gain. Without the gain limitation imposed by a feedback resistor, the op amp operates at its maximum ("open-loop") gain. A small input voltage will then cause the output of the op amp to change state immediately. The resulting voltage swing is so dramatic that the comparator can be considered a switching circuit.

The operation of a noninverting analog comparator is shown in Fig. 1. A known reference voltage is applied to the comparator's inverting (−) input, and an unknown voltage to its noninverting (+) input. The LED indicates the status of the comparator's output.

In operation, the output of the comparator is at −V when the input voltage is more negative than the reference voltage which in this case is ground. The LED indicates this by glowing. When the input voltage is more positive than the reference voltage, the comparator output switches from −V to +V and the LED is extinguished. Because the reference is ground, a very small positive voltage will trigger the comparator. In both cases, the voltage difference is measured in millivolts.

Comparator Demonstration Circuit. Unless you have previously worked with analog comparators, you will probably want to take a few minutes to breadboard the simple demonstration circuit shown in Fig. 2 before trying any of the circuits that will be described later.

The comparator in this circuit is a 741 op amp without a feedback resistor. A variable input voltage is provided by *R1*, a potentiometer operated as a voltage divider. Resistors *R2* and *R3* form a fixed voltage divider that provides a reference at half the supply voltage.

When the input voltage is below the reference voltage, the LED glows to indicate that the comparator's output is low (at ground). The LED switches off to indicate the comparator's output is high (at +9V) as soon as the input voltage exceeds the reference voltage. With the values shown in Figure 2, *R1*'s wiper will be at the center of its rotation when the comparator switches, assuming that *R1* is a linear potentiometer.

Sine- to Square-Wave Converter. One of the simplest applications for a comparator is the sine- to square-wave converter shown in Fig. 3. The reference voltage is ground so the comparator switches its output to its maximum positive value when the sine-wave voltage exceeds ground potential. Similarly, the comparator output switches to its maximum negative value when the sine-wave voltage is at or below ground potential. The result is a square wave with the same period as the sine wave.

Peak Detector. Another simple but useful comparator application is the peak detector. As its name implies, the peak detector retains the maximum amplitude of a fluctuating input voltage for subsequent readout and analysis. Suitable transducers connected to the input of a peak detector permit the determina-

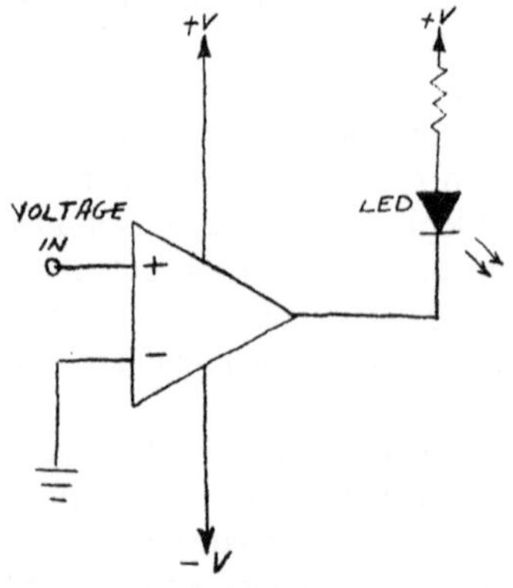

Fig. 1. Operation of a basic comparator circuit.

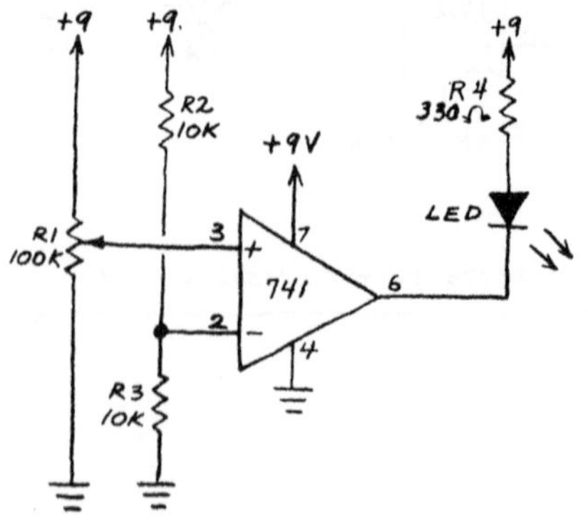

Fig. 2. Schematic of a demonstration comparator circuit.

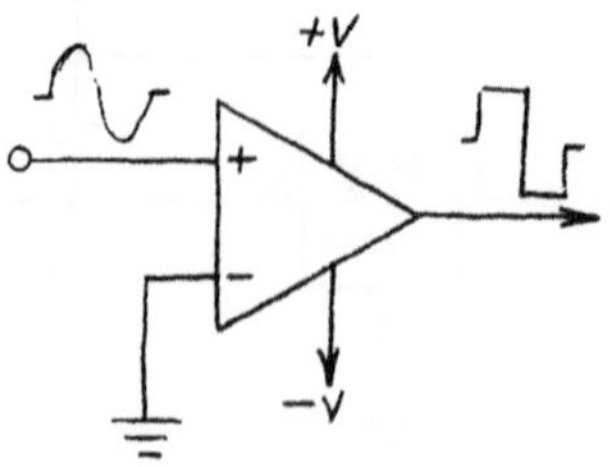

Fig. 3. Comparator as sine-wave to square-wave converter.

tion of such parameters as maximum wind velocity, temperature, light intensity, vehicle speed, and many others.

Figure 4 shows a basic peak detector circuit that you can easily assemble. To understand its operation, assume that *C1* is initially *discharged* (i.e., the RESET switch has been momentarily closed). This means that the reference voltage at the inverting input of the comparator is 0 and that a positive input voltage will immediately switch the output of the comparator to +9 volts. The comparator output will then begin to charge *C1* until the voltage across the capacitor equals the input voltage. As soon as the two voltages are equal, the comparator output immediately drops to ground potential and *C1* stops charging.

If a subsequent input voltage exceeds the charge stored in *C1*, the comparator output will again go high and allow *C1* to charge to the new peak voltage. This tracking process ensures that *C1* always retains the peak voltage applied to the input. When you want to track a new (lower) peak voltage, close the RESET switch to discharge *C1*.

The peak detector circuit is subject to drift because *C1* will gradually lose its charge. Diode *D1* prevents discharge through the comparator, but discharge can take place through the output circuitry or through the dielectric leakage of the capacitor. For these reasons, it is important to use a low-loss polystyrene or Mylar capacitor for *C1* and a high-impedance monitoring circuit.

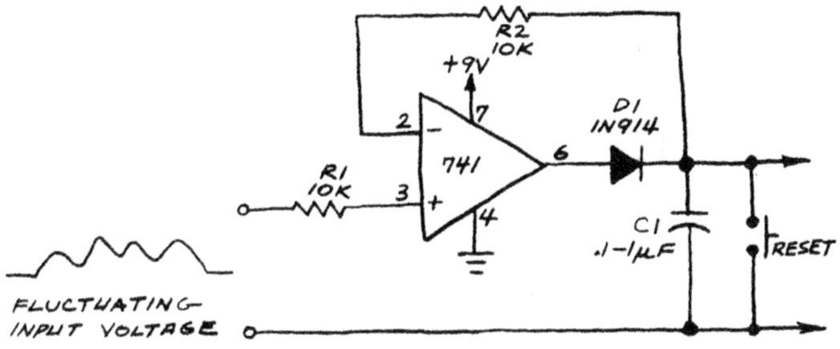

Fig. 4. Schematic of a simple peak detector.

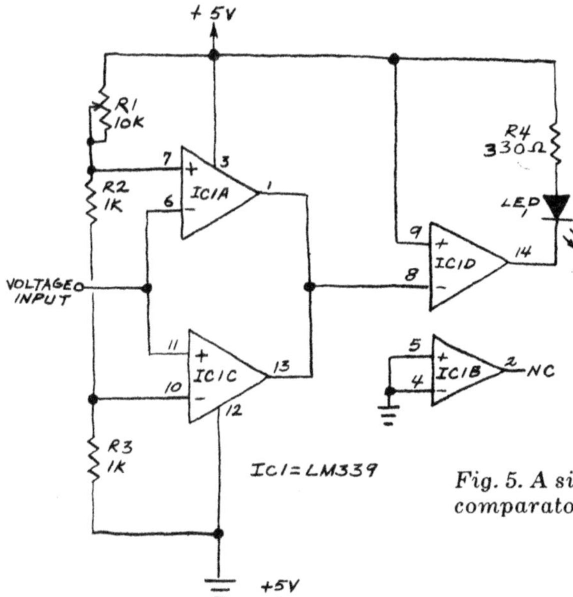

Fig. 5. A simple window comparator circuit.

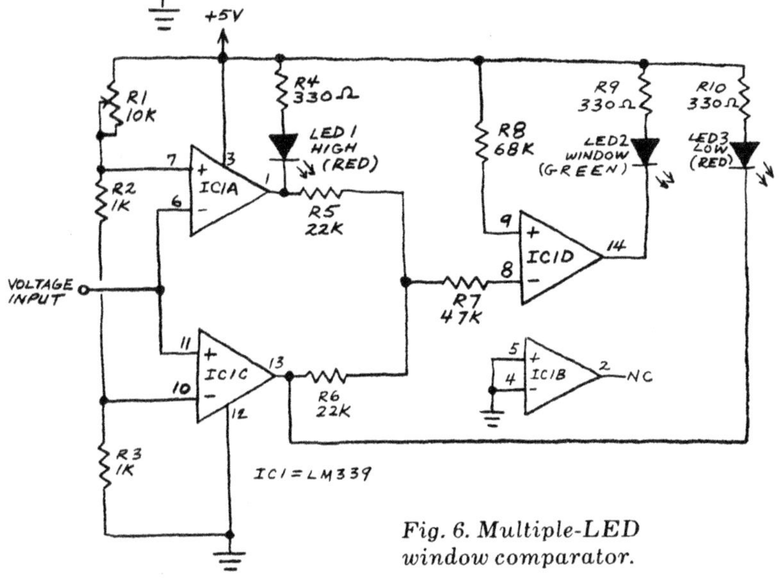

Fig. 6. Multiple-LED window comparator.

The Window Comparator. The comparator circuits described thus far operate in the noninverting mode. That is, they generate an output identical in polarity to the input voltage. However, a comparator can be operated in the inverting mode by simply reversing the two inputs. This makes possible many additional applications, one of which is called the limit or window comparator.

A window comparator can be made from three-fourths of an LM339 quad comparator as shown in Fig. 5. This chip was the subject of the January 1977 Experimenter's Corner. Unlike the 741, the LM339 is specifically designed to operate with a single-polarity power supply.

In operation, *IC1C* functions as a noninverting comparator, but *IC1A* operates as an inverting comparator. Potentiometer *R1* and fixed resistors *R2* and *R3* form a divider chain that delivers slightly different voltages to the two comparators. These voltages define the upper and lower limits of the circuit's switching "window," which can be changed easily by varying *R2* and *R3*.

The output of each comparator in the LM339 is an uncommitted collector. This means two or more outputs can be tied together to achieve a logic OR function without using diodes or a logic gate.

When the input voltage is less positive than *IC1C*'s reference voltage, the output collector of this comparator is low. When the input voltage is more positive than *IC1A*'s reference voltage, its output collector is low. When either output is low, the other is pulled low, causing a LED connected between the two outputs and the positive power supply to glow.

If the input voltage falls in the window region between the two reference voltages, the output of each comparator is high. This will cause a LED connected to the outputs to be darkened.

It's usually desirable for an indicator to light when a desired condition is met. The third comparator in Fig. 5 serves this purpose by inverting the output of the window comparator. The LED then glows only when the input voltage falls within the window region.

An even more useful version of the circuit is shown in Fig. 6. Here, the third comparator is employed as a NAND gate. Three LED's connected to the outputs of all three comparators provide a HIGH/WINDOW/LOW indication. For best results, use a green LED for the WINDOW indicator and red LED's for the HIGH and LOW indicators. The green LED will glow when the input voltage is within the window. The red LED's will indicate that the input voltage is either above or below the window. The LED's should be mounted in a vertical row with the HIGH LED on top, the WINDOW LED in the middle, and the LOW LED on the bottom.

If you use three different colors for the LED's, the circuit will tell you whether

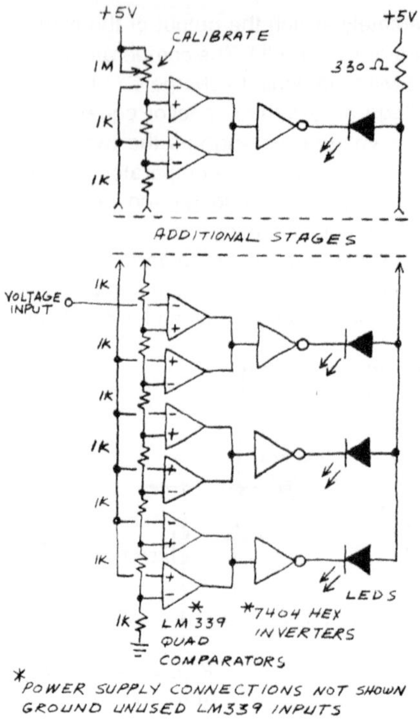

Fig. 7. A moving-dot voltage indicator.

the input voltage is above, below, or in the window no matter how the LED's are mounted. A red LED connected to the output of *IC1A*, for instance, would indicate a HIGH voltage. A yellow LED at *IC1C* would indicate a LOW voltage. Finally, a green LED at *IC1D* would indicate an input voltage within the WINDOW.

Incidentally, the comparator used as a NAND gate in Fig. 6 can be replaced by a conventional TTL 7400 NAND gate. In fact, the first breadboard version of the circuit I assembled used a 7400. Similarly, the third comparator in Fig. 5 can be replaced by one of the inverters in a 7404 hex inverter or an npn transistor and a 10,000-ohm base resistor. Keep this in mind when building a window comparator in a complex circuit that includes digital logic chips. A 7400 with an unused gate will allow you to eliminate the extra resistors required by the comparator NAND gate. ■

4. Voltage-to-Frequency Converters

MANY INTERESTING circuit applications have been made possible by some relatively new monolithic ICs that convert voltages applied to their inputs into pulse trains whose frequencies vary in step with changes in the input voltages. In the past, voltage-to-frequency or simply V/F converters were available only as expensive hybrid modules or do-it-yourself patchwork versions made from IC timers and op amps. This month, we'll look at several straightforward applications for two new V/F chips.

V/F Converter Basics. Figure 1 is a simplified block diagram of a basic V/F converter. The circuit functions as a relaxation oscillator whose frequency is determined by the voltage applied to the noninverting input of the comparator. If capacitor *C1* is initially discharged, the output of the comparator will switch to the positive supply voltage as soon as the input voltage becomes positive. This triggers a one-shot timer that closes a switch to connect a constant current source to *C1* for a fixed time interval determined by the values of timing components R_T and C_T. Depending on many factors, this one injection of charge might develop a voltage across *C1* that is more positive than the input voltage. If this happens, the one-shot will not be triggered again and will remain in its "off" state. The comparator will continue to monitor the input.

This charging cycle will be repeated any time the input voltage becomes more positive than that across *C1*. In the meantime, *C1* is gradually discharged by *R1*. Should the voltage across *C1* fall below the input voltage, the charge sequence will be repeated—even if the input voltage has not changed.

This automatic tracking process, known as *charge balancing*, enables the circuit to generate a pulse train whose frequency is precisely proportional to the input voltage. The output pulses developed by the one-shot timer are buffered by transistor *Q1*.

This is a highly simplified description of how most V/F converters work. For more details, see Walter G. Jung's "IC Timer Cookbook" (Howard W. Sams and Co., 1977, pp. 184-192). The data sheets for the various

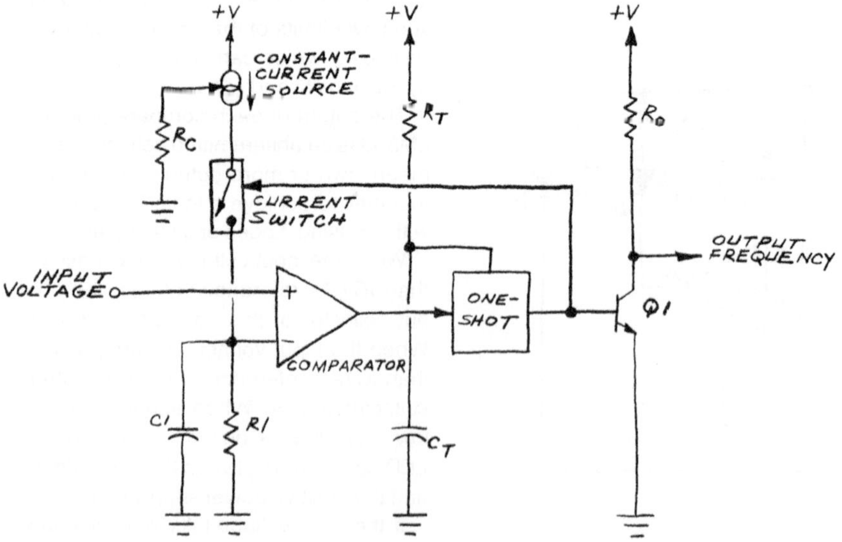

Fig. 1. Functional diagram of a typical voltage/frequency converter.

V/F ICs also include good explanations of how they operate.

Teledyne 9400 V/F Converter. This 14-pin DIP incorporates both CMOS and bipolar circuitry on a single silicon substrate. The result is very low current comsumption, typically 3.5 mA when the IC is powered by a single 9-volt battery. The chip can, however, be powered by either a dual- or single-polarity supply. Figure 2 is the schematic of a V/F converter made with a 9400 and some external parts. The circuit, which is powered by a single-ended supply, was adapted from one appearing in the manufacturer's data sheet.

A breadboard version that I assembled began to emit an output signal with a frequency of 0.3 Hz when the input voltage reached 0.25 volt. The maximum input voltage to which the circuit would respond was exactly 8 volts when the circuit was powered by a 9-volt alkaline battery. The output frequency corresponding to this input voltage was 13.53 kHz. A plot of the output frequency versus the input voltage at half-volt intervals for the prototype circuit is shown in Fig. 3. The striking linearity of this chip's output-frequency/input-voltage characteristic, which in this case extends over a five-decade frequency range, is characteristic of V/F ICs.

The output frequency of the circuit in Fig. 2 can be increased to a maximum of 100 kHz by reducing the values of *C1*, *C2* and *R2*. The 9400 data sheet gives detailed information.

9400 Digital Data Transmission. Frequency shift keying (FSK) is a complicated name for a very simple way to transmit digital data. In most digital electronic circuits, the logic 0's and 1's of a binary signal are represented by two voltage levels. FSK data transmission assigns one audio-frequency tone to logic 0 and a second (usually higher) frequency to logic 1.

This permits a stream of bits to be transmitted over a pair of wires, by radio, or by light. At the receiver, a frequency-to-voltage (F/V) converter transforms the received tones back into two distinct voltage levels.

A block diagram of a basic FSK data transmission system is shown in Fig. 4. Potentiometer *R1* in the transmitter input network permits the quiescent output frequency to be preset to any convenient value. This permits many different FSK transmitters, each with different 0 and 1 frequencies, to share a common transmission channel. Of course, each information (*not* transmission) channel will require a separate FSK receiver.

9400 Frequency Modulator. Several articles in this magazine have described ways of transmitting information over a pulse/frequency-modulated beam of infrared radiation emitted by a LED or injection laser. This method of light-beam modulation is superior to amplitude modulation because each pulse transmitted has the same amplitude, usually the maximum signal power the transmitter can radiate. The received signal is not as subject to fading as that in an AM system when propagation conditions change or when the transmitter-to-receiver distance changes.

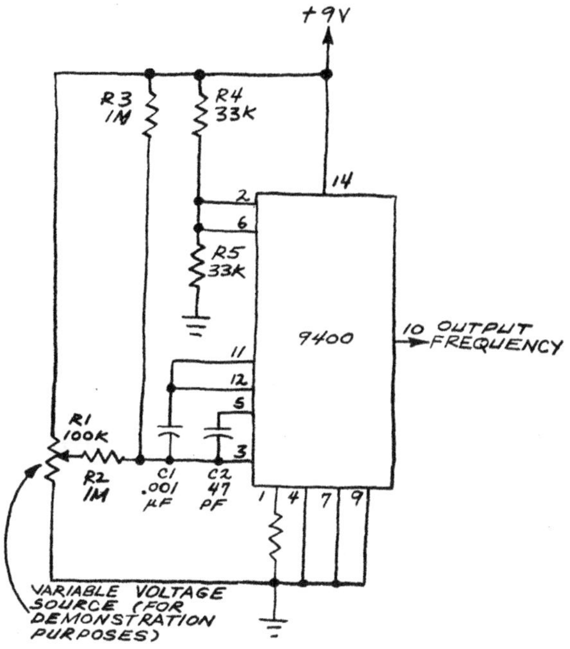

Fig. 2. Voltage-to-frequency converter using a 9400 IC.

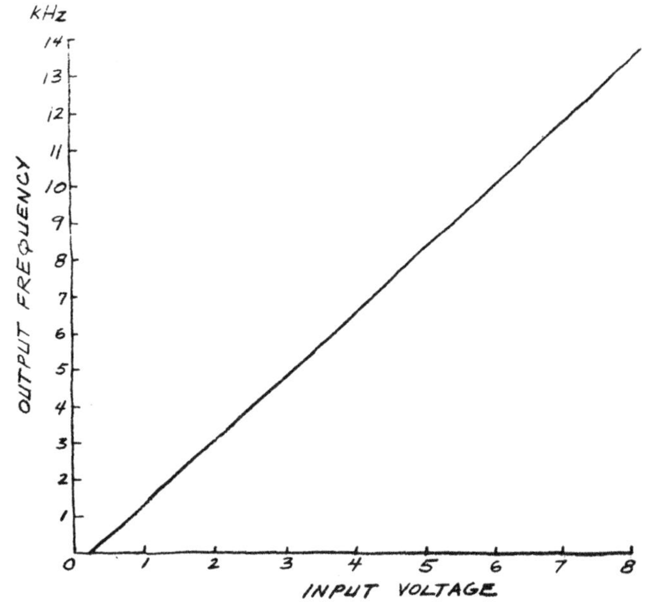

Fig. 3. Output frequency vs. input voltage for 9400 V/F converter circuits.

TYPICAL TRANSMITTER OPERATION		
C2 (FIG. 2)	OUTPUT FREQUENCY	
	LOGIC 0 IN	LOGIC 1 IN
47 pF	3943 Hz	17,671 Hz
0.1 µF	1000 Hz	1665 Hz

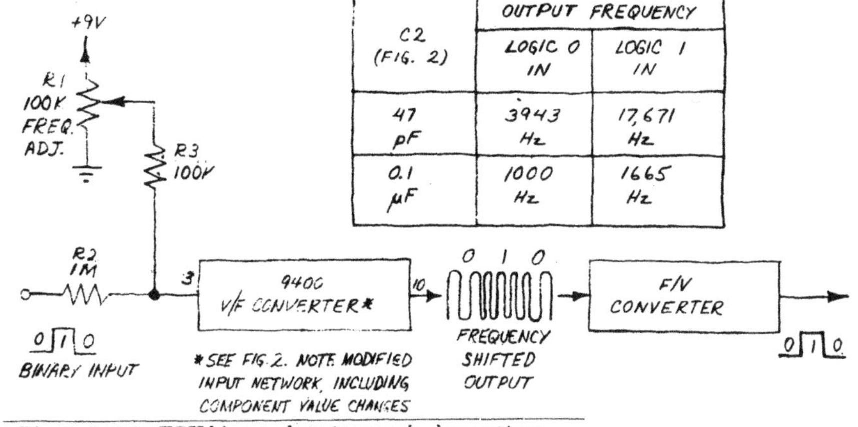

Fig. 4. A 9400 FSK binary data transmission system.

The 9400 and other V/F converters can be used as exceptionally linear frequency modulators. Figure 5, for example, shows a basic FM transmitter that will transform an audio signal such as voice into a train of variable-frequency pulses suitable for driving a LED or modulating a radio transmitter.

Note that the duration of the pulses in the output signal is variable. For several reasons, it's desirable to drive a LED with pulses of uniform duration, especially in a long-range voice-communication system in which the LED is driven by ampere-level current pulses.

The circuit shown in Fig. 5 can easily be modified to accomplish this purpose. One possibility is to connect its output to a one-shot that delivers a pulse of uniform duration to the LED each time a pulse is generated by the 9400. Another is to trigger the gate of an SCR which, in turn, dumps the charge that has accumulated in a capacitor through the LED. Still another method is to trigger a transistor which then dumps charge from a capacitor through the LED. Whichever means you select, it's important to make sure that the pulses from the one-shot are not too wide. Otherwise, some of the closely spaced pulses generated by the 9400 will be missed, resulting in distortion.

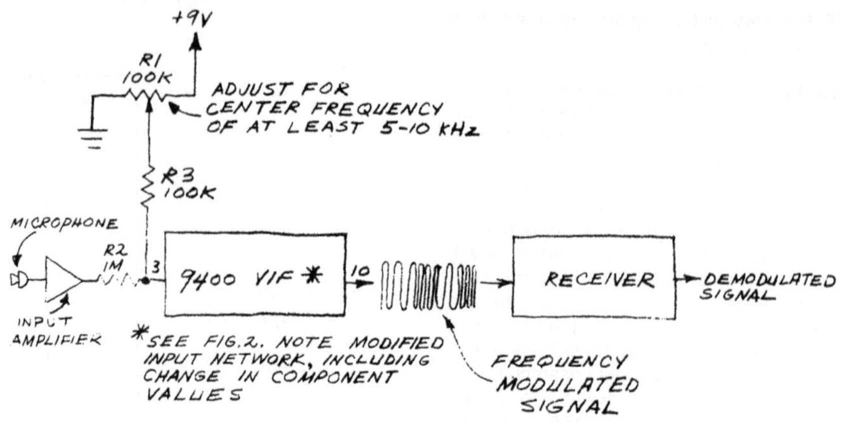

Fig. 5. A 9400 frequency-modulated transmitter.

The frequency modulated signal must be demodulated after it is received. One way to accomplish demodulation is to connect a one-shot to the receiver output, the method employed in the P/FM laser receiver described in the *1979 Electronic Experimenter's Handbook* (F. Mims, "Semiconductor Laser Communications System," pp. 64-73). Another demodulation system employs a phase-locked loop, a method covered in a prior installment of "Experimenter's Corner" ("Pulse Modulation and Phase-Locked Loops," May 1976, pp. 101-102). That column also described a simple two-transistor P/FM LED transmitter.

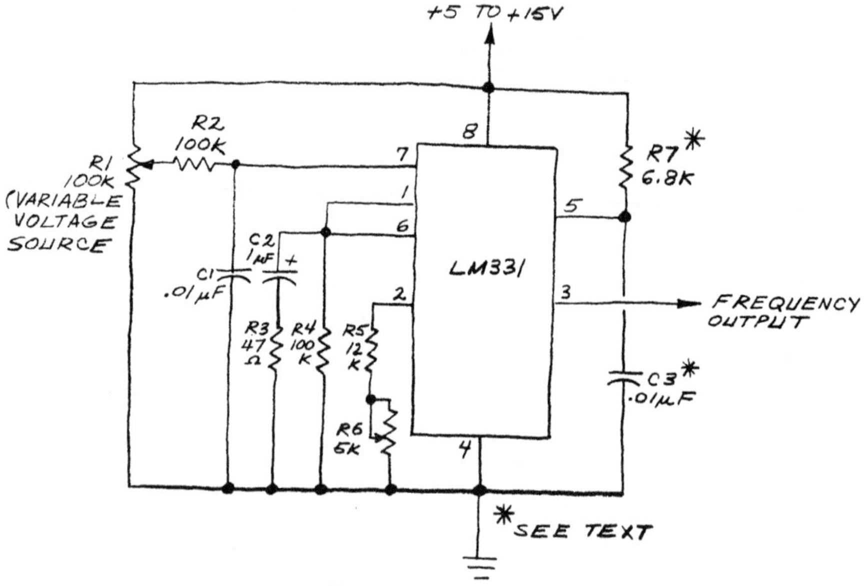

Fig. 6. Voltage-to-frequency converter using an LM331.

National LM331 V/F Converter. After spending a good deal of time experimenting with the 9400, I received a few sample LM331 V/F converters from Robert A. Pease, a staff scientist for National Semiconductor. The LM331 has a guaranteed linearity of at least 0.01 percent when connected in the V/F mode. Like the Teledyne Semiconductor 9400, it can be operated from a single-ended or dual-polarity supply, and can generate an output frequency of up to 100 kHz.

Figure 6 shows a basic V/F converter adapted from the LM331 data sheet. Potentiometer *R1* serves as a voltage divider that delivers a variable input voltage to the V/F circuit. A breadboard version of this circuit yielded the plot of voltage versus frequency shown in Fig. 7. The increasing nonlinearity in V/F operation when the input voltage exceeded 8 volts is probably due to my use of standard-tolerance components. For ±0.03% linearity (typical), use 1% tolerance resistors for *R4* and *R7* and a low-temperature-coefficient capacitor for *C3*.

One of the simplest applications for the LM331 is the ultra-stable oscillator shown in Fig. 8. This circuit, which Don Pease of Na-

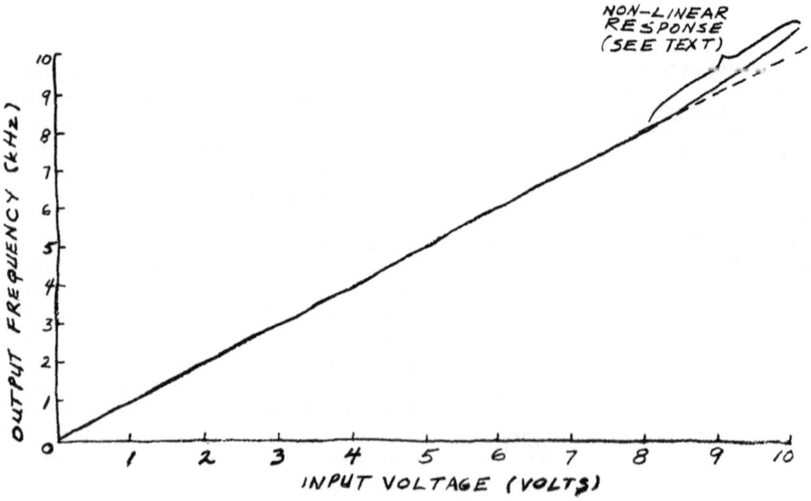

Fig. 7. Output frequency vs.input voltage for LM331 V/F converter.

tional Semiconductor described in *Electronic Design* (December 6, 1978, pp. 70-76), has a frequency stability of ±25 parts per million per degree Centigrade (ppm/C) if low-temperature-coefficient parts are used for *R3, R4* and *C2.* These components determine the output frequency of the oscillator.

Note that *R1* is composed of *two* 15,000-ohm resistors in series. Don recommends that these resistors and the one used for *R2* be from the same production batch. This makes the circuit from five to ten times more immune to temperature changes than it would be if *R1* were a single 30,000-ohm resistor. Incidentally, although these resistors should have a tolerance of no more than 1% for best results, the circuit *will* operate (but with less accuracy) if standard 10% tolerance resistors are used.

Like the 9400, the LM331 is not yet readily available from many of the hobby distributors who advertize in this magazine, but it will be as soon as the demand exists. Until then you can get the LM331 from Hamilton/Avnet, Schweber, Hall Mark, Sterling or any of the dozens of major industrial distributors who handle National Semiconductor parts. ■

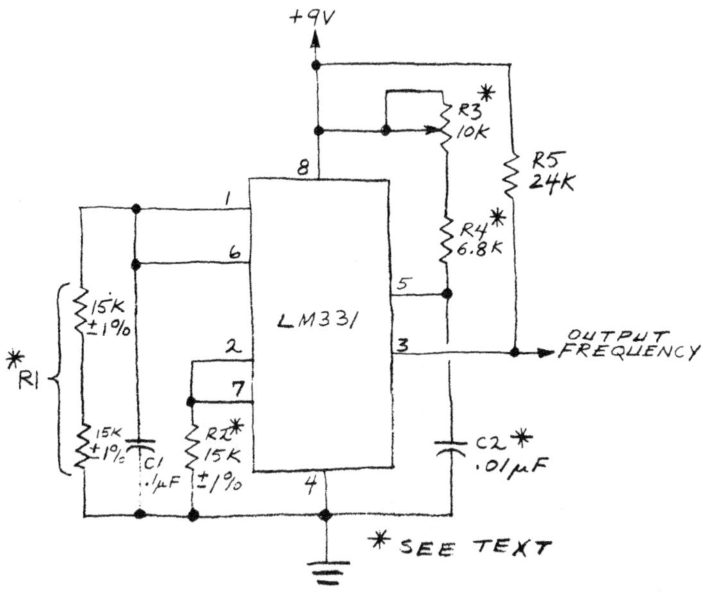

Fig. 8. LM331 operated as stable frequency oscillator.

5. Frequency-to-Voltage Converters

Important features of the 9400 and LM331 are that both chips can also be used for frequency-to-voltage (F/V) conversion applications. Whether operated in conjunction with a V/F converter or alone in a straightforward F/V mode, both the 9400 and LM331 have many interesting applications. We'll look at several, but first a brief explanation of how a typical F/V converter works.

F/V Conversion. The F/V operation of a V/F converter chip is very straightforward and is even easier to understand than operation of the same IC in its V/F mode. Figure 1 is a simplified functional diagram of the LM331 connected as a F/V converter.

In operation, the incoming signal is applied directly to the noninverting input of the comparator. The inverting input of the comparator is biased at a voltage determined by the values of divider resistors *R1* and *R2*. The comparator output switches states each time the amplitude of the incoming frequency exceeds or drops below the reference voltage.

The one-shot is triggered by a positive transition at the output of the comparator. This in turn, closes the current switch and allows the current source to charge output filter capacitor *C2* for a period determined by the time constant *R3C1*. Bleeder resistor *R4* continually discharges *C2* so that the charge stored in this capacitor at any instant approximates the average charge available from the current source. In short, the charge stored in *C2* (and thus the voltage across it) is directly proportional to the input frequency.

Incidentally, can you think of a drawback to the basic circuit in Fig. 1? (Hint: Consider the time constant *R4C2*. How will this affect the circuit's ability to respond to an input signal whose frequency is rapidly changing?)

LM331 F/V Converter. In last month's column, I described some straightforward V/F applications for the LM331. Shown in Fig. 2 is a working F/V converter whose operation is similar to that of the functional diagram in Fig. 1. This schematic shows how to use the LM331 as a simple F/V converter.

In operation, the incoming signal is coupled to the comparator in the LM331 via *C1*. Resistors *R2* and *R3* provide the reference voltage to the comparator. The operation of the

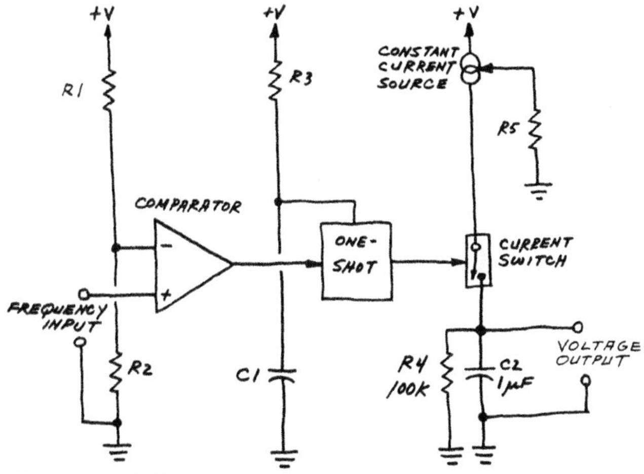

Fig. 1. A typical frequency-to-voltage converter.

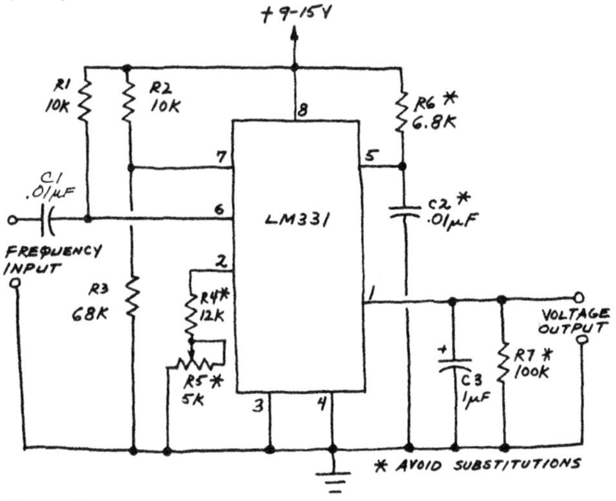

Fig. 2. Frequency-to-voltage (F/V) converter using LM332.

one-shot in the LM331 is determined by the time constant *R6C2*, and the output signal is filtered by *C3* and *R7*. Potentiometer *R5* provides a very useful calibration feature. Together with *R4*, it controls the current that charges *C3*. In short, *R4* and *R5* in Fig. 2 are equivalent to *R5* in Fig. 1.

Figure 3 shows the highly linear response of a breadboard version of the circuit shown in Fig. 2. When the circuit was powered by a 15-volt supply, its response was linear beyond 10,000 Hz. When I powered the circuit with a single 9-volt battery, however, its response was linear only up to about 6500 Hz. This happened because the largest possible output voltage the circuit can provide is 6.5 volts when powered by a 9-volt supply.

For both tests, *R5* was used to calibrate the circuit so that a 3000-Hz input would result in a 3.00-volt output and that a 1000-Hz change in the input frequency would cause the output to change by precisely 1.00 volt. This very simple alignment procedure yielded excellent results, as Fig. 3 indicates.

When the F/V measurements were made for a third time with the circuit calibrated so that a 5000-Hz input signal yielded a 5.00-volt output, the following strikingly linear results were obtained:

Input Frequency (Hz)	**Output** (volts)
0	0.00
100	0.10
500	0.50
1000	1.00
2000	2.00
3000	3.01
4000	4.00
5000*	5.00
6000	6.00
7000	7.00
8000	7.97
9000	8.94
10000	9.91

* Calibration point.

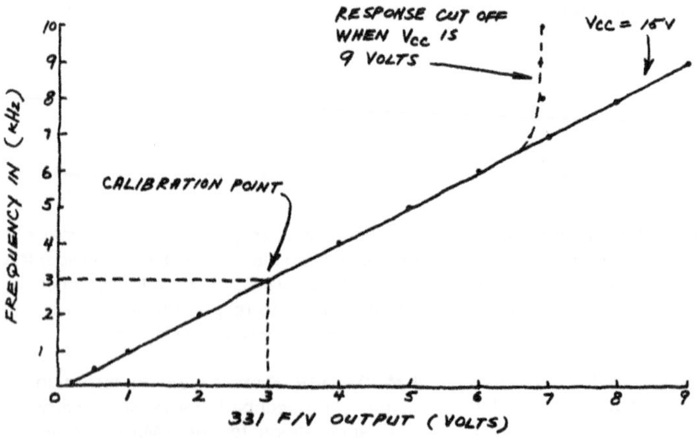

Fig. 3. Plot of output voltage vs. input frequency for a frequency-to-voltage converter using an LM331.

These measurements were made with the help of a DVM and a digital frequency counter. This excellent linearity is typical of the results obtained from breadboard V/F and F/V converter circuits employing the LM331 or 9400.

9400 F/V Converter. Last month, we experimented with a very linear V/F converter built around a 9400 IC. Figure 4 shows a 9400 connected as a F/V converter to decode data transmitted by an infrared LED driven by a 9400 V/F converter. This pair of circuits, which is adapted from a design by Michael O. Paiva, Teledyne Semiconductor's Product Marketing Manager, makes an excellent analog data transmission system.

The transmitter portion of the circuit is essentially identical to Fig. 2 in last month's column. The only significant change is the addition of *R7*, a miniature 8-ohm speaker and an infrared LED. Resistor *R7* is required to limit current through the LED to a safe value. The speaker is entirely optional. Because the frequency of the pulse train generated by the transmitter's V/F converter remains within the audio range, the speaker provides a convenient way of monitoring the circuit's operation, particularly during preliminary testing and evaluation. The speaker can be removed later if desired.

The receiver consumes only about 3.5 mA when powered by a 9-volt battery. Its output voltage appears at pin 12, so a DVM can be used for a readout. It's more convenient, however, to connect a small 0-to-1 or 0-to-10-mA panel meter directly to the output as shown in Fig. 4.

A DMM operating in its current-reading mode can be connected in place of the meter for more accurate measurements. I experimented with both methods and found the conventional meter best for initial adjustments and tests, while the DVM proved to be superior for taking data such as that used to plot graphs.

The transmitter LED and receiver phototransistor can be replaced by an optoisolator if electrical isolation is the only reason for using this circuit. If this is done, it might be necessary to increase the value of *R7* in the transmitter to reduce the LED's forward current and prevent the receiver from responding in a nonlinear fashion. In other applications, the signal from the transmitter LED can be sent through the air or through a plastic or glass optical fiber.

I spent most of one day experimenting with this analog transmission system and made a few observations you might find useful should you decide to duplicate it. First, the receiver operates erratically when the optical signal at *Q1* is excessive. If the signal is too weak, the meter indicates zero output current. But when the signal is too strong, the meter needle may slam or swing wildly back and forth. These same effects can occur if the value of *R12* is reduced significantly.

While experimenting with the system, I experienced considerable trouble when the frequency of the incoming signal exceeded a few kilohertz. Although the receiver would readily respond to frequencies in excess of several kilohertz when the signal was coupled directly into pin 11 of the 9400, the re-

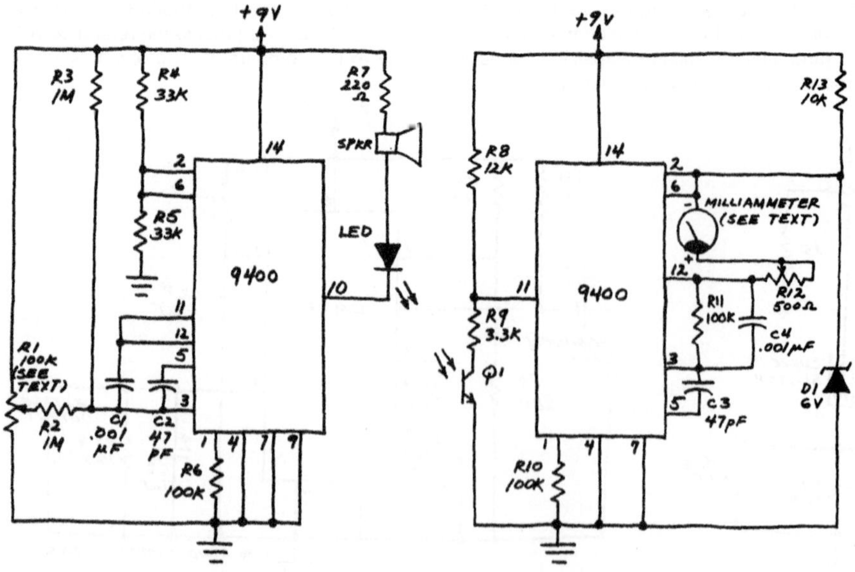

Fig. 4. Analog data transmission system designed by Michael Paiva.

ceiver meter would indicate zero when the signal was received from the transmitter LED via the phototransistor.

The problem was traced to the integrating effect of *Q1*, *R8* and *R9* in the receiver. When the incoming pulse train has a low to moderate frequency, the individual pulses are sufficiently wide to produce a signal at pin 11 having enough amplitude to trigger the receiver. As the transmitter frequency is increased, however, the pulses become correspondingly narrower. The *Q1/R1/R2* combination then produces an output pulse having insufficient amplitude to trigger the receiver. This effect is illustrated in Fig. 5.

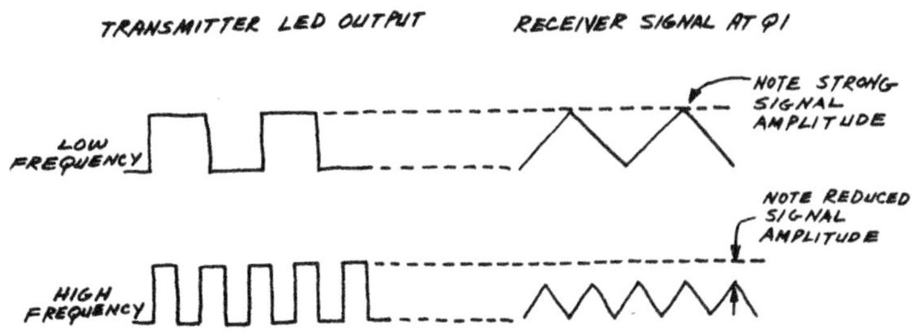

Fig. 5. Reduction in receiver signal of 9400 F/V circuit at a high input frequency.

With the component values specified in Fig. 4, the analog system can transmit an input ranging from 0.25 to 8 volts. The received signal, as displayed on a milliammeter, will range from a minimum of 0.2 mA to a maximum of about 6 mA. Figure 6 is a plot of the transmitted frequency versus the receiver output current. Increasing *C2* in the transmitter to 220 pF will lower the maximum frequency and restrict the maximum receiver output current to about 0.8 mA, thus allowing use of a 0-to-1-mA meter. Shown in Fig. 7 is a plot of the transmitted frequency versus the receiver output current when *C2* is 220 pF.

Before attempting any practical application of the analog transmission system it is necessary to make a graph that plots the transmitter's input voltage versus the receiver's output current. Last month's column included graphs showing the operation of a V/F converter, and we've just looked at graphs describing the operation of a F/V converter. It's very easy to make graphs like these, so I'll leave to you the preparation of a graph showing the operation of the analog transmission system. All you have to do is record the receiver output current for each 1-volt increase in transmitter input and plot the results. You don't have to measure the transmitter's output frequency.

Other F/V Applications. After you've gained some experience with the basic F/V circuits described in this column and the V/F circuits featured last month, you should be well prepared to move ahead on your own. Try to obtain copies of the 9400 and LM331 data sheets. Besides providing valuable operating tips, they suggest many interesting applications.

Also, remember that you can use either the 9400 *or* LM331 in most F/V (and V/F) applications. The chips are not directly interchangeable and they're not even functionally identical. But their operation in both V/F and F/V modes is very similar. For example, with the help of the National Semiconductor data sheet, you should be able to adapt the analog transmission system shown in Fig. 4 for use with LM311 instead of 9400 ICs. ■

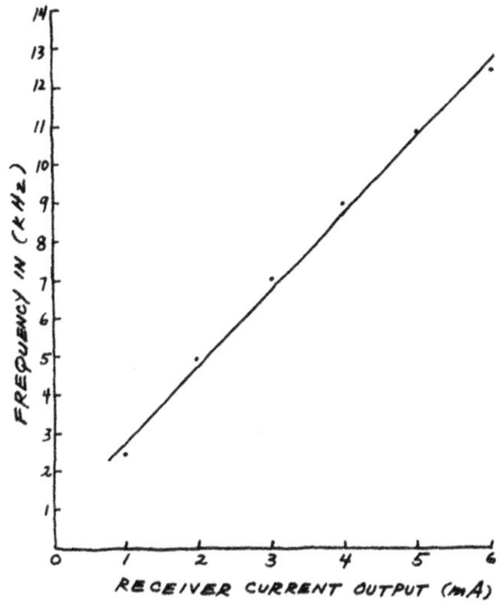

Fig. 6. Receiver output current vs. input frequency for 9400 analog transmission system.

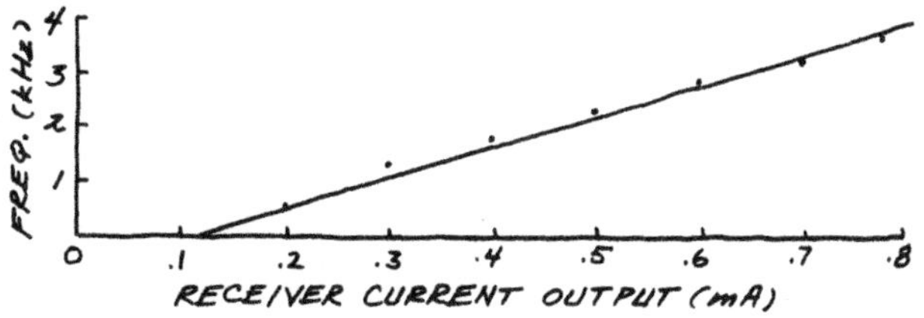

Fig. 7. Analog transmission system with 0-1-mA output.

6. Dark/Light Detector

HERE IS a simple but useful circuit that can function as either a light detector or a dark detector. The circuit's photosensor is a standard cadmium-sulfide (CdS) light-dependent resistor. When the project is operating in its light-detection mode and the photosensor is dark, there is no output. When light strikes the sensitive surface of the LDR, the speaker emits a tone. When the circuit is in its dark-detection mode and the LDR is illuminated, the speaker is quiet. It emits a tone when the photosensor is dark.

The circuit is actually an astable oscillator operating as a tone generator. The oscillator is designed around a 555 timer chip whose reset input (pin 4) is the key to the project's two modes of operation. When pin 4 is at or close to $+V_{cc}$, the circuit will oscillate. When pin 4 is grounded, however, *C1* is discharged and the circuit ceases oscillation.

In both the light- and dark-detection modes, the light-dependent resistor and *R3* form a voltage divider whose center node is connected to pin 4 of the timer IC. When *S1*, a dpdt toggle switch, is placed in position L, the photosensor is connected between pin 4 of the IC and $+V_{cc}$. When the level of ambient light increases sufficiently, the resistance of the photosensor decreases to a low value, pin 4 approaches $+V_{cc}$ and the circuit oscillates. This is the circuit's light-detection mode.

When *S1* is placed in position D, the photosensor is between pin 4 and ground and fixed resistor *R3* is between pin 4 and $+V_{cc}$. Now, when sufficient light strikes the photosensor, pin 4 approaches ground potential and the circuit ceases to oscillate. The project thus functions as a dark detector because removing light from the LDR permits the 555 to oscillate.

The circuit is easily modified. For example, increasing the value of *C1* will decrease the frequency of oscillation. Reducing the capacitance of *C1* will increase the frequency. For more volume, the speaker can be driven by an audio amplifier whose input is capacitively coupled to pin 3 of the timer IC. If only light (or dark) detection is desired, *S1* can be eliminated. The photosensor and *R3* should then be permanently in the positions corresponding to the desired operating mode.

This project has many useful applications. In its light-detection mode, for example, it can be used as an open-door alarm for a refrigerator or freezer or an open-drawer alarm for a cash register. The circuit makes a simple annunciator when used in its dark-detection mode. A source of steady light (artificial or sunlight) beamed at the photosensor inhibits the tone. An interruption of the light beam, such as occurs when a physical object passes between the light source and the sensor, stimulates oscillation.

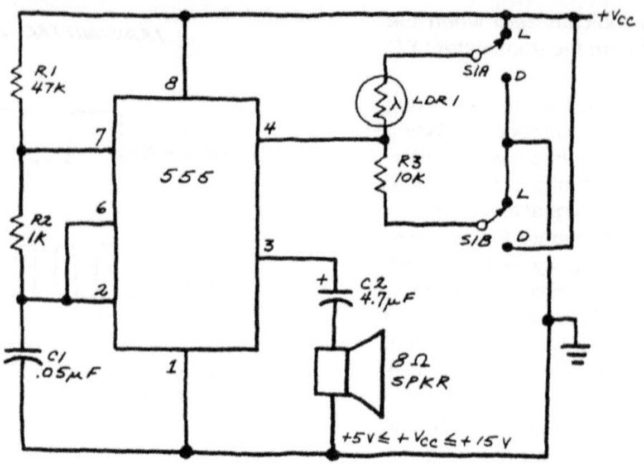

Both operating modes make interesting day/night indicators. In the light-detection mode, the speaker will sound when the sun rises; and in the dark-detection mode, it will sound when the sun sets. ■

7. Experimenting with Noise

THOSE of us who appreciate good-quality sound reproduction might disagree about the definition of good music, but it is safe to say that all audio enthusiasts share a common opinion of noise—the less the better! Noise is equally unpopular among radio astronomers, biomedical engineers, radio communications users, and others who work with low-level electrical and electromagnetic signals.

So much engineering effort is devoted to the suppression of noise (it can *never* be entirely eliminated) that it might come as something of a surprise that there are many useful applications for noise. These include acoustics measurements, instrument calibration, antenna tuning, signal jamming, data encryption, electronic music and even applied psychology! This month, we'll examine several methods of generating noise and explore a few of its uses. First, let's define a few basic terms.

Noise can loosely be called an electronic or electromagnetic weed. More precisely, noise is an undesired electronic or electromagnetic signal having frequency components within the frequency range of interest which tends to interfere with the reception or detection of desired signals. By definition it excludes crosstalk and interference from other information-carrying signals within the frequency range of interest.

There are many kinds of noise. We are primarily interested in *white noise* and *pink noise*. White noise is a complex waveform with a Gaussian amplitude probability characteristic. It is formed by contributions from all frequencies over a theoretically infinite but, in practice, broad and specified bandwidth. White noise has a flat (constant) spectral power density. Thus, it contains equal energy per unit of frequency (Hertz). Transduced, audible white noise contains equal contributions from all audio frequencies perceptible to the human ear. It is thus analogous to white light, which comprises all wavelengths (colors) perceptible to the human eye.

Pink noise is also a complex waveform with a Gaussian amplitude probability characteristic and is also formed by contributions from all frequencies over a theoretically infinite but, in practice, broad and specified bandwidth. Pink noise contains equal energy levels in each octave of its spectrum. Because each next-higher octave possesses twice the number of discrete frequencies (in hertz) as compared to the octave immediately below it, the low-frequency components of pink noise have higher amplitudes than the high-frequency components. This is necessary if pink noise is to contain equal amounts of energy in each octave of its spectrum. Audible pink noise, therefore, has more bass content than white noise and sounds "warmer."

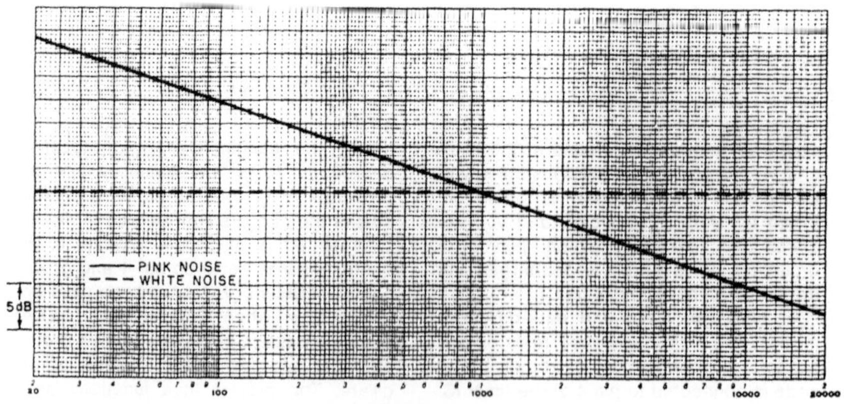

Fig. 1. Plots of amplitude versus frequency for white noise and pink noise.

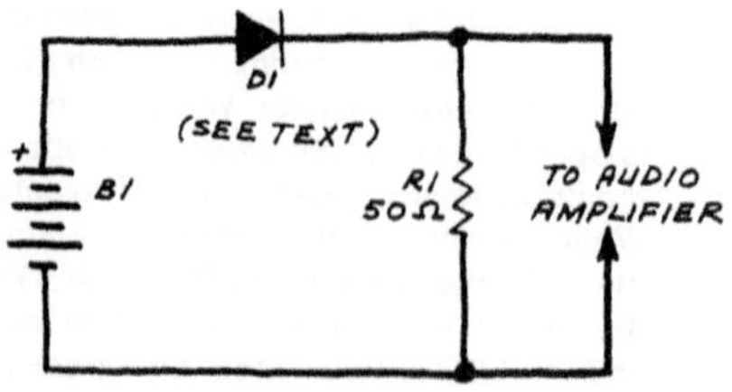

Fig. 2. Simple schematic of basic diode noise generator.

Plots of amplitude versus frequency for white noise (dashed line) and pink noise (solid line) appear in Fig. 1. Note that pink noise displays a −3/dB octave slope. If white noise is routed through a low-pass filter having a −3-dB/octave response, the filtered signal will be pink noise. Pink noise is commonly used as a test signal in audio work because many audio spectrum analyzers are "constant percentage bandwidth" instruments. That is, the passband of each bandpass filter in these analyzers is an unchanging percentage of its center frequency. Therefore, the higher the center frequency of the filter, the broader its bandpass. If white noise is applied to such an analyzer, a rising 3-dB/octave characteristic will be displayed. If pink noise is applied to the analyzer input, a flat amplitude-versus-frequency characteristic will be indicated. The most common audio applications for a pink noise source and such a spectrum analyzer is in frequency-response testing of audio preamplifiers and amplifiers and in the equalization of an audio system in a listening room.

Now that we have examined some basic ideas about noise, let's see some circuits that generate and employ it.

Diode Noise Generators. The simplest noise generator is a forward-biased diode. Figure 2 shows a basic diode noise generator that you can quickly assemble. Connect the circuit to an audio amplifier (capacitive coupling might be necessary) and the speaker will produce a continuous rushing or hissing sound.

A circuit like this can be used to adjust a radio receiver for optimum noise figure. With a suitable diode such as the 1N21 or 1N23 and short, point-to-point wiring, the generator will produce wideband noise with components extending as high as 148 MHz. The ARRL *Radio Amateur's Handbook*, which describes how to make receiver noise adjustments, suggests adding a 500-pF capacitor between the anode of *D1* and ground. It also suggests inserting a 50,000-ohm potentiometer, preferrably one with a logarithmic taper, in series with the anode of *D1* and the positive power supply terminal to permit adjustment of the noise amplitude.

If you like to experiment, try various kinds of diodes for *D1*. A red LED, for example, produces both light *and* noise. Of course, you must increase the value of *R1* to protect the LED from excessive current levels. Use 470 ohms when the voltage of *B1* is +6 volts and 820 ohms when *B1* is a 9-volt battery.

Transistor Noise Generators. When reverse-biased beyond the avalanche point, the emitter-base junction of a bipolar junction transistor generates noise. In the winter edition of the 1975 *Electronic Experimenter's Handbook*, John S. Simonton, Jr. described how to assemble a pocket-size sonic noise generator based upon this effect. John noted in his article that pink noise is an excellent mechanism for masking and thereby concealing low-level sound such as a confidential conversation. John also noted that audible pink noise can help produce a feeling d relaxation and can in some cases block pain stimuli.

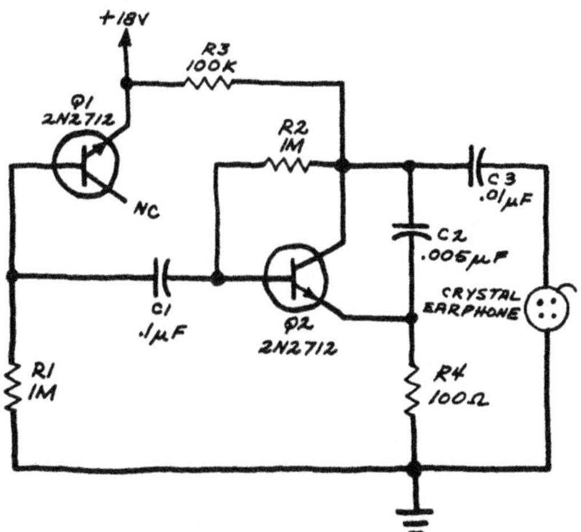

Fig. 3. Noise generator designed by John Simonton.

Figure 3 is the schematic diagram of John's circuit. In operation, *R1* limits the current through the reverse-biased emitter-base junction of *Q1* to a safe value. The noise signal is coupled to amplifier *Q2* via *D1*. After the noise has been amplified, *C2* shunts some of the high-frequency components to ground. The resulting output signal, which is transduced by a high-impedance earphone, is a reasonable approximation of pink noise.

John points out that most 2N2712s will produce noise, but some will not. Should you want to try other transistor types, make sure they have an emitter-base breakdown voltages of less than 18 volts.

Shift Register Digital Noise Generators. Figure 4 shows a simple 7-stage shift-register pseudorandom bit generator which produces a sequence of 127 bits before recycling. Other shift-register/exclusive-OR gate arrangements can be used to produce shorter or longer sequences.

White noise is synthesized when a pseudorandom bit generator such as the one in Fig. 4 is clocked at a sufficiently fast rate. Shift-register generated noise is not necessarily as random as that produced by a diode, especially if a relatively small number of stages is involved. But the noise level is more uniform and of a much higher amplitude than that from a diode.

Fig. 4. Basic pseudorandom bit generator.

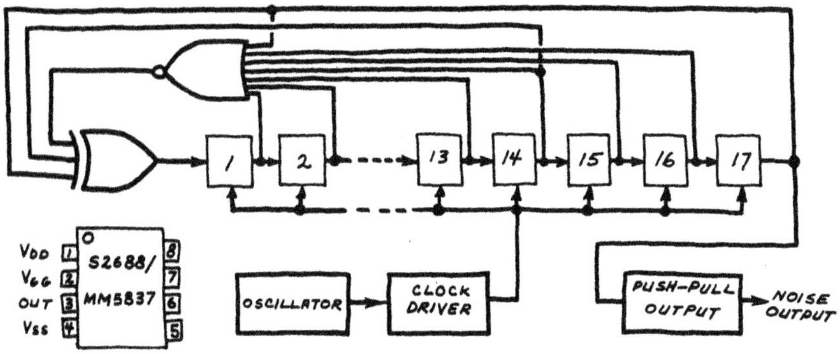

Fig. 5. Block diagram of S2688/MM5837 digital noise generator.

If you would like to experiment with digital noise generators of this type, see pages 277 to 283 of *TTL Cookbook*, by Don Lancaster (published by Howard W. Sams). Don describes several interesting applications, including a scrambler for encoding computer data, and he also gives schematics of several shift-register pseudorandom sequencers.

S2688/MM5837 Digital Noise Generator. The S2688/MM5837 (National Semiconductor) is a complete PMOS digital noise generator in an 8-pin mini-DIP. The internal circuit, shown in Figure 5, consists of a 17-stage shift register, some gates and a clock. Pseudorandom bit patterns are produced by connecting the outputs of the 14th and 17th stages of the shift register to an exclusive-OR gate whose output is applied to the input of the first stage in the shift register.

A 17-input NOR gate monitors the output of each stage of the shift register. Should the outputs of all 17 stages simultaneously go low, the NOR gate prevents a lockup condition (a continuous output of all 0's) by automatically applying a logic 1 to the third input of the exclusive-OR gate. This, in turn, applies a logic 1 to the first stage of the shift register.

The S2688/MM5837 is exceptionally easy to use. If the output is connected to an op amp or other high-impedance circuit, the chip can be powered by a single supply (V_{SS} = 0 V and V_{DD} = −14 V ±1 V). If the chip must drive a low impedance, V_{GG} should be connected to −27 V ± 2 V.

Though it is recommended that V_{DD} be within a volt of −14 V, I've found that the internal clock speed can be altered by varying V_{DD}. Here's what I measured:

V_{DD}	Approximate Clock Frequency (Hz)
−5	0
−6	0.7
−7	2267
−8	8731
−9	16,382
−10	23,531
−11	32,564
−12	38,347
−13	40,010
−14	37,800
−15	33,173

Because broadband noise is best suited for most audio applications, it's evident that a supply voltage of −12 to −14 volts gives the best noise quality. However, the lower frequency noise generated when lower supply voltages are used has several possible applications. For example, when V_{DD} is between −6 and −7 volts and the noise generator is coupled to an audio amplifier, the random clicks of a radiation counter can be simulated.

S2688/MM5837 Pink Noise Generator. Pink noise, which is required for room equalizing and other acoustical applications, can be produced by following an S2688/MM5837 with a −3 dB/octave low-pass filter. One such filter appears in National Semiconductor's *Audio Handbook* (Fig. 2.17.6, p. 2-56) and is shown connected to an S2688/MM5837 in Fig. 6. The pink noise produced by this generator contains equal amounts of energy in each octave of the audio spectrum from 20 Hz to 20 kHz. The output is about 1 volt of pink noise superimposed on an 8.5-volt dc level.

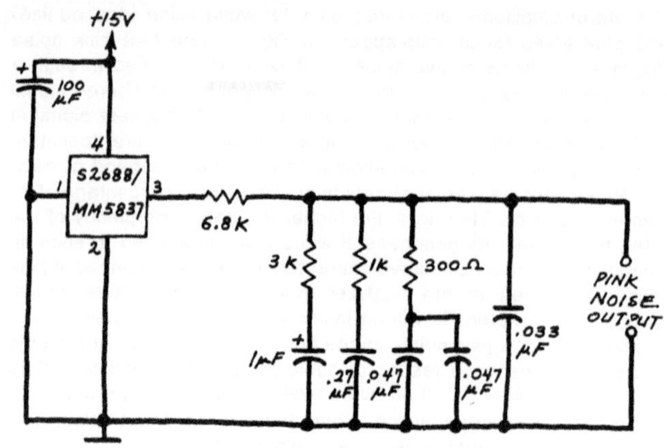

Fig. 6. Pink noise generator using S2688/MM5837.

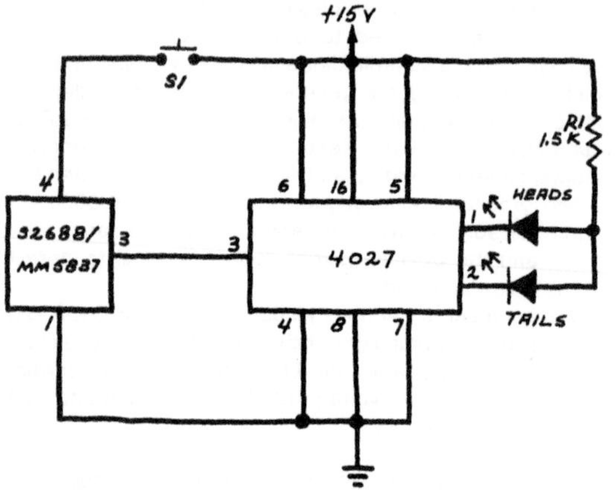

Fig. 7. Noise generator coin tosser.

Coin Tosser. Circuits that produce a completely random binary output are much in demand. Figure 7 shows a simple random-output circuit made from a noise generator and a 4027 flip-flop operated in its toggle mode.

Pressing *S1* applies power to the noise-generator chip and causes noise pulses to be applied to the flip-flop. The output LEDs appear to glow continuously even though they are rapidly being switched on and off by the noise pulses. Releasing *S1* turns off the noise chip. The logic states at the outputs of the 4027 then reflect the input status at the time that the noise is cutoff. Therefore, only one of the two LEDs glows.

Ideally, the output of the tosser should be completely random. With my circuit, however, in 100 tosses, green came up 56 times and red 44 times. I tried another 100 tosses and this time green came up only 43 times while red came up 57 times. These results seemed contradictory and not very random until I added them together. The result: In 200 tosses green glowed 99 times and red glowed 101 times. ■

8. An Op-Amp AM Radio

THE CIRCUIT shown schematically in Fig. 1 is a simple AM radio receiver. You can assemble it and start to receive radio stations within minutes. This receiver does not generate enough output to drive a loudspeaker, but that's a small price to pay for a circuit which derives *all* of its operating power from the radio signal it receives.

The major factor that limits the sensitivity of the receiver shown in Fig. 1 (which is an updated version of the old-fashioned crystal radio) is the barrier potential across the pn junction of the diode detector. For the circuit to respond, the received signal must exceed about 300 millivolts if *D1* is a germanium diode such as the 1N34 and a hefty 600 millivolts if *D1* is a silicon device like the 1N914. To maximize the input signal level, a good antenna and earth ground should be used. The antenna should be a length of copper wire at least 10 feet long positioned as high and in the clear as possible. A low-resistance connection to a coldwater pipe that extends deeply into the earth or to some other good ground will help the receiver gather as much signal power as possible.

The threshold effect imposed by the diode restricts reception to relatively powerful stations. If the forward voltage drop across the diode could be eliminated, the receiver would be able to demodulate any strength signal.

Russel Quong of Palos Verdes, CA, has found a simple way to reduce the voltage drop of a standard silicon diode from 600 millivolts to only one millivolt or so. Russel's idea, which was described in a brief note on page 148 of the July 20, 1978 issue of *Electronics*, is to substitute an op-amp precision half-wave rectifier for the standard diode. I've tried several versions of this basic idea, one of which is shown in Fig. 2. They all work well.

The circuit in Fig. 2 can be divided into four sections, the first comprising antenna coil *L1* and tuning capacitor *C1*. These components form a simple tunable filter which enables individual stations to be selected from a broad band of received frequencies.

The received signal is detected or demodulated by the second section, a half-wave rectifier formed by *IC1A*, *D1*, *D2*, and *R1*. The demodulated signal is then amplified by the third section, a high-gain driver amplifier consisting of *IC1B*, *C3*, *R2*, and *R3*. Potentiometer *R3* governs the gain of this amplifier and therefore serves as a volume control.

In the output stage, transistor *Q1* functions as a simple power amplifier for driving a small 8-ohm speaker or an earphone. Resistors *R4* and *R5* set the base bias for *Q1*, and *R6* limits current through the speaker.

You can assemble most of the circuit on a small solderless breadboard. Variable inductance *L1* is a standard loopstick antenna coil with an adjustable ferrite core, and *C1* is a miniature 0-to-365-pF variable capacitor. Both these components used to be widely available, but you might now have trouble finding them because the demand for radio parts has seemingly been reduced to a trickle. None of my catalogs list either part, but I've seen them in some shops.

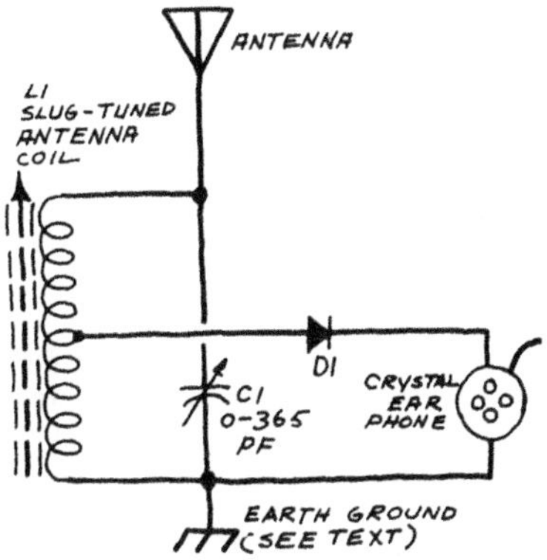

Fig. 1. Super-simple diode AM radio.

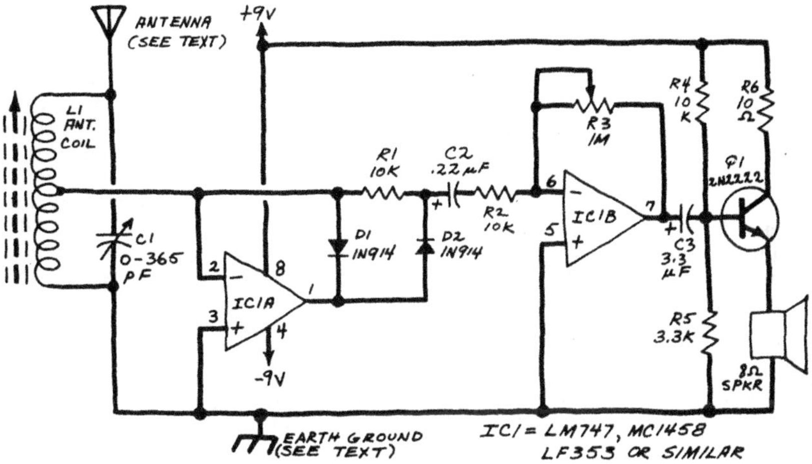

Fig. 2. An AM radio using an operational amplifier.

If you can't find *L1* and *C1* as new commercial items, salvage them from a transistor radio. The leads from the coil will be very fragile, so use care when disconnecting them from the radio's circuit board. One or more of the leads probably go directly to the tuning capacitor. If the coil has more than one tap, try all of them and use the one that gives best results.

If you can find an adjustable loopstick, substitute a fixed capacitor for *C1* if you prefer. Try values ranging from 100 to 250 pF. Higher values of capacitance will favor the low end of the AM broadcast band and lower values the high end.

In most areas, this radio will require an external antenna. If you live fairly close to several stations, you might find that a few feet of dangling copper wire will suffice. I live in a rural area 35 miles from each of several cities, and have had excellent results by clipping a short antenna lead to the dial stop on a rotary-dial telephone. This antenna allows my radio to pull in five stations with plenty of volume, and several others at somewhat lower levels. A good earth ground will help this receiver perform as well as it can. It is not as important to provide this receiver with a good earth ground as it is in the case of the "crystal" receiver described earlier, but this should be done if possible. ■

9. Programmable-Gain Amplifiers

HOW would you like to control the volume of a radio or gain of an amplifier with a series of switches instead of a knob? Or, better still, how would you like to perform this and other, similar functions digitally, perhaps under microprocessor or computer control? If so, read on!

The Operational Amplifier. Figure 1 shows a straightforward inverting amplifier built around a standard operational amplifier. The voltage gain of this circuit is the quotient of the value of the feedback resistor(R_F) divided by the value of the input resistor (R_{IN}).

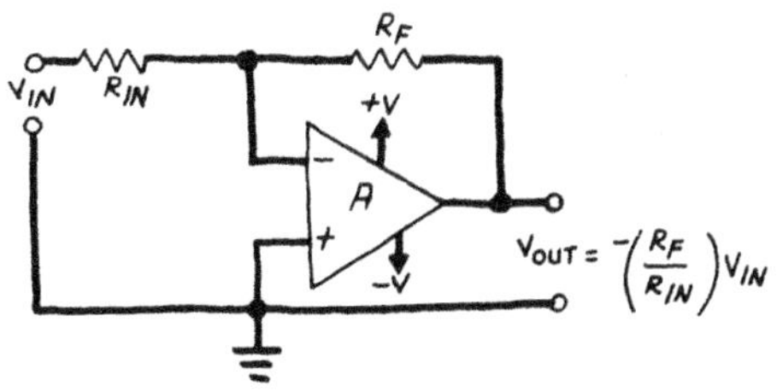

Fig. 1. An inverting operational amplifier.

The gain of the amplifier can be altered by varying the values of R_{IN} or R_F or both. In practice, it's usually best to keep R_{IN} fixed because changing it alters the amplifier's input impedance.

Potentiometers are customarily used when it's necessary to make either R_{IN} or R_F variable. Unless expensive ten-turn or

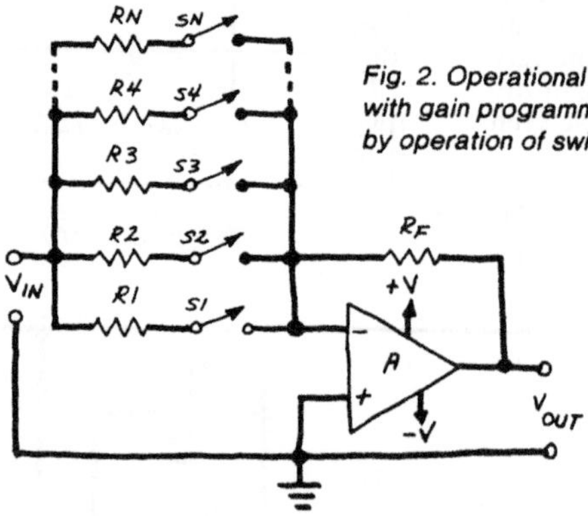

Fig. 2. Operational amplifer with gain programmable by operation of switches.

detented potentiometers are used, this makes accurate changes in gain difficult. The output of the amplifier must be monitored with an oscilloscope or digital voltmeter while the adjustment is made.

One way to provide selectable fixed, and therefore repeatable, gain settings is to use a parallel network of resistor/switch pairs as shown in Fig. 2. By preselecting the resistor values, any desired gain setting can be achieved by simply closing the appropriate switch. If R_F and *R3* in Fig. 2 are

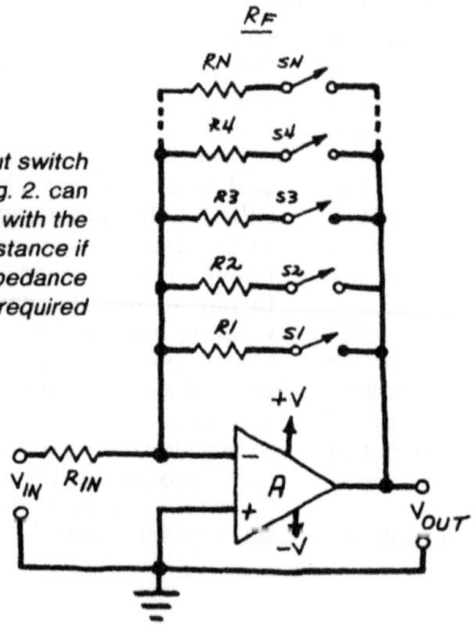

Fig. 3. The input switch network of Fig. 2. can be interchanged with the feedback resistance if constant input impedance is required

10,000 and 1,000 ohms respectively, then the voltage gain will be −10 when only *R3*'s switch is closed. The polarity of the gain figure is negative due to the fact that the amplifier operates in the *inverting* mode.

The input network and R_F can be interchanged (Fig. 3) if a constant input impedance is required. In this case, if R_{IN} is fixed at 10,000 ohms and *R4* is one megohm, the voltage gain will be −1,000 when only *R4's* switch is closed.

Simplified Programming. As more gain settings are required, the circuits just presented become impractical because more and more resistors and switches become necessary. One way to reduce circuit complexity while greatly increasing the number of available gain steps is to replace the custom-selected resistor network with a binary-weighted resistor network. Figure 4 shows a four-level network connected in place of R_{IN}

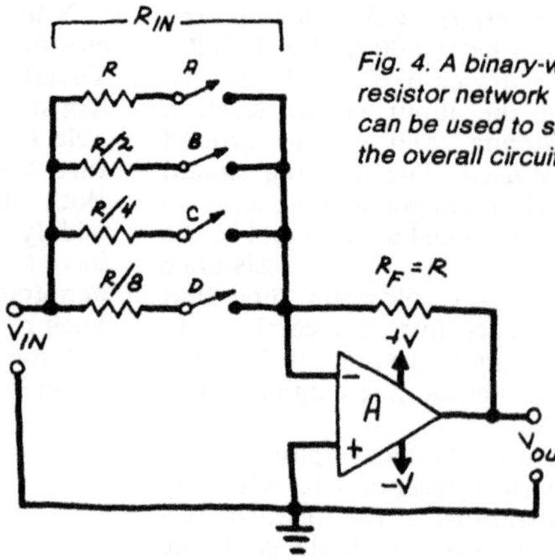

Fig. 4. A binary-weighted resistor network can be used to simplify the overall circuit.

in an op-amp circuit. Closing various combinations of switches will provide gains of each integer from 1 to 15. The previous circuits would require fifteen resistor/switch pairs to provide this same number of gain steps.

More resistors and switches can be added to provide progressively more gain steps. Eight resistor/switch pairs, for example, will provide 255 gain steps!

It's helpful to understand the operation of the binary-weighted resistor network in Fig. 4. Although its input network has four levels, we can understand how it works by referring to the simpler, three-level version shown in Fig. 5.

First, remember that the total resistance of two or more resistors in parallel is the reciprocal of the sum of the reciprocals of the individual resistors, or $R_{TOTAL} = 1/(1/R1 + 1/R2 + 1/R3 \ldots + 1/RN)$. Knowing this, we can make a table that shows the resistance given by each of the eight possible switch combinations. In this table, a zero represents an open switch and a one a closed switch.

Switch C	B	A	Resistance
0	0	0	----
0	0	1	R
0	1	0	R/2
0	1	1	1/(1/R + 2/R) = R/3
1	0	0	R/4
1	0	1	1/(1/R + 4/R) = R/5
1	1	0	1/(2/R + 4/R) = R/6
1	1	1	1/(1/R + 2/R + 4/R) = R/7

As you can readily see, the seven switch combinations give seven different resistances of descending value. We can make practical use of this table by assigning a value to *R*. If, for instance, *R* is 40,000 ohms, we obtain:

Switch C	B	A	Resistance
0	0	0	----
0	0	1	R = 40,000
0	1	0	R/2 = 20,000
0	1	1	R/3 = 13,333
1	0	0	R/4 = 10,000
1	0	1	R/5 = 8,000
1	1	0	R/6 = 6,666
1	1	1	R/7 = 5,714

You can verify the accuracy of this table by actually calculating the value for each step. You should arrive at the same result, whether you divide each of the integers 1 through 7 into 40,000 or use the formula for the total resistance of resistors in parallel.

Now that you know how the binary-weighted resistor network works, let's assume that it's connected as the input resistance in a straightforward inverting amplifier like the one shown in Fig. 1. If R_F is made equal to *R*, then the amplifier will have switch-programmable gains ranging from −1 to −7 in equally spaced increments.

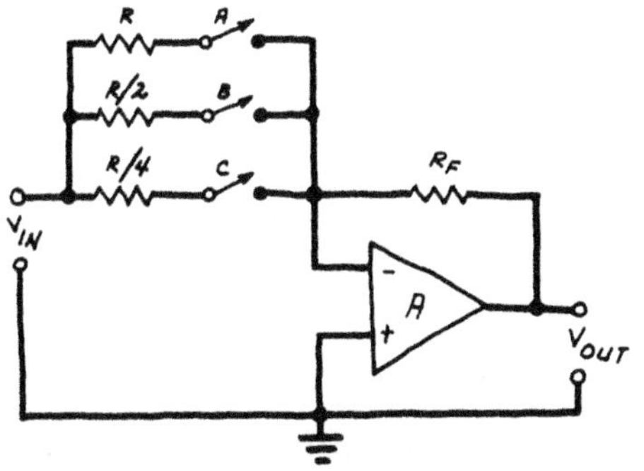

Fig. 5. Three-level binary-weighted input network.

It should now be obvious to you how the circuit shown in Fig. 4 provides a switch-programmable gain of from 1 to 15. Add another resistor-switch pair to the network and the available gain will range from −1 to −31. The use of eight resistor-switch pairs results in switch-programmable gains from −1 to −255.

This method of programming the gain of an op amp has a few disadvantages of which you should be aware. One is that the value of the resistor in the most significant switch position becomes progressively smaller as the number of resistor/switch pairs is increased. This places a practical limit on the maximum number of resistors that can be added since the maximum available (open-loop) gain of the op amp cannot be exceeded.

The second disadvantage also concerns the most significant switch position and is of more importance. Since the resistance of the most significant resistor is much less than that of the least significant switch position, its tolerance is much more critical. For best results, one-percent or closer tolerance resistors should be used.

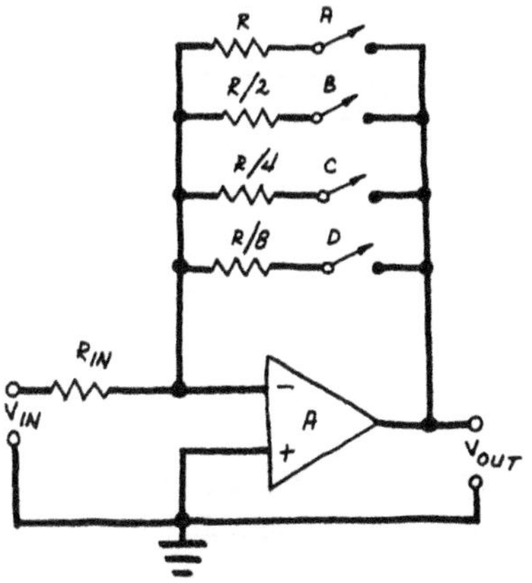

Fig. 6. Using a binary-weighted resistor network in the feedback circuit.

Thus far, we've considered the use of the binary-weighted network as the input resistance. Of course, the network can also be used as the feedback resistance as shown in Fig. 6. The advantage of this configuration is that the amplifier's input impedance remains constant for the various gain settings. However, the disadvantage is that the uniform, stepped gains of the previous circuit are not available. If R_{IN} is 1 ohm, for example, the gains range from −5,714 to −40,000. This circuit thus does not provide uniformly incremental gains, but it does give plenty of range for such an application as audio amplification.

Adding Digital Control. A straightforward way to provide digital control of the circuits that have been presented is to replace the conventional, manual switches with CMOS analog switches. Figure 7 shows how to use a CD4066 to replace all four mechanical switches in the circuit of Fig. 4. Keep in mind that the "on" resistances of the four analog switches in the CD4066 should be subtracted from the values of the input-network resistors to calculate the actual gains such a circuit would provide.

Barry B. Woo of Fluke Automated Systems, Inc. has written an excellent paper on various methods of adding digital control to a programmable-gain amplifier. His paper, "Digitally Programmable Gain Amplifiers with Arbitrary Range of Integer Values," appears in the *Proceedings of the IEEE* (July 1980, pp. 935-936). You can find this journal at most well-stocked technical libraries.

Other Applications. The emphasis in this column has been amplification, but the techniques which have been presented can be used to implement such interesting applications as a digitally controlled potentiometer with no moving parts. Other applications include analog-to-digital and digital-to-analog conversion.

You might want to explore some of these applications on your own. If so, you might also want to experiment with an interesting alternative to the binary-weighted resistor networks described here. The *R-2R* network requires more resistors, but it uses only two resistor values and is more tolerant of resistance variations.

The *R-2R* network was the subject of the July 1978 "Experimenter's Corner." It is also covered in a very useful book by Sol Libes entitled *Fundamentals and Applications of Digital Logic Circuits* (Hayden, 1975). Mr. Libes gives a concise treatment of the subject, complete with circuits, on page 136.

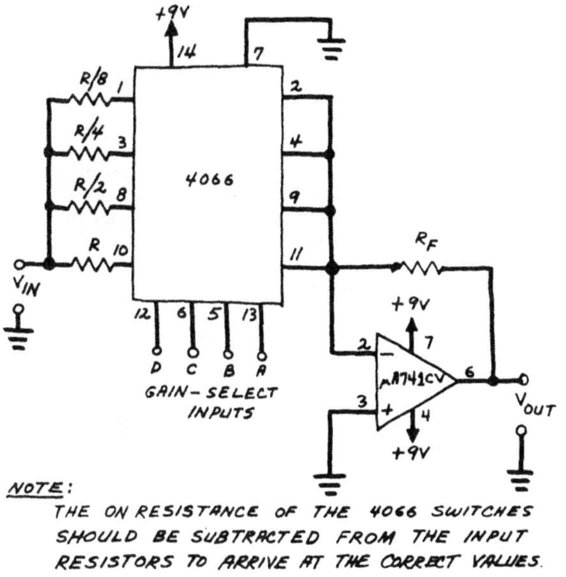

Fig. 7. How to use a CD4066 to replace mechanical switches.

10. Remote Sensing, Part 1

ALL matter emits, absorbs, and reflects different wavelengths of electromagnetic radiation, each type in its own unique fashion. The combination of these properties provides an electromagnetic "signature" which permits various kinds of sensors to identify unknown matter from afar.

The study of electromagnetic signatures has given rise to the method of observation and measurement called *remote sensing*. In this two-part discussion, we will consider several types of remote sensing and describe the assembly of some circuits which can identify a portion of the electromagnetic signatures of various man-made and natural objects.

Elements of Remote Sensing. In its broadest sense, remote sensing is the perception of an object from a distance by means of a suitable sensing device. By this definition, observation devices ranging from the human eye to telescopes, cameras, and spectroradiometers are remote sensors.

Effective remote sensing usually requires that the sensor be capable of distinguishing a range of wavelengths emitted by, reflected from or transmitted through the object or matter being sensed. Photography, an early and still important remote-sensing method, provides a good illustration of the importance of spectral sensitivity.

A black-and-white aerial photograph, for example, displays as various shades of gray all the wavelengths to which the film is sensitive and which are emitted from or reflected by the matter within the camera's field of view. Such a black-and-white photograph can convey a considerable amount of information, but a *color* aerial photograph simplifies the location of man-made structures and can even permit the identification of various kinds of vegetation.

The use of black-and-white film in remote sensing can be made more productive by exposing the emulsion through a narrow-bandpass optical filter. Several such black-and-white photos of the same scene, each exposed through a different filter, can be superimposed to provide as much or even more information than a color photograph.

Another important form of remote sensing was developed in the last century when astronomers used glass prisms and diffraction gratings to break up the light from distant stars into its component parts. Hot gasses, whether on earth or in a star, emit characteristic wavelengths of radiation. Astronomers learned how to determine the composition of stars by analyzing the spectra of their emissions. *Spectrometry*, as this research tool is called, has become far more sophisticated in recent years due to the development of more sensitive equipment. The basic technique, however, remains unchanged.

Remote-Sensing Methods. Figure 1 illustrates four important remote-sensing methods. The *passive reflection* method is probably the most widespread. In this method, the object being observed, called the *target*, reflects ambient radiation emitted by the sun or a nearby source of artificial light. The *active reflection* method is more specialized, because in it the target is illuminated by a source of artificial light designed specifically for this purpose. In both of these methods, target characteristics can be determined by the way various wavelengths are reflected.

The *emission* method is totally passive and relies only upon radiation emitted by the target. So long as the detector is sufficiently sensitive, this method will detect *anything* from ice cubes to stars, for all matter at a temperature greater than absolute zero emits electromagnetic radiation. It permits the temperature of a target to be determined from afar. Also, if the target is heated to incandescence, the characteristic spectral lines it emits permit its constituents to be identified.

The *transmission* method, which is commonly used to detect dust, gases, precipitation and other matter in the earth's atmosphere, requires a separate source and detector. Sometimes, the source can be the sun or some other natural or artificial light source suitably placed with respect to the target. Generally, however, the source is designed specifically for the purpose. In any case, the target can often be identified by the way it absorbs and scatters different wavelengths of light.

Many variations on these basic methods are possible. For example, the active-reflection method may employ a wide-band, "white" light source and a single detector, before which various narrow-bandpass filters are placed. While the detector is pointed at a fixed target (which is illuminated by the source), detector-output measurements are taken each time a different filter is moved into position.

Alternatively, the filters can be placed before the light source. Another variation is to eliminate the filters entirely and illuminate the target with a narrow-band source such as a laser. If necessary, sunlight or artificial ambient light can be blocked by a suitable narrow-bandpass filter placed in front of the detector.

As you can see, there are many ways to implement remote sensing. We will first consider active- and passive-reflection remote sensing, two methods which are also known as *reflection spectroscopy*.

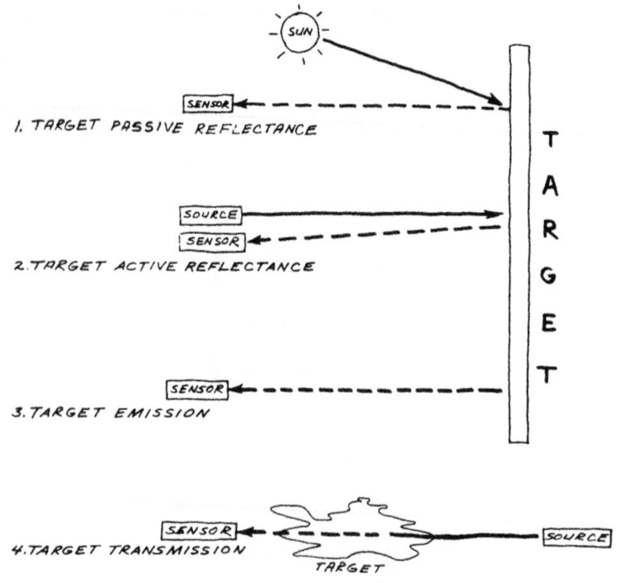

Fig. 1. The principal kinds of remote sensing.

Reflection Spectroscopy. I first became interested in reflection spectroscopy while evaluating the performance of various kinds of infrared travel aids for the blind. To predict the range of such a device, it is necessary to know the optical reflectance of many different materials at the wavelengths employed by the aid. The most common wavelengths are 880 and 940 nanometers in the near infrared. Very efficient, powerful LEDs which emit radiation at these wavelengths are readily available.

Figure 2, for example, shows the spectral reflectance of a typical green leaf. This plot nicely illustrates how well such leaves reflect incident radiation at near-infrared wavelengths. It also shows that a leaf has a distinctive *reflection signature*. Chlorophyll, the key chemical constituent of green plants, readily absorbs blue and red wavelengths. The small peak in reflectance at 550 nanometers produces the characteristic coloration of photosynthetic plant life.

The much larger peak beyond 700 nanometers is in the near infrared, and is therefore invisible to the unaided eye.

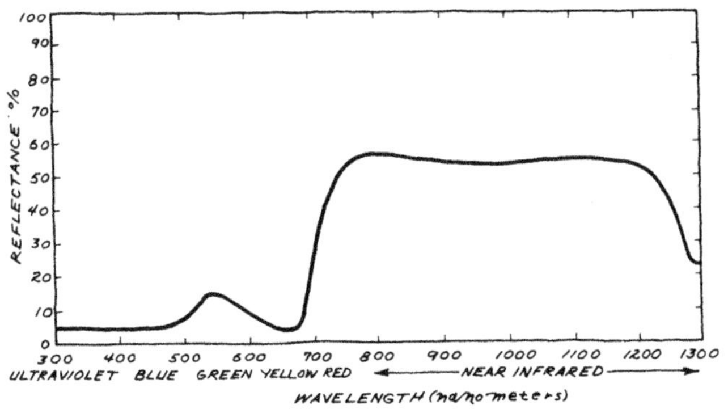

Fig. 2. Spectral reflectance of a typical green leaf.

This spectral region, however, is readily detectable by means of infrared film or an infrared image converter such as a starlight scope. The peak explains why vegetation appears bright white on infrared film or when viewed through an infrared-to-visible-light image converter.

The large difference in reflectance values at 650 nanometers and in the 750-to-1200-nanometer band means that a GaAsP red LED and an (AlGa)As or GaAs:Si near-infrared emitter can be used in their light *detector* (either reverse-biased or photovoltaic) mode as the heart of a detector circuit which indicates the presence or absence of green vegetation As you may recall from previous columns, LEDs make excellent narrow-band detectors.

The same result can be achieved with two silicon detectors, one covered with a 600-to-670-nanometer bandpass filter, the other covered with a near-infrared bandpass filter. This approach, however, is more expensive—suitable filters may cost as much as $50 or more.

Using LEDs to Detect Vegetation. To determine if a GaAsP LED could be teamed with a GaAs:Si LED to detect green vegetation, I tried an experiment that you might want to duplicate. In this experiment, I used the variable-gain operational amplifier shown in Fig. 3.

First, a GaAsP LED enclosed in a clear (*not* diffuse) red encapsulant was connected to the input of the amplifier. The diode was then pointed at the white side of a Kodak Neutral Test Card (available at most camera stores) which was illuminated by a single incandescent lamp. All other lights were extinguished.

The Kodak test card has a reflectance of 90 percent in the visible and near-infrared regions of the spectrum. Therefore, I adjusted *R1 and* the distance of the light from the card to achieve a meter reading of 0.9 milliamperes. I then removed the card and placed a fresh leaf from a Japanese ligustrum in its place. The meter indicated 0.05 milliamperes, which signifies a reflectance of 5 percent.

I then repeated this procedure with the GaAs:Si LED. This time, I measured a reflectance of 52 percent. Both reflectance measurements coincide well with published values. And though GaAsP LEDs are less sensitive than GaAs:Si LEDs, this simple experiment proved that *both* LEDs can be used in tandem to measure the reflectance of green leaves.

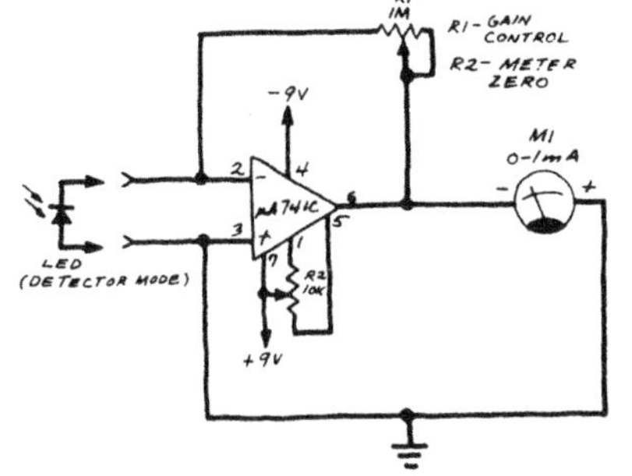

Fig. 3. Simple green detector circuit as described in text.

A Practical Green-Leaf Detector. The "truth table" for a dual-wavelength, leaf-signature detector is:

Reflectance

A 650 nm (red)	B 940 nm (near infrared)	Leaf present?
Low	Low	No
Low	High	Yes
High	Low	No
High	High	No

The Boolean function for this table for an active low output is

$$(\overline{A \cdot \overline{B}})$$

Figure 4 shows a practical green-leaf detector circuit. As in the previous experiment, two LEDs function as narrow-band wavelength detectors.

In operation, each LED is reverse-biased and connected to a series resistor (*R1* and *R5*) across which a voltage drop appears when light striking the LED generates a photocurrent. The series resistor for the GaAsP LED (*R1*) has a much larger resistance than that for the GaAs:Si LED (*R5*) because the GaAsP LED is not as sensitive to light.

The output voltages generated by the two LEDs are applied to the inverting inputs of two comparators (*IC1A* and *IC1B*). When the light level is sufficiently high, the outputs of the two comparators go low. Otherwise the outputs remain high.

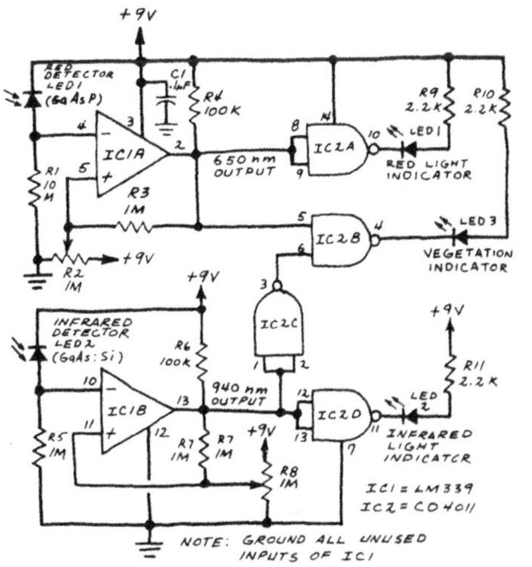

Fig. 4. Circuit for remote sensing of green vegetation.

The required Boolean function for the truth table is implemented by two of the four NAND gates in a 4011 (*IC2B* and *IC2C*). When both detectors receive reflected light from a leaf illuminated by sunlight or a bright incandescent lamp, the output of comparator *IC1A* stays high and that of *IC2B* goes low. This combination is decoded by NAND gates *IC2B* and *IC2C*, and VEGETATION INDICATOR *LED3* glows. The LED is dark for any other combination of inputs.

For preliminary work, assemble the circuit in Fig. 4 on a solderless breadboard. The two detector LEDs should be mounted next to one another and installed in an opaque housing such as a short length of heat-shrinkable tubing.

Before it can be used, the circuit must be calibrated. The calibration procedure is greatly simplified by the three LEDs. Begin by rotating the wiper of trimmer potentiometer *R2* until *LED1* just begins to glow. Then rotate the wiper of *R8* until *LED2* just begins to glow. Indicator *LED3* should now be dark.

Now place a white card a few centimeters from the two detector LEDs, and illuminate the card with a bright incandescent lamp or sunlight. Both *LED1* and *LED2* should darken, and *LED3* should remain dark. If either *LED1* or *LED2* remains on, adjust the appropriate potentiometer until the LED functions properly. If this fails to solve the problem, make sure the fields of view of the chips in each LED are not blocked by the edge of the opaque tube. Also, make sure light is not entering the rear of the opaque tube and illuminating the LEDs from behind.

The circuit is now ready for use. Leave the light source in place and remove the white card. Both *LED1* and *LED2* should glow. Then place a fresh green leaf where the card was located. Indicator *LED1* should continue to glow, and *LED2* should darken. Diode *LED3* will glow to indicate the presence of a leaf.

With careful adjustments, the circuit will detect a single leaf up to 10 centimeters away. Under the proper conditions, trees and shrubs illuminated by bright sunlight can be detected over much greater distance. ■

Remote Sensing, Part 2

Leaves, as you might recall, reflect red light poorly but reflect near-infrared radiation very well. This generates a characteristic reflectance signature which makes it possible to use a red LED and a near-infrared LED as a pair of narrow-band radiation *detectors*. This is done in the leaf-detector circuit described last month in Part 1 of this series.

NASA's Image Classification Circuit. An expanded version of the leaf-detector circuit has been developed for NASA's Langley Research Center by Roland L. Hulstrom, Roger T. Schappell and John C. Tietz of the Martin Marietta Corporation. Like the circuit I described, NASA's circuit also teams a red sensor and a separate near-infrared sensor to detect green vegetation. Moreover, these two detectors also permit the detection of water, bare land, clouds and snow.

Figure 1 is the schematic for this new circuit as given in a recent NASA Tech Brief. The circuit, an expanded version of which is slated to be flight-tested aboard one or more Space Shuttle missions, is designed to automatically reduce the quantity of unwanted imagery transmitted to earth from camera-carrying earth satellites.

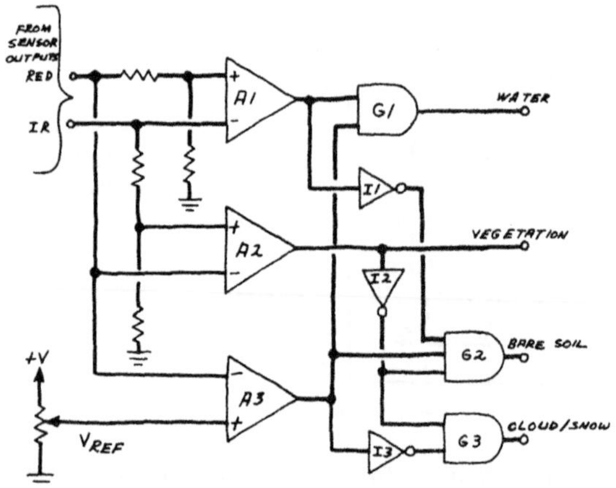

Fig. 1. Earth satellite picture classification circuit.

NASA explains the objective behind the design of the new circuit as follows. Earth-observation satellites generally do not make decisions about the usefulness of the data being sent to earth. As a result, a significant amount of time and money is spent in sorting out the useful data. A great saving could be realized if circuits aboard the satellite could recognize useless imagery or actually look for specific features. The circuits do not have to be very smart to be useful. For example, about 70% of the earth's surface is water. Of the 30% that is not water, about one-third to one-half will be obscured by clouds at any given time.

This means that a satellite might get one clear picture of land out of perhaps five or six observations. The amount of unwanted, more or less useless data that is stored, processed and indexed could, therefore, be greatly reduced by a circuit that simply blocked transmission of the 80% of the images that is of water and clouds.

A simple circuit has been developed to classify picture elements by spectral signature alone. No pattern recognition is required. Computer simulations and field measurements have confirmed that the four basic features—vegetation, bare land, water and clouds or snow—can be separated by radiance measurements at two discrete wavelengths: 650 and 850 nm.

It's very significant that the reflectance signatures of four key topographic features can be classified by examining only two wavelengths of their reflected radiation. From last month, you already know that green vegetation has a very low reflectance at 650 nanometers—typically less than 5 percent. At 850 nanometers in the near infrared, the reflectance of vegetation is typically from 45 to 55 percent.

Soil usually has a higher reflectance at near-infrared wavelengths than in the visible portion of the spectrum. The transition between low and high reflectance is more gradual than for vegetation, and occurs in the visible region. This means that the difference in soil reflectance at 650 and 850 nanometers is not as dramatic as it is for green leaves.

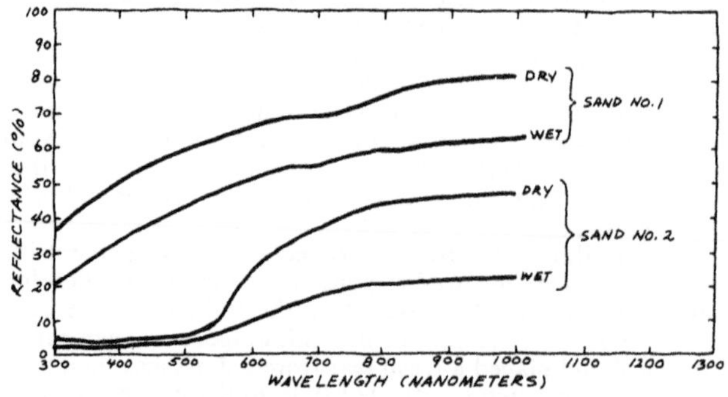

Fig. 2. Spectral reflectance of two different sands.

Figure 2 shows the reflectance curves of two highly reflective soils (actually, sands). Sand number 1 is white beach sand from Ft. Walton Beach, Florida. Sand number 2 is a darker sand from Monument Valley, Utah. Note that both sands, like all other soils, reflect less light when they are wet. These reflectance curves, and many others, can be found in "The Spectral Reflectance of American Soils" by H. R. Condit (*Photogrammetric Engineering*, Sept. 1979).

Water's reflectance at 650 and 850 nanometers is the reverse of that of leaves, because water reflects red light but absorbs near-infrared wavelengths. Clouds and snow have much higher reflectances than soil, but the differences in reflectance at 650 and 850 nanometers are similar to that of some soils.

Remarkably, the two wavelengths selected by NASA for its Image Classification Circuit are very close to the optimal detection regions of the GaAsP LED (650 nanometers) and the new (AlGa)As "super" LED (880 nanometers). A practical version of NASA's circuit can be made by using two such LEDs as detectors. The green-leaf detector circuit described last month shows how LEDs can be coupled to a circuit like the one shown in Fig. 1.

A detailed report on NASA's image-classification circuit is available for $6.00 (paid in advance) from the National Technical Information Service, Springfield, VA 22161. The report is entitled "Experimental and Simulation Study Results for Video Landmark Acquisition and Tracking Technology" (NASA CR-158997). Request the publication by name and by the identification number LAR-12589.

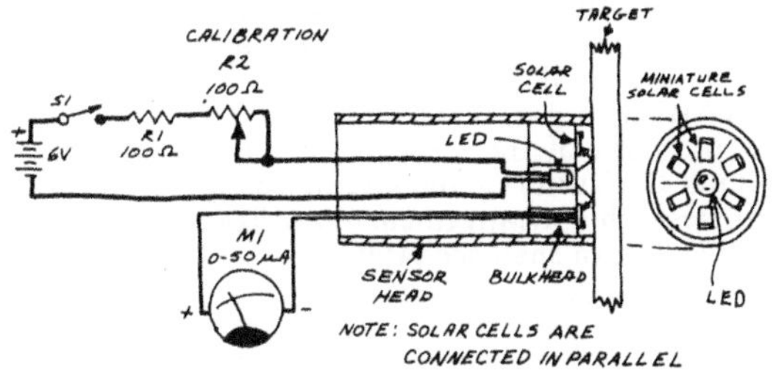

Fig. 3. Construction of a simple, low-cost reflectometer.

An Inexpensive Narrow-Band Reflectometer. Sometimes, it is important to know the reflectance of an object at only one wavelength. Since 1970, I have measured the reflectance at 940 nanometers of scores of different objects. These measurements make possible the accurate prediction of the detection range of various infrared travel aids for the blind.

Soon, I plan to repeat many of these measurements at the 880-nanometer wavelength emitted by the new, (AlGa)As high-power emitters. I will use the simple reflectometer illustrated in Fig. 3. This ultrasimple system requires no sophisticated electronics. The reading on the 0-to-50-μA meter is doubled to obtain the target's reflectance in percent.

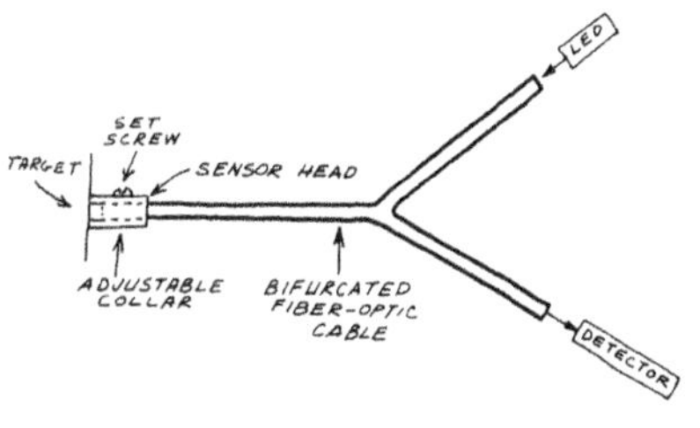

Fig. 4. How to build a fiber-optic reflectometer.

To make a reflectance reading, the sensor head is first placed against a Kodak photographic test card or a similar target with a known reflectance. The CALIBRATION potentiometer is adjusted until the reference target's reflectance is indicated on the meter. If, for example, the reference target has a reflectance of 90 percent (as does the Kodak test card), the CALIBRATION control should be adjusted for a meter reading of 45 μA.

You need not duplicate exactly the arrangement shown in Fig. 3 to make a working reflectometer. To reduce erroneous readings to a minimum, the solar cells should be mounted in a ring around the source LED. It is important that ambient light be kept away from both the target and the solar cells when measurements are being taken.

When making a measurement, place the sensor head firmly against the target. The output of the LED will fluctuate with changes in temperature and battery voltage, so the circuit should be recalibrated just before each reading is taken.

L. A. Lott and D. L. Cash have described a more sophisticated reflectometer in a paper entitled "Spectral Reflectivity Measurements Using Fiber Optics," which appeared in the April 1973 issue of *Applied Optics* (pp. 837-840). In their device, one branch of a "Y"-configured, *bifurcated* fiber-optic cable carries light to the target. The reflected light is carried through the second branch of the cable to a detector. The low light levels involved necessitate the use of a detector amplifier.

I've assembled such a fiber-optic reflectometer, and it works quite well. The small size of the sensor head means that the reflectance of very small objects, or different parts of the same object, can easily be measured. Figure 4 is a simplified diagram of such a device. See Lott and Cash's paper for more detailed information.

Remote Sensing of Water Vapor. If you read the "Solid-State Developments" in the February 1981 issue of this magazine, you may recall that 940-nanometer radiation is strongly absorbed by water vapor in the atmosphere, but that absorption at 880 nanometers is negligible. This provides a characteristic signature which permits the remote sensing of water vapor by dual-wavelength *transmission spectroscopy*.

Figure 5 is a simple circuit I've designed to demonstrate this method of detecting water vapor. It is a dual-wavelength transmission spectrometer with an audio output.

In operation, a GaAs:Si 940-nanometer emitter and an (AlGa)As 880-nanometer emitter are both pointed at a silicon phototransistor that drives an amplifier. The two LEDs are alternately driven by pulses with a duty cycle of 50 percent that are generated by an astable multivibrator made from two of the four NAND gates in a 7400.

Initially, the receiver will generate a tone coinciding with the pulse rate at which the LEDs are driven. The position of the silicon detector is then adjusted to null out the tone.

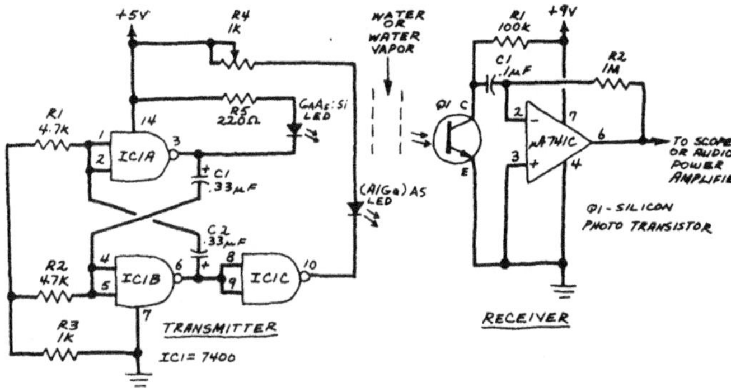

Fig. 5. Dual-wavelength water-vapor detector.

Operation of the circuit can be demonstrated by placing a small, transparent container between the two LEDs and the detector. If necessary, align the detector to cancel any tone output from the receiver. When water is poured into the container, radiation from the 940-nanometer LED will be suppressed, but that from the 880-nanometer emitter will be largely unaffected. Consequently, the null condition will be disturbed and a tone will be emitted by the receiver.

This simple circuit proves that an 880-nanometer LED can be teamed up with a 940-nanometer LED to detect water. Detecting water *vapor* is more difficult, but it can be done. One way to demonstrate the detection of water vapor is to allow steam to pass between the two LEDs and the silicon detector. More sophisticated versions of this dual-wavelength circuit are possible, but I will leave their design to those of you interested in remote sensing. ■

11. Experimenting with a Joystick, Part 1: Basic Concepts and Applications

JOYSTICKS are used to provide an interface between an operator and radio-controlled airplanes, video games, computers, audio systems and many automated industrial systems. In the past, joysticks were rather expensive, and only a small number was sufficient to supply those hobbyists and experimenters able to afford them. Increased production to meet the demand of video-game makers and the use of more plastic have brought the single quantity price as low as $5.00 for some models. Inexpensive units are now available from several of the electronic parts suppliers that advertise in POPULAR ELECTRONICS.

Basic Concepts. A typical joystick consists of two potentiometers installed in a boxlike assembly as shown in Fig. 1. A two-axis mechanical linkage allows the rotation of each potentiometer to be controlled by a single, movable rod (the stick). Some joysticks include four potentiometers that operate in two ganged pairs.

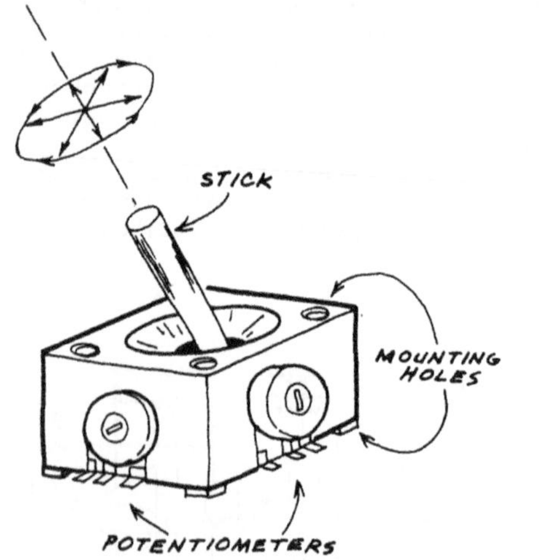

Fig. 1. Three-dimensional representation of a typical two-axis joystick.

As you can see in Fig. 2, moving the control stick up and down rotates only *R1*'s shaft. Moving the stick left and right rotates only *R2*'s shaft. When the stick is moved in any other direction, the shafts of both potentiometers are rotated. In short, the resistances of *R1* and *R2* are functions of the position of the stick.

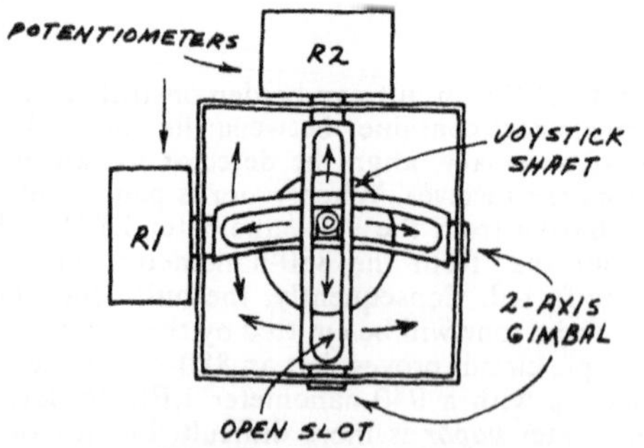

Fig. 2. Diagram of internal construction of a typical low-cost joystick.

Depending upon the application, various methods may be employed to connect the potentiometers of a joystick into a working circuit. Figure 3, for example, shows how the two potentiometers can be connected to form a two-stage voltage divider. A single potentiometer is essentially a one-stage voltage divider. Cascading two potentiometers as shown provides a wide range of output voltages for various positions of the stick.

Figure 4 shows some of the voltages I measured for various positions of a low-cost joystick (Radio Shack 271-1705). Each of the potentiometers in this joystick has a resistance of 100,000 ohms. The input voltage was 5.5 volts. So long as the resistances of the two potentiometers are equal, other joysticks will give similar results.

Examination of Fig. 4 shows that the output voltages for various positions of the stick are not necessarily exclusive. This limits the utility of the circuit configuration in Fig. 3. Nevertheless, this arrangement does have some interesting applications as we shall see later.

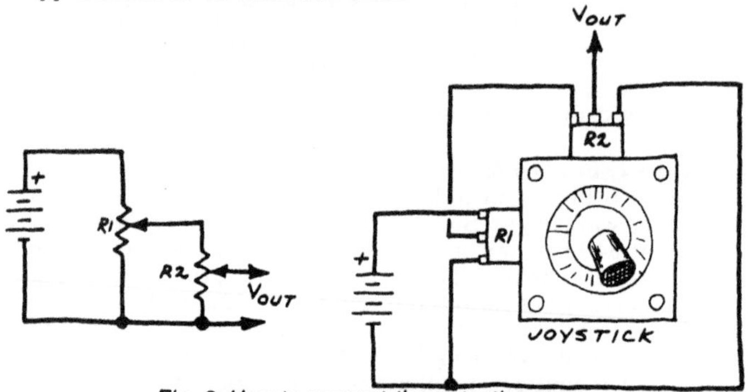

Fig. 3. How to connect the potentiometers in a joystick as a two-stage voltage divider.

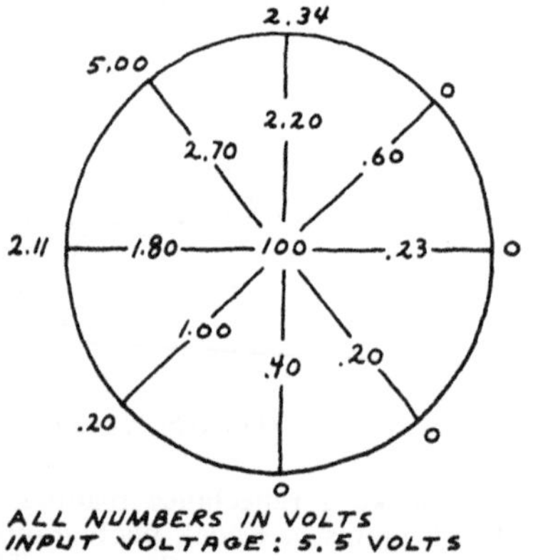

Fig. 4. Output voltage versus stick position for 2-axis joystick.

A Single-Axis Joystick. Even if you do not yet have a joystick, you can begin experimenting with circuit techniques by converting a standard potentiometer into a single-axis joystick. Figure 5 shows a simple way to accomplish the transformation with the help of two short lengths of wood dowel and a single 6-32 set screw.

Figure 6 shows a 1-of-10 controller circuit ideally suited for use with a single-axis joystick. The circuit consists of an LM3914 LED dot/bar generator. The joystick (*R1*) provides a variable voltage input to the LM3914.

In operation, the outputs of the LM3914 are sequentially enabled as the voltage at pin 5 is increased. This voltage, of course, is a function of the position of *R1*'s shaft.

The circuit in Fig. 6 includes output indicator LEDs. Resistor *R2* provides current limiting for all the LEDs. The outputs can drive SCRs, TRIACs, small relays or external logic.

If *R1* is a linear taper potentiometer, it can be effectively converted to a logarithmic taper potentiometer by substituting an LM3915 for the LM3914. The LM3915 is functionally identical to the LM3914 except it provides a logarithmic output in which each output level is separated from its adjacent levels by 3 dB.

If you want to convert the 1-of-10 output from the LM3914 or LM3915 into binary coded decimal (BCD), connect the outputs to a 74147 priority encoder. This will provide true analog-to-digital conversion for the joystick.

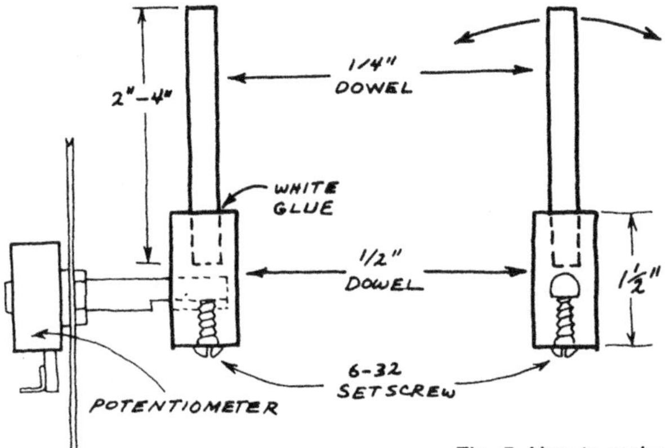

Fig. 5. How to make a one-axis joystick.

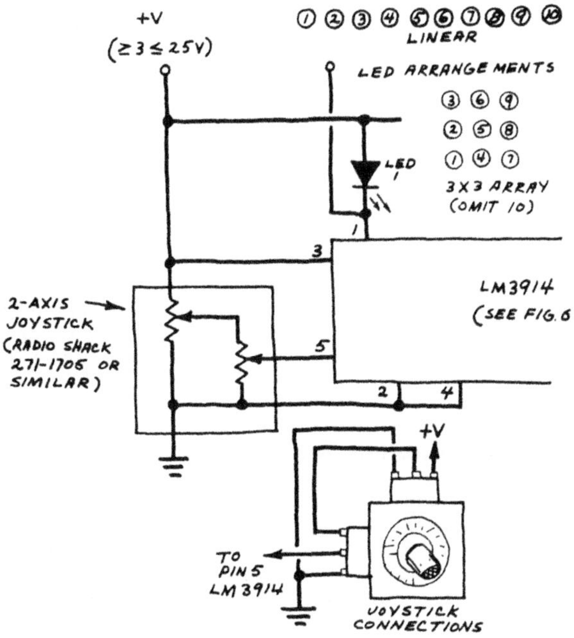

Fig. 7. How to add a two-axis joystick to the circuit shown in Fig. 6.

A Two-Axis Joystick Controller. Figure 7 shows how a two-axis joystick can be connected to the one-axis controller shown in Fig. 6. The result is a highly flexible, combination narrow- and wide-range controller system which is fully adjustable with a two-axis joystick.

This application utilizes the cascaded voltage divider circuit shown in Fig. 3. It requires that the LM3914 range potentiometer (*R3* in Fig. 6) be adjusted so that the lowest and highest order LEDs are off when the joystick is at its extreme lower left and upper right positions.

In this application the output LEDs provide important visual feedback which informs the operator exactly what is happening. The LEDs can be arranged in a row or in a 3 x 3 square array as shown in Fig. 7. In either case, moving the stick from full lower left to full upper left sequentially activates LEDs 1-3. Moving the stick from lower center to upper center sequentially activates LEDs 1-6. And moving the stick from lower right to upper right sequentially activates LEDs 1-9. Depending upon *R3*'s adjustment (see Fig. 6), these results may vary by one or two LEDs.

For best results, a square template that restricts the movements of the stick may be necessary. You can make a template by cutting a square aperture in a piece of plastic or card stock. The joystick I used has four mounting holes to which the template could be attached with self-threading screws supplied with the joystick. You can determine the approximate dimensions of the aperture in the template by using strips of tape to restrict the movement of the stick while monitoring the results.

As in the single-axis joystick controller in Fig. 6, the LM3914 outputs can control external logic, various solid-state switches or relays. An interesting possibility is to use a 74147 priority encoder to convert the results to BCD for digital processing. ■

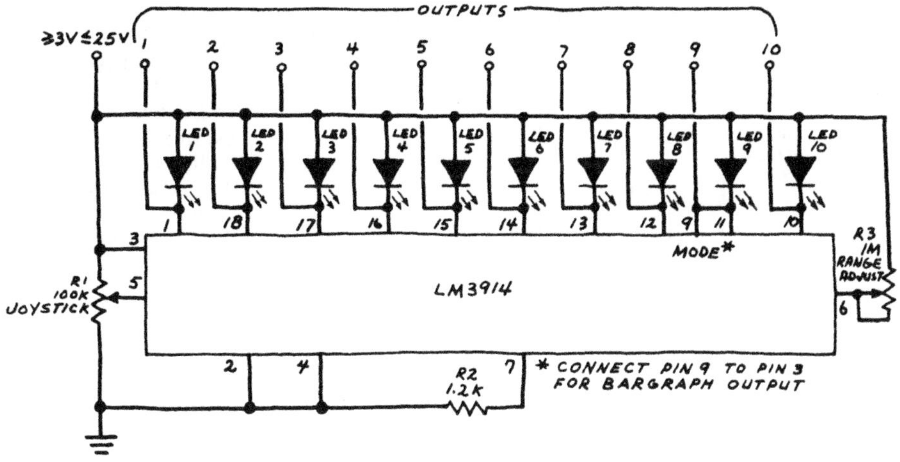

Fig. 6. Schematic diagram of a single-axis joystick controller using an LM3914 LED dot/bar generator chip.

Experimenting with a Joystick, Part 2: Some Typical Applications

The joystick I used with the experimental ciruits we'll be discussing, each of which I assembled and tested, contains two 100-kΩ linear-taper potentiometers. It's available from Radio Shack for $4.95 (catalog number 271-1705). Other joysticks may also be used. Those containing logarithmic-taper potentiometers, for example, are usually better suited for audio control applications. Other sources for joysticks include some of the mail order electronic parts supply companies that advertise in this magazine. You can also salvage used joysticks from discarded video games and remote-control transmitter systems.

A Joystick-Controlled Mixer. A *summing amplifier* allows two or more signals to be simultaneously amplified. When used for audio applications, summing amplifiers are usually called *mixers*.

A typical mixer is a relatively simple circuit that allows two or more microphones to be connected to a single power amplifier. The mixer may or may not include a stage of preamplification. More elaborate mixers have half a dozen or more inputs, each having its own gain and, perhaps, other signal controls.

You can make a simple, but functional mixer by connecting several resistors, one for each channel, to one of the inputs of an operational amplifier. If potentiometers are used instead of fixed resistors, it is possible to control the amplitude of the signal from each channel.

This is where the joystick comes in. Normally, a separate potentiometer is required for each channel. Since the joystick incorporates two or four pots in a single structure, the signal amplitude of two to four channels can be controlled by moving a single control.

Figure 1 shows a very simple joystick-controlled mixer. The simplicity of this circuit will allow you to appreciate the versatility of a dual-channel mixer having a single mixing control. For example, I fed two audio channels (music and a tone) into the two inputs and was able to control the mixture with one hand while making other changes with the other hand.

I used a joystick containing 100-kΩ linear-taper potentiometers in this circuit. Logarithmic-taper pots would be better. You can try the basic resistive summer with many different op amps and audio amplifier chips. Joysticks containing four pots are hard to find, but they permit four channels to be simultaneously controlled.

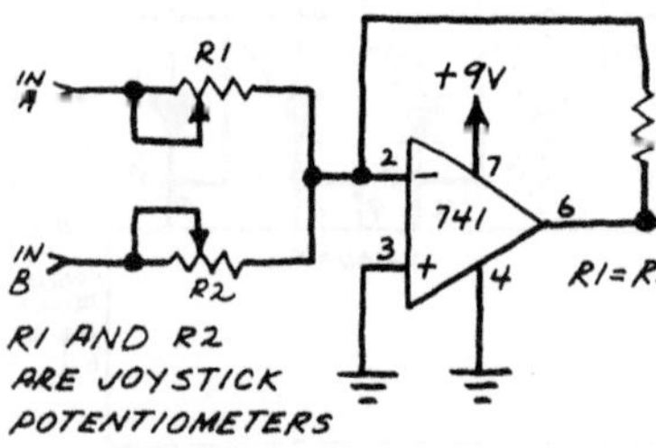

Fig. 1. A joystick-controlled mixer amplifier.

Joystick-Controlled Tone and Amplitude. Figure 2 shows an experimental joystick tone- and amplitude-controlled amplifier. In operation, *R1* controls the amplitude of the signal while *C1* and *R2* serve as a simple, but adjustable, tone-control filter.

To experiment with this circuit, connect an audio-frequency tone source to the input. A sine wave works better than fast-rising square waves. Move the stick so that *R1* is rotated and the amplitude of the tone from the speaker will be altered.

Move the stick so that *R2* is rotated and the amplitude of the tone will also be altered; but in this case, the change in amplitude is dependent upon the frequency of the incoming signal.

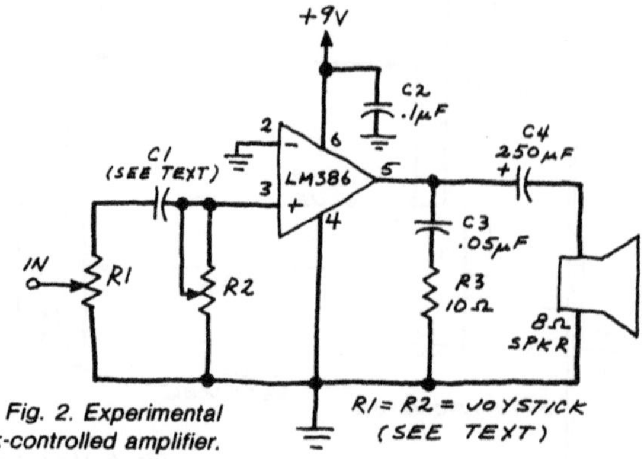

Fig. 2. Experimental joystick-controlled amplifier.

The first-order, high-pass Butterworth filter formed by *C1* and *R2* has a half-power frequency cutoff that is the reciprocal of 6.28*R2C1* or 0.159/*R2C1*. The actual cutoff frequency you obtain may differ from that predicted by the formula. For example, when *R2* is 50,000 ohms and *C1* is 0.0001μF, the formula predicts a half-power frequency cutoff of 3,185 Hz. I measured a half-power frequency cutoff of about 1,000 Hz.

In any event, this simple circuit demonstrates how a single control can alter both the gain and frequency response of an amplifier. If you wish to design your own joystick-controlled amplifiers, there are several excellent books which contain helpful design guidelines. Among them are: *The Design of Active Filters, With Experiments* by Howard Berlin (E&L Instruments, 1977); *Handbook of Operational Amplifier Circuit Design* by David Stout and Milton Kaufman (McGraw-Hill, 1976); and Don Lancester's *Active-Filter Cookbook* (Sams, 1975). You can find these and other books on the design of suitable filters in better libraries. You can also find some design information in application notes published by various op-amp manufacturers. See, for example, National's *Audio/Radio Handbook* and *Linear Applications Handbook*.

Dual-Tone Mixer. Figure 3 is an experimental joystick-controlled dual-tone mixer with which I've been experimenting. The circuit is designed around a 556 dual timer chip.

The 556 is functionally identical to two 555 timers on a single chip. Each half of the circuit in Fig. 3 is connected as an astable multivibrator whose frequency of oscillation is controlled by one of the two 100-kΩ potentiometers in a joystick. The frequency is also controlled by *C1* and *C2*.

Both halves of the circuit will oscillate without *C3* and *C4*; but the capacitors are necessary to prevent uncontrolled interactions between the two oscillators.

The output from each tone generator is coupled into a simple audio mixer made from *R5*, *C6* and a 741 op amp. Potentiometer *R5* also serves as a balance control.

Capacitor *C5* is not part of the mixer, but serves to stretch the fast rising and falling pulses from the two tone generators. This causes the sounds produced by this circuit to be more tolerable than they normally are.

For initial experiments, connect an external amplifier to pin 6 of the 741. The values of *C1* and *C2* should be similar or identical. Move the joystick off center, apply power and adjust *R5* until the two simultaneous tones have approximately the same amplitude. You can then move the joystick in various directions to produce a wide range of unusual tone combinations.

The tones produced by this circuit are for the most part very unpleasant. Nevertheless, the circuit nicely demonstrates how a single control device can regulate two independent circuits. If you would like to follow up with circuit ideas of your own, consider substituting the 4046 CMOS phase-locked loop for the 556 (see the July and August 1980 installments of this column). The following circuit shows just one way the 4046 can be used in conjunction with a joystick.

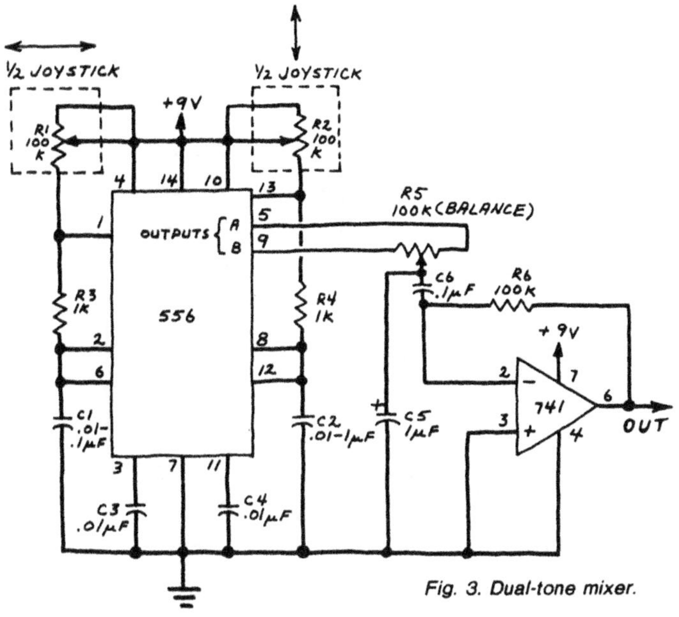

Fig. 3. Dual-tone mixer.

Percussion Synthesizer. The sound produced by percussion instruments such as drums and wood blocks can be simulated by a damped op-amp oscillator. Changing one or two component values permits a variety of percussion sounds to be synthesized.

The circuit in Fig. 4 permits many different percussion sounds to be generated under complete manual control. Percussion synthesizers described previously in this column operate when a switch is closed and released. The percussion effects produced by the circuit in Fig. 4 can be controlled in real time simply by moving a joystick back and forth. Wide, circular movements of the joystick produce sounds ranging from that emitted by a plucked violin string to a tapped glass. Various bell-like sounds can also be produced.

Referring to Fig. 4, note that *R1* is the horizontal-axis pot in a two-pot joystick and is connected as a voltage divider

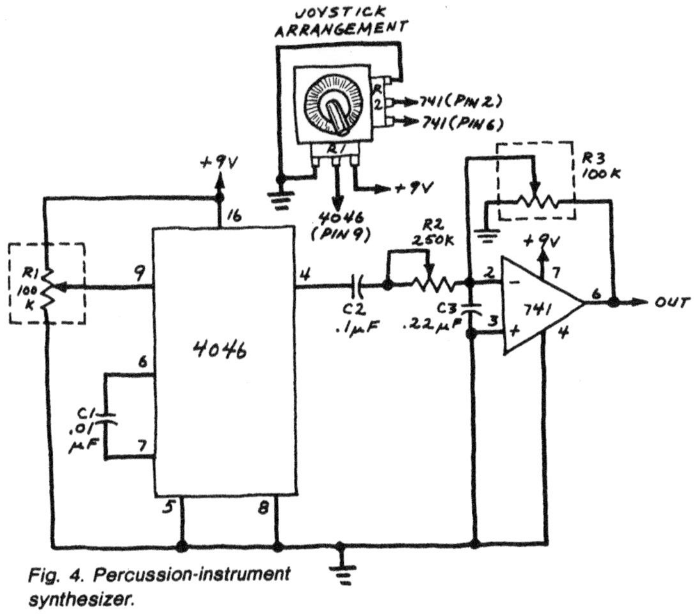

Fig. 4. Percussion-instrument synthesizer.

which provides a variable voltage to the input of the voltage-controlled oscillator (vco) section of a 4046 phase-locked loop. The vco output is coupled into a 741 op amp through gain control *R2*. Capacitor *C3* enhances the percussion sounds by stretching the incoming signal from the 4046.

The vertical-axis pot in the joystick, *R3*, can be adjusted along with *R2* to alter the gain of the 741. Experimenting with the resistance of *R2* will help provide a full range of percussion sounds. The sounds themselves are produced when the joystick is moved in various patterns.

To create percussion sounds, for example, orient the joystick so that back and forth movements control the amplitude pot (*R3*). Push the stick *away* to its outermost limit, and the sound will cease as the noninverting input of the 741 is brought to ground through *R2*. Pull the stick straight back to achieve the percussion effect. You'll hear a sudden tone followed by rapid damping as the stick reduces the amplitude of the signal. For different tone frequencies, move the stick to the left or right before pulling it back. This will let you quickly change from plucking sounds to bells, etc.

If the joysticks are connected as shown in Fig. 4, the sound will cease when the stick is pulled back and rotated to the *left*. This allows you to fully recover and prepare for a new sound cycle without the presence of an intermediate (and highly distracting) tone.

For tinkling bell or plucked string effects, rapidly rotate the stick in a *clockwise* direction. You can experiment with the values of *C1* and *C3* and of *R2* for different effects.

Digitizing the Output from a Joystick. Joysticks provide a very important interface between computers and their operators. The practical implementation of this interface generally requires some form of analog-to-digital conversion.

Some joystick interfaces utilize software to detect the various tones produced by a joystick-controlled tone generator.

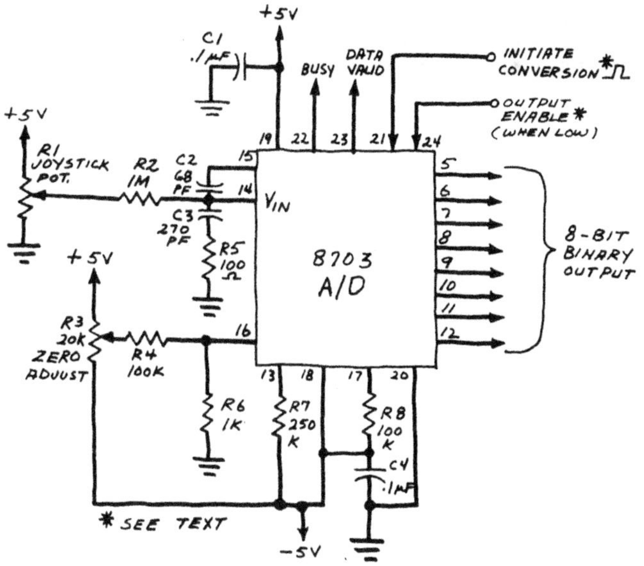

Fig. 5. Digitizing the output from a joystick with an A/D converter.

Figure 5 shows a straightforward hardware approach. This circuit digitizes the status of one of the pots in a joystick. The joystick pot is connected as a voltage divider which provides a variable voltage to the input of the analog-to-digital (A/D) converter chip. Many different A/D chips can be used in this application. The 8703, a product of Teledyne Semiconductor (1300 Terra Bella Ave., Mountain View, CA 94043), is a 24-pin DIP which provides 8 bits of conversion. Since the chip is a CMOS device, its power dissipation is a very low 20 mW.

The 8703 has three-state outputs and is thus ideally suited for connection to microcomputer buses. Pin 24 controls the outputs; when it is low, the outputs are enabled. When pin 24 is high, the outputs assume the high impedance (off) state.

Eight bits per joystick pot is probably all or even more than you will need. If you need even higher resolution, you can select the 8704 (10-bit) or 8705 (12-bit) chips. These chips are essentially identical to the 8703 with the addition of more outputs.

When the circuit in Fig. 5 is connected as shown, the output will indicate an 8-bit binary count of 0000 0000 to 1111 1111 as the joystick pot is rotated from one extreme to another. Since joystick pots are usually limited by a plastic or metal template to only a portion of their maximum rotation, you may not obtain the full range of conversion. However, you can connect the joystick pot to a higher voltage to enhance the overall range.

Having spent some time experimenting with the circuit in Fig. 5, I would urge you to obtain the 8703 data sheet before using the chip in an actual circuit. ADC chips are sometimes a little tricky to use (certainly more so than op amps and unclocked logic), and the data sheet will help you understand their operation.

For example, you can vary the rate at which the 8703 performs conversions by connecting an external clock to pin 21. Each time the clock sends a positive pulse to pin 21, the 8703 will begin a new conversion cycle unless it is busy performing an existing cycle. When pin 21 is connected to V_{DD} (pin 19), the 8703 goes into a free-running mode in which conversions are automatically performed at a rate of approximately 800 per second.

Decoding the Output from an ADC. The binary output from a joystick with an ADC can be connected directly to a digital data bus. Or it can be decoded for a real-time indication of the position of the joystick. Figure 6 shows one way to decode half the output from the 8703 A/D converter.

The 4049 buffers provide CMOS-to-TTL interfacing for the 8703. (The 8703 will interface directly with C... LS chips.) The 74154 is a 4-to-16-line decod... forms a 4-bit input into its hexadecim... joystick is swept from one extreme to another, ...

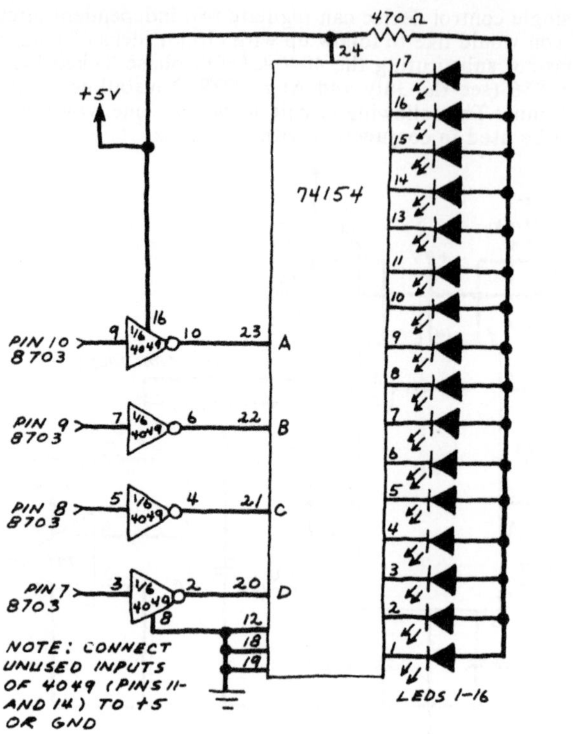

Fig. 6. How to decode the four lowest-order bits from the 8703 A/D.

in succession. By placing the axis of the joystick pot perpendicular to the row of LEDs, the position of the glowing LED will track the position of the joystick. ■

Two

Light and Light Communications

1. Phototransistor Receiver Module

It's easy to squeeze miniaturized LED transistors and phototransistor receivers into 16-pin DIP modules with the help of 8-pin MINIDIP ICs. Figure A is a photo of an

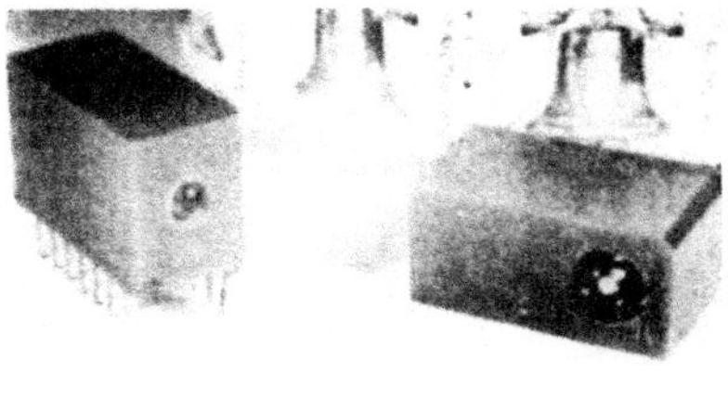

Fig. A. Miniature IR modules

infrared transmitter and receiver assembled in this fashion.

Figure B is a circuit diagram for the receiver. In operation, photons impinging upon the phototransistor cause a small photocurrent to flow. This signal is passed by *C1* to the 741 op amp which has a gain (determined by *R2* and *R3*) of 1000. The amplified signal appears at pin 6 of the 741

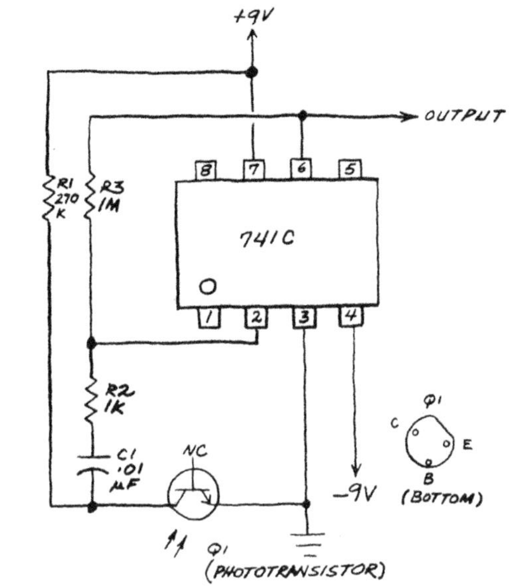

Fig. B. Schematic of an op amp phototransistor receiver.

Fig. C. Phototransistor module.

where it can be coupled to another circuit or used to energize a small relay or drive a small speaker.

Figure C is a photo of the interior of the receiver module, and Fig. D shows the assembly details. Begin assembly by installing all the resistors in the bottom of the module header and inserting their leads deep in each pin slot. Next, install *C1* and solder it and the resistors in place. Avoid using too much solder.

Next, clip off the base lead of the phototransistor and install it on the module header as shown in the figures. Make sure the collector and emitter leads are properly oriented before soldering them in place.

Place the pins of the IC adjacent to or inside the slots in the appropriate module header pins. Make sure they don't protrude too far or the module cover will not fit. Carefully solder the pins in place. Then remove excess solder from the outside edges of the header pins with a file. Finally, drill a 3/16-inch (4.8-mm) hole in the module cover directly over the location of the phototransistor and snap the cover in place.

Test the module by inserting it in a solderless breadboard and applying power from two 9-volt batteries via jumper leads. A small speaker or earphone connected to the receiver output through a 1000-ohm series resistor will emit a loud buzzing sound when the phototransistor is pointed toward a flourescent lamp. If you use an earphone instead of a speaker, use caution when conducting this test! The sound from the earphone can be *uncomfortably* loud. It's best to hold the phone near rather than inserting it in your ear until you've had some experience with the receiver.

After the module is working properly, try listening to the pulsating tone from multiplexed LED displays in digital watches, clocks and calculators with the receiver. You will also be able to "hear" lightning, vibrating car headlights, flickering candle flames and other modulated light sources. ■

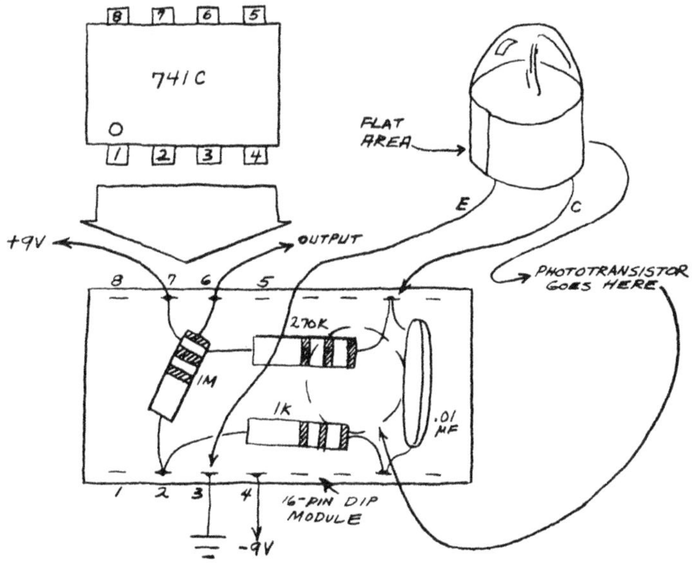

Fig. D. Assembly details of phototransistor receiver module.

2. LED Transmitter Module

Figure A is a complete circuit diagram of the transmitter. The circuit uses a 3909 LED flasher IC for the utmost in simplicity. This IC is designed to drive red LEDs and *not* IR (infrared) LEDs. IR LEDs have a lower forward voltage drop (about 1.2 volts) than red LEDs (about 1.7 volts). This means you can fool the 3909 into driving an IR LED by adding an ordinary silicon diode in series with the LED. A diode like the 1N914 has a drop of 0.6 volt and this gives a total drop of about 1.8 volts when connected in series with an IR LED.

For best results, use a GaAs:Si LED instead of a GaAs LED. Both types are available from companies that advertise in the Electronics Marketplace section of this magazine. GaAs LEDs emit at a peak wavelength of 900 nanometers (nm) while GaAs:Si LEDs have a peak wavelength of about 935 nm. Visible light ranges from about 400 nm to 700 nm.

Most GaAs:Si LEDs are at least twice as efficient as GaAs units, and that's why they will work better in this project. GaAs units have a much faster rise time, but this is irrelevant because the rise time of the transmitter is not fast enough to tax a GaAs:Si diode.

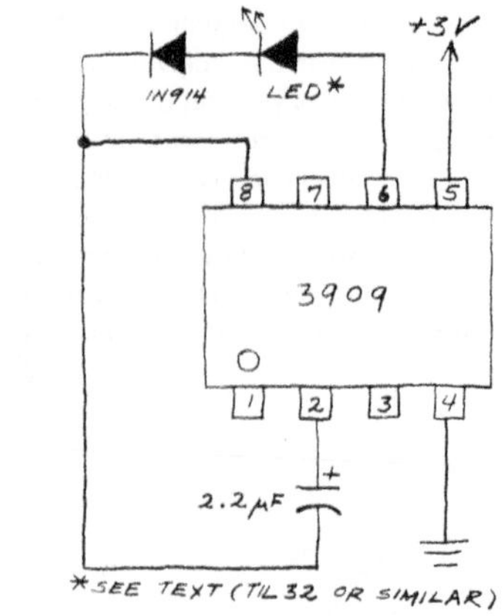

Fig. A. An infrared light emitting diode transmitter.

Fig. B. LED transmitter module.

Figure B is a photo of the interior view of the transmitter module and Fig. C shows its assembly details. Begin assembly by installing the capacitor and diode in the bottom of the module header and inserting their leads deep in the indicated pin slots. Then install the LED as shown in Fig. C, making sure its leads are oriented properly and that it doesn't protrude too far over the edge of the header. Secure all the leads in place with a small amount of solder and clip off the excess lead lengths close to the header pins.

Next, place the pins of the IC adjacent to or inside the slots in the appropriate header pins. Make sure they don't protrude too far or the module cover will not fit. Then carefully solder the pins in place. Use a small file to remove excess solder from the outside edges of the header pins so that the module cover will fit. Then bore a hole (⅛-inch if you use a TIL32 LED) in the module cover and snap the cover in place.

Unless you use a red LED, you'll need a receiver such as the phototransistor receiver module described in the previous Project of the Month to test the transmitter. Insert the module in a solderless breadboard and connect a 3-to-6-volt supply to the power connections. With a 3-volt supply, the transmitter LED will flash on and off at 360 Hz. If you connect an earphone to the receiver module and point the LED toward the phototransistor, you will hear a loud tone. Block the path between the two modules and the tone will stop.

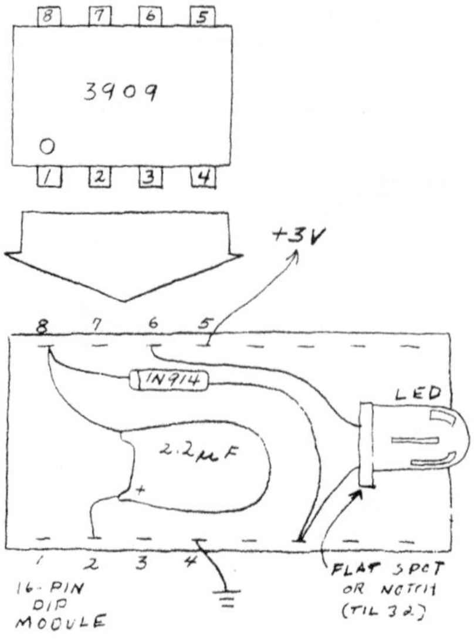

Fig. C. Assembly details of LED transmitter module.

Remember to be careful if you connect an earphone to the output of the receiver module instead of a small speaker. The sound generated by the earphone can be *very* loud.

Try experimenting with the two modules to see how far you can separate the receiver module from the transmitter module and still recover a usable signal. Also, try using the two modules as an object detector by pointing both units at a white card and seeing how far away the card can be placed without losing the signal. ■

3. Eavesdropping on Light

WE ARE literally surrounded by modulated sources of light, both natural and artificial. Seeking them out can be an enlightening and entertaining experience.

We ordinarily discuss circuits that are *not* available as preassembled commercial products. This column, however, marks a departure from our general practice in that it calls for the use of a commercially available, battery-powered audio amplifier. Of course, you can use a home-brew audio amplifier that you have on hand, or you can build one using an audio IC or a few transistors. You can then begin tracking down modulated light sources within minutes of reading this column, assuming you already have a few common components.

Suitable Detectors and Amplifiers. Silicon solar cells, photodiodes, phototransistors and other photovoltaic devices can all be employed as sensors in the detection of modulated light sources. Whatever sensor is employed can usually be directly connected to the input of the audio amplifier. In some cases, however, a transformer or other impedance-matching device or circuit will be required.

Although the high-fidelity amplifier found in any home audio system can be used with excellent results, a portable amplifier is best suited for this application because it can be readily used outdoors and in automobiles. Shown in Fig. 1 is a Realistic Micro-Sonic battery-powered amplifier I have used with suitable sensors to detect many different modulated light sources over the past several years.

Fig. 1. Portable battery powered amplifier suitable for monitoring modulated light. Plug inserted into microphone jack incorporates a miniature silicon photodiode.

Notice the miniature plug inserted into the amplifier's microphone jack. This plug contains a small silicon photodiode whose two leads are soldered directly to the plug's terminals. The opening in the plastic cap intended for the connecting cable was enlarged slightly with a reamer so that as much light as possible could strike the photodiode.

You might be able to save a little money by using one of the transistorized amplifier modules sold by some electronic parts suppliers. Mount the amplifier in a plastic case along with a battery, volume-control potentiometer and speaker. Incidentally, defective portable tape recorders are a good source of amplifier modules.

Many different kinds of light detectors can be connected to the audio amplifier. For very low light levels, I've found that a large-area silicon solar cell works best. However, this type of cell is easily broken so you will need to attach the cell you select to a rigid substrate of plastic, metal or wood. A few drops of cement will secure it in place. You can give additional protection to the cell as well as provide a directional detection capability by installing it at one end of a (10 to 30 cm) plastic, aluminum or cardboard tube. A lens is not necessary if the surface area of the cell is about the same as that of the tube's aperture. Use a long tube and paint its inside surface flat black for best results.

Most inexpensive, large-area silicon solar cells available on the surplus market are *not* supplied with connection leads. It is very important to use care when soldering connection leads to these cells because improper soldering procedures will cause the fragile electrodes to peel away from the cell.

The thin upper electrode is more difficult to solder than the large electrode that covers the entire bottom of the cell. For best results, heat a portion of the upper electrode near a corner of the cell if it is rectangular or near the perimeter if it is circular. Apply heat for only a few seconds with a low-power iron and then apply a small amount of solder. Next remove ⅛" (3.2 mm) of insulation from one end of a length of Wire-Wrap wire and place the exposed conductor along the electrode adjacent to the solder. Reheat the solder for a moment. It will suddenly flow over and around the wire to provide a perfect solder connection. Use this same procedure to solder a wire to the cell's bottom electrode.

You will have to provide a means for protecting the wire leads after the cell is mounted on a card or in a tube. I prefer to attach a shielded phono cable to the tube or card and then solder the cell's leads to the cable. This prevents the leads attached to the cell from being broken by a sudden jerk. The shielded cable reduces unwanted noise from nearby ac power lines and other sources.

For special-purpose detectors, try light-emitting diodes instead of solar cells. The peak response of a LED is confined to a much narrower group of wavelengths than that of a solar cell, and roughly corresponds to the wavelength emitted by the diode when forward biased. For example, a high-efficiency, GaAs:Si near-infrared emitter has a peak spectral response at about 940 nanometers.

Visible LEDs work as detectors also, but they are not as efficient as near-infrared LEDs. Figure 2 shows a GaAs infrared emitter soldered to a miniature plug that can be inserted directly into a modular amplifier's input jack.

Whatever detector you select, tune in as many different light sources as possible. Many LED clock, watch and calculator displays are multiplexed at relatively low-frequencies and will usually produce a buzz or hum. Fluorescent lamps and neon lights produce a very strong 120-Hz buzz. Several years ago, a long-range light-beam communications experiment I was conducting was interrupted by a persistent buzz originating from a large neon advertising sign more than two miles away! Flickering candles, matches, lighters, campfires and fireplaces produce a variety of interesting sounds.

Electrical storms are particularly fascinating to monitor, especially at night. Lightning flashes produce the same crackling and popping sounds as those heard over a radio during a storm. The light detector, however, finds line-of-sight discharges which makes it possible to identify areas of peak activity.

Although the photodetector's sensitivity is reduced in daylight due to the unwanted dc bias which is produced, lightning can still be detected. Often, in fact, you'll detect with a solar cell lightning that you cannot see with your eyes.

Fig. 2. Infrared emitting diode connected to miniature phone plug.

Steady light originating from the sun and dc-powered lamps normally produces only a hiss. Movement, however, adds a new dimension to steady light sources. You will discover its effect the first time you "hear" light from the sun interrupted by a picket fence or overhead branches. You will even be able to pick up the hum of a flying insect by capturing the sunlight reflected from its oscillating wings. Similarly, you will detect the wing beats of a hummingbird when you position your detector so as to form a straight line with the sun and a bird hovering at a feeder.

To liven up the otherwise uninspiring hiss produced by a flashlight, tap its reflector with a pencil. This will cause a pleasant chime-like sound as the filament vibrates in and out of the reflector's focal point.

Calvin R. Graf, an acquaintance who shares my interest in monitoring modulated light sources, has described some of these and many other observations in a recently published book entitled *Listen to Radio Energy, Light and Sound* (Howard W. Sams & Co., Inc., 1978). Calvin's book reports on many of his personal observations and suggests experiments that can be conducted easily.

Op Amp Preamplifier. The circuit shown in Fig. 3 will serve as a crude but effective preamplifier for a battery-powered portable amplifier. The preamplifier can be assembled on a small perforated board. Insert the leads from a pair of 9-volt battery connector clips through a ¼" hole drilled in the board and tie a knot in the leads to keep them from pulling loose. Then solder them to the appropriate circuit nodes. (Red is positive and black is negative.) Connect the preamp to the amplifier with shielded cable.

The voltage gain of the preamp is the quotient of *R2* divided by *R1*. With the values shown on Figure 3, its gain is 1000. This should be more than adequate for most sensors. Too high an input signal will overdrive the audio amplifier, so keep the volume control set to a low level when using the preamp.

A word of caution is in order for those who want to eavesdrop on light in noisy areas. An earphone will prove very helpful when the ambient sound level is high, but be sure the volume is turned to a low level until you have focused in on a light source you wish to monitor. Unexpected flashes of light can produce very *loud* sounds! ■

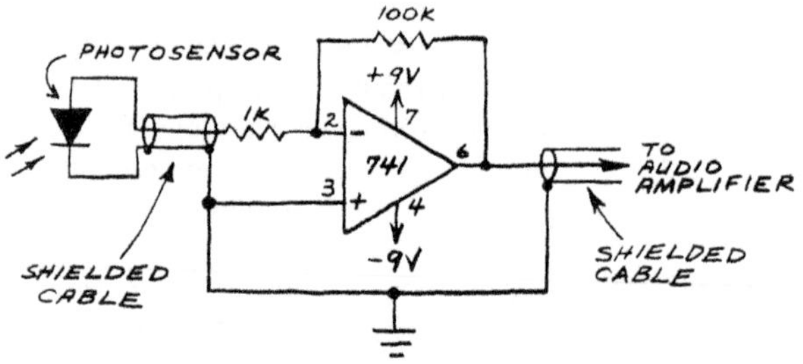

Fig. 3. Simple op-amp preamplifier for monitoring modulated light sources.

4. Optical Fiber Communications, Part 1

OPTICAL FIBER communications is one of the fastest growing areas of modern electronics. Here's why.

- A hair-thin glass fiber can carry more information than 900 pairs of copper wires comprising a cable as thick as your fist.
- Glass and plastic fibers are immune to electromagnetic interference and do not attract lightning strokes.
- Because fibers are insulators and not conductors, they don't generate sparks or present a shock hazard, nor can they be short-circuited.
- Some glass fibers can transmit a modulated beam of light more than ten kilometers before it is necessary to employ a repeater to strengthen the signal.
- Communications-grade fiber is already cheaper than coaxial cable, and even greater price advantages are in the offing. The raw material for glass fiber is sand, but coax is manufactured using copper (for conductors) and petroleum-derived plastic (for the dielectric).
- Fiber systems are impossible to jam and difficult to intercept.
- Glass fiber has a higher tensile strength than a steel wire of the same (small) diameter.

In this first installment of a two-part series, we'll find out how fibers transmit light and learn something about their idiosyncrasies. In Part 2, we'll put fibers to work in some practical communication systems that you can easily build.

How Fibers Transmit Light. Figure 1 shows how a ray of light travels through an optical fiber by making multiple reflections from the fiber's core/cladding boundary. The core and cladding are both transparent but the index of refraction of the core is slightly higher than that of the cladding. Just as the boundary between air and water is highly reflective, the core/cladding boundary behaves as a mirror to light waves striking it within the fiber's *acceptance angle*.

The material in Fig. 1 is known as a *step-index* fiber because of the sharply defined transition between its core and cladding. Step-index fibers are easily manufactured, but have one major disadvantage. Light waves entering one end of a fiber at the same time can arrive at the opposite end at slightly different times due to the different travel paths or modes they can follow. This causes narrow optical pulses to be stretched, and places an upper limit of a few tens of megahertz on the rate at which data can be sent through the fiber.

Graded-index fibers are manufactured to reduce the delay problem associated with step-index fibers. Instead of a well defined core/cladding interface, this type of fiber merges the core with the cladding to form a gradual change in refractive index. This causes light rays to curve through the fiber as shown in Fig. 2. Because the light near the cladding travels faster than light in the core, there is considerably less pulse broadening. Accordingly, data rates as high as hundreds of megahertz are possible.

Attenuation of Optical Fibers. The first question most people ask when they learn about optical fiber communications is, "How clear are they?" The answer is *incredibly* clear. If ocean water were as clear as typical communications-grade glass fiber, it would be possible to see clearly the bottom of the deepest depths of the sea.

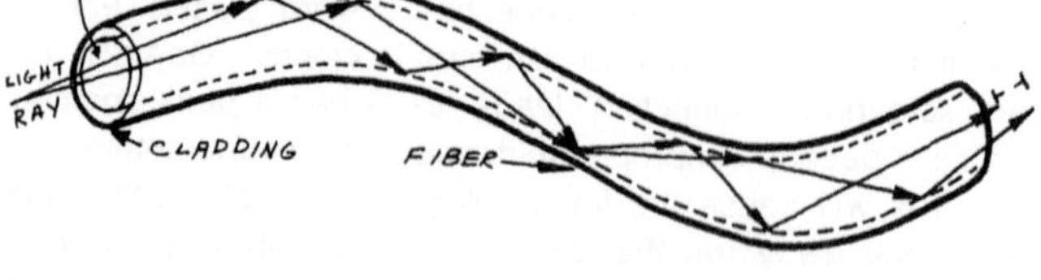

Fig. 1. Light rays in ordinary optical fibre take multiple paths causing narrow pulses to be stretched.

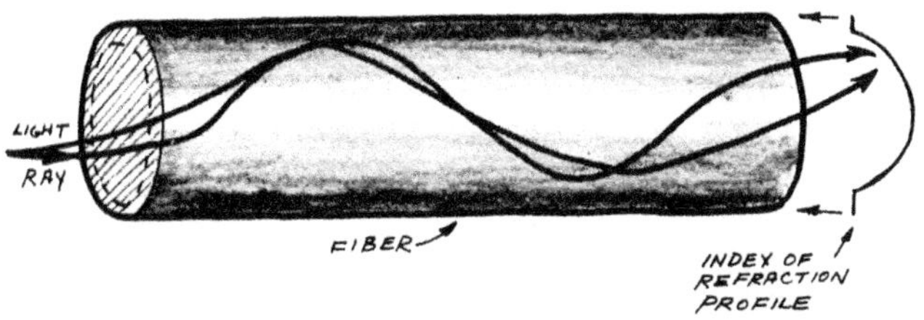

Fig. 2. A graded-index fiber is made so that pulse stretching is considerably less severe, allowing faster data transmission.

The attenuation of optical fibers is given in decibels per kilometer (dB/km). Inexpensive plastic fibers such as those used to make optical art displays often have an attenuation of hundreds or thousands of dB/km, so they're practical only for very short-range communications links of a few meters or less. Communications-grade glass fibers have attenuations ranging from approximately 20 to as little as one dB/km!

A 3-dB/km glass fiber one kilometer long attenuates only half the light injected into one end. In other words, apply one milliwatt of radiation into one end of a 1-km fiber and you'll receive half a milliwatt at the other end. Solid-state detectors can work with signal levels of a few tens of nanowatts, so it's possible to transmit high-quality data for 10 km or more over a 3-dB/km fiber without the need for a repeater.

It's important to note that optional fibers do not transmit all wavelengths of light equally well. Figure 3, for example, graphically shows the amplitude-versus-wavelength response of two different fibers. Because of the variations in response over a range of wavelengths, it is important to match optical sources with wavelength-compatible fibers.

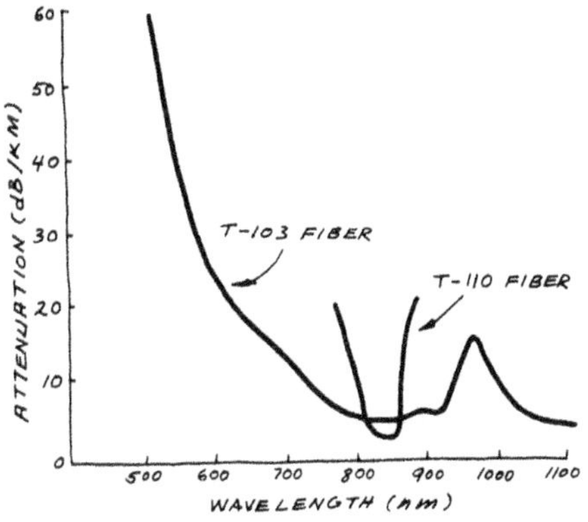

Fig. 3. Spectral response of two ITT communications-grade glass fibers.

Where to Buy Fiber. More than a dozen companies have entered the optical-fiber field, and communications-grade fiber should soon become available to experimenters for considerably less than a dollar a meter. In the meantime, you can purchase fiber from Edmund Scientific Company (Edscorp Bldg., Barrington, NJ 08007). Their catalog lists high-attenuation plastic fibers and 40-dB/km silica fiber. Another source for optical fibers is Math Associates (376 Great Neck Road, Great Neck, NY 11021), which sells communications-grade, low-loss fibers in lengths as short as one meter.

If you have the money and are serious about fiber communications, you can buy unjacketed fiber in minimum lengths of 500 or 1000 meters at prices ranging from 50 cents to $1.00 per meter. Here are some manufacturers to whom you can write for detailed specifications, prices and shipping information:

ITT Electro-Optical Products Division
7635 Plantation Rd.
Roanoke, VA 24019

Corning
Telecommunication Products Dept.
Corning Glass Works
Corning, NY 14830

Valtec Corporation
West Boylston, MA 01583

Siecor Optical Cables, Inc.
631 Miracle Mile
Horseheads, NY 14845

E.I. Du Pont De Nemours & Co.
Plastic Products and Resins Dept.
Wilmington, DE 19898

Quartz Products Corp.
688 Somerset Street
Plainfield, NJ 07061

Before ordering large reels of fiber, be sure you know exactly what your application is and how you intend to implement it. You should also have spent some time beforehand experimenting with short lengths of fiber to determine if you can work with the material without extraordinary difficulty, and if it can solve your communications problems better than an ordinary wire link.

Cutting Fibers. You can cut plastic fibers with a razor blade or hobby knife, but glass fibers require a more elaborate procedure.
Here's how I cut them. First, I carefully strip off any protective coating(s) with a hobby knife. Some fibers are coated with acetate lacquer which can be removed with acetone. (Use acetone only in a well ventilated area and avoid contact with skin.) Next, tape one end of the fiber to your work surface and pull the exposed portion of the fiber over your index finger. Finally, lightly score the fiber over your finger with a carbide glass cutter while applying a small amount of tension to the fiber.

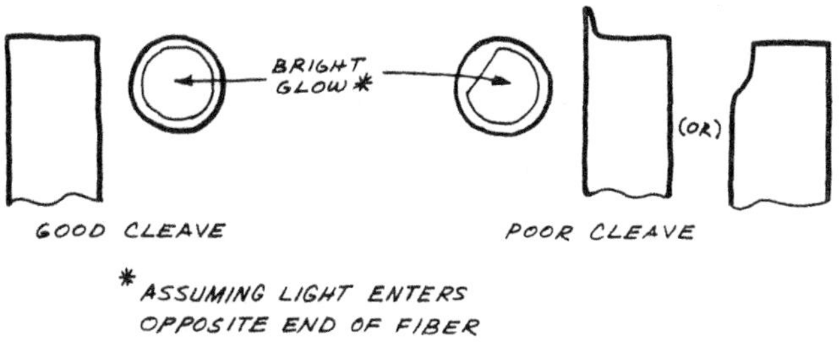

Fig. 4. Typical appearances of well and poorly cleaved fibers when viewed under 50-power magnification.

This procedure should result in an almost perfectly flat cleavage plane, but you must confirm this by examining the exposed end of the fiber with a 50-power phono-stylus microscope. Figure 4 shows what you will see when light is passing through the fiber. If the first cut is unsatisfactory, try again. You might even want to experiment with other methods of cleaving the fiber, such as scoring the fiber while simply pulling on the portion not taped to your work table.

Incidentally, be sure to carefully discard bits of fiber removed during cutting procedures. Small-diameter fibers can easily penetrate a finger or a bare foot!

Attaching Fibers to LEDs and Photodetectors. The two principal methods of attaching fibers to LEDs and photodetectors are removable connectors and fiber pigtails. Removable connectors are expensive, but AMP, Inc. (Harrisburg, PA 17105) has introduced a moderately priced connector which is finding widespread popularity. This connector will probably become available to experimenters in the near future.

Laser diodes and LEDs with factory-installed fiber pigtails cost hundreds of dollars. Motorola's solution to this problem is a new series of emitters and detectors with integral light pipes which mate with AMP connectors. One of the emitters (MFOD402F) includes a built-in integrated preamplifier. For more information, you can request data sheets for the MFOE102F LED and the MFOD102F and MFOD302F detectors from Motorola (P.O. Box 20912, Phoenix, AZ 85036).

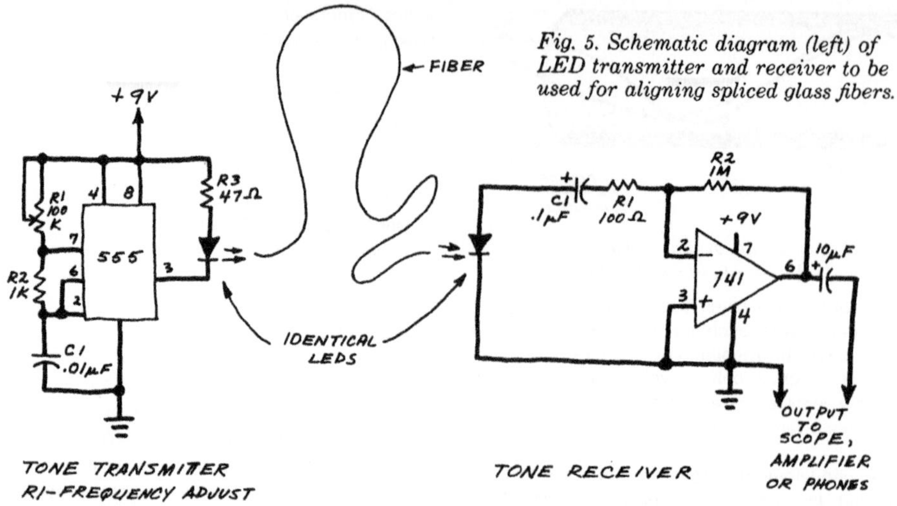

Fig. 5. Schematic diagram (left) of LED transmitter and receiver to be used for aligning spliced glass fibers.

Although I've found AMP connectors the best solution for coupling fibers to LEDs and detectors, I often attach fibers directly to epoxy-encapsulated LEDs. The easiest way to do this is to heat a small awl in a flame and push its hot point through the epoxy all the way to the semiconductor chip that emits the light. You should then test the LED to make sure it has not been damaged and that a bright point of light is visible at the exposed surface of the chip.

For temporary experiments, you can anchor a fiber in the hole using cyanoacrylate adhesive such as Eastman 910. For more permanence, insert the LED in a short length of heat-shrinkable tubing, insert the fiber into the LED (making sure the end has been cleaved properly) and surround the LED and fiber with epoxy. You'll need to hold everything together with tape, clothespins or clamps until the epoxy hardens.

For best results, pulse-modulate the LED and monitor the amplitude of the signal emerging from the opposite end of the fiber while slightly moving the end being cemented to the LED until maximum signal is received. This procedure is very much like tweaking the catwhisker in an old-fashioned crystal radio.

Because LEDs can also function as detectors, you can reverse this procedure as long as the source LED is made from the same type of semiconductor material as the receiver LED. Figure 5 shows a transmitter and receiver circuit you can use to align the fiber.

The procedure outlined above works best with GaAsP red LEDs encapsulated in clear epoxy. The 650-nm wavelength emitted by these diodes transmits well through most glass and plastic fibers, and the clear epoxy makes hole formation easier. Litronix RL-50 and RL-55 and Monsanto MV-50 miniature LEDs make excellent sources and detectors. Figure 6 shows a homebrew pigtailed RL-50 LED epoxied in a short length of tubing. Note how the leads are bent back and connected to lengths of wrapping wire.

If you find this procedure too time consuming, you can always try the AMP connectors mentioned earlier. The AMP Optimate single-position, fiber-optic connector is designed for single plastic fibers or bundles of glass fibers. It attaches quickly to an input/output bushing containing receptacle for an LED, photodiode or phototransistor. ■

Fig. 6. One way (below) of mating a glass fiber to a common light-emitting diode.

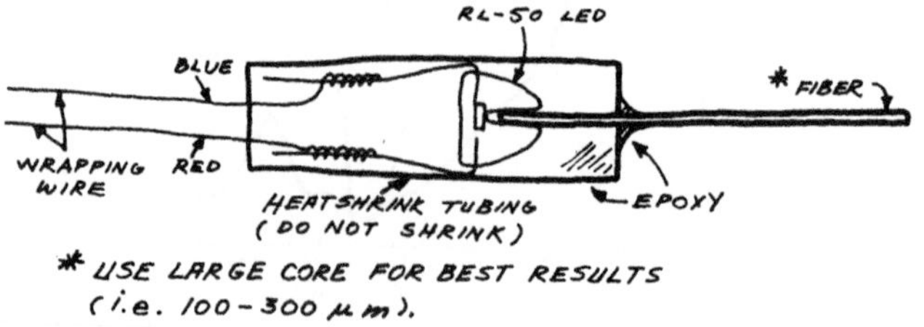

Optical Fiber Communications, Part 2

I have built and tested all these circuits. Nevertheless, I suggest that you assemble versions on solderless breadboards before making permanent units. This will allow you to make gain and modulation adjustments and to perform preliminary operating tests to determine whether or not a particular circuit is suitable for your application.

Selecting Emitters and Detectors. Double-heterostructure injection lasers generate higher power levels over greater bandwidths than light-emitting diodes, but their high cost rules out their use for all but the most affluent experimenters. Fortunately for the rest of us, LEDs provide enough power for fiber links of a kilometer or more. GaAs: Si LEDs are very powerful, but the 940-nm wavelength they radiate is readily absorbed by most fibers. Therefore, GaAs (900 nm), GaAlS (780 to 900 nm) and even common GaAsP (650 nm) red LEDs are better choices.

Among the devices that are suitable detectors are phototransistors, solar cells, photodiodes, PIN photodiodes and even LEDs. Phototransistors are excellent for low-bandwidth links, but they must be shielded from ambient light. PIN photodiodes are the best choice for high-bandwidth applications. For 2-way communications over a single fiber, use a LED as a dual-function emitter/detector.

Operating Tips. Some of the circuits we'll be using employ operational amplifiers with large feedback resistances to provide very high gain. These circuits will probably oscillate violently (at least mine did) unless you take the following precautions. Connect a 0.1-μF disc ceramic capacitor directly across the power-supply pins of the op amp. The capacitor should have short leads. Avoid long component leads and interconnection wires. Use miniature shielded cable (Radio Shack 278-752 or similar) to connect components such as detectors and microphones to the input of an op amp if the distance involved is more than a few centimeters. *Never* use an earphone to monitor the output of an untested receiver! If the circuit oscillates, the resulting sound pressure level can quite easily exceed the threshold of pain.

PFM Transmitter. Transmitting voice and other analog information as a stream of light pulses rather than an amplitude- (intensity-)

modulated continuous light carrier offers several important advantages. Perhaps the most important is noise immunity. Unlike the signal transmitted by an AM light-wave system, all bursts of light in a pulsed system have the same amplitude. This means a threshold circuit can be connected to the receiver to automatically block noise pulses having an amplitude smaller than that of the information-carrying pulses.

Another important advantage is the fact that most LEDs and certain types of injection lasers emit far more power when driven by brief current pulses than when operated more or less continuously as in an AM system. Other advantages of pulse communications include increased bandwidth, reduced (continous) operating power and interesting data encryption and multiplexing possibilities.

A pulse frequency-modulated (PFM) transmitter is fairly simple. In quiescent operation, the circuit produces a continuous stream of pulses at a specified center frequency, usually above the audio range. Audio signals applied to the input of the modulator cause the center frequency to vary in direct proportion to both the amplitude and frequency of the input signal.

Figure 1 is the schematic diagram of a simple unijunction-transistor PFM transmitter I first described in this column in May 1976. (Back issues of POPULAR ELECTRONICS are available at many libraries.) Although this circuit works very well, the pulses delivered to the LED do not have enough duration and amplitude for maximum optical power generation.

Figure 2 is an even simpler circuit designed around a 555 timer IC. This modulator delivers more current to the LED. When powered by an 8-volt rechargeable battery, the peak current as measured across a 1-ohm resistor in series with the LED is 320 milliamperes. This is from three to six times the maximum allowable LED current in an amplitude-modulated continuous-carrier light-wave system.

The pulse duration is a brief 400 nanoseconds. This keeps both the duty cycle of the LED and average power consumption of the circuit very low, but reduces the effectiveness of a phototransistor detector because its response time is slower than that of a PIN photodiode.

For best results, *R1* should be adjusted to give a center frequency between 20 and 30 kHz. You don't need a frequency counter to make this adjustment. You can monitor a suitable light-wave receiver (see below) while adjusting *R1* for optimum sound quality.

Initial tests and adjustments are simplified if you connect the output of a transistor radio (via its earphone jack) to the input of the modulator. The modulator works best when the amplitude of the input is 2 to 4 volts peak-to-peak. For voice operation, connect any standard audio amplifier to the modulator's input. Figure 3 is the schematic of a microphone preamplifier that I like to use.

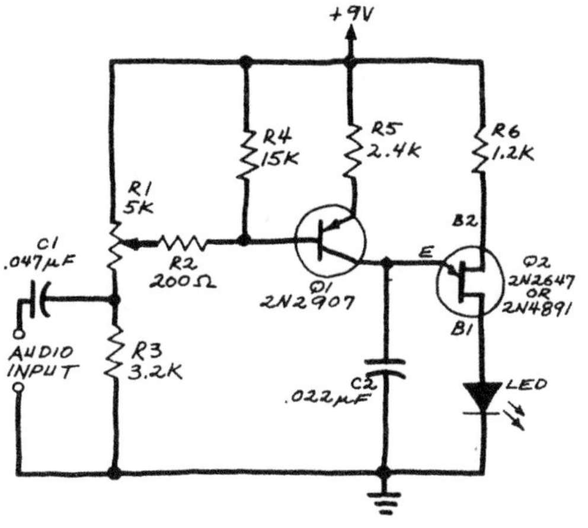

Fig. 1. Simple unijunction transistor PFM transmitter.

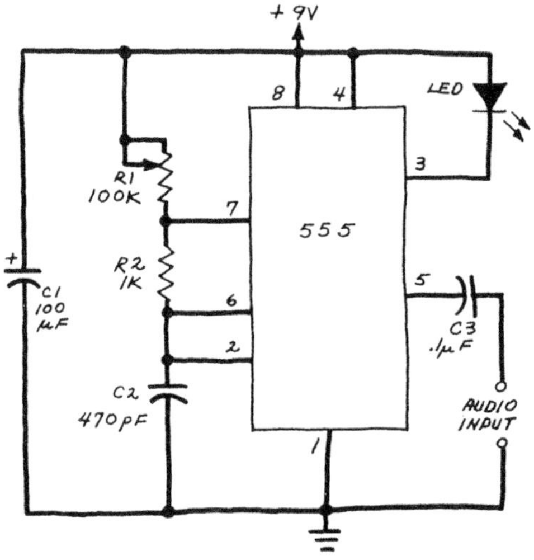

Fig. 2. A light-wave transmitter designed around a 555.

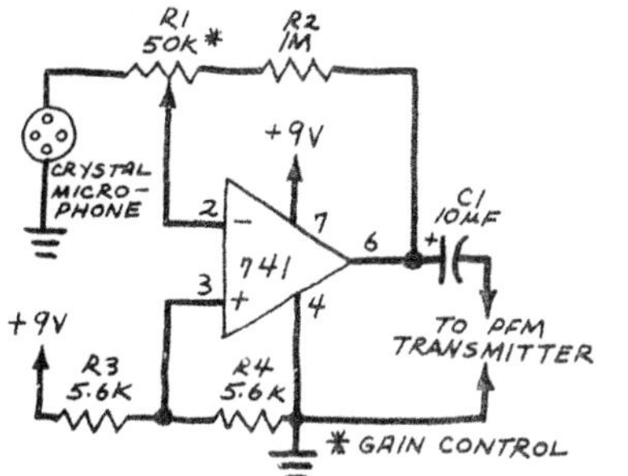

Fig. 3. Microphone preamplifier for PFM transmitter.

PFM Receiver. You can receive signals from a PFM transmitter with any light-wave receiver if the center frequency of the transmitter is higher than approximately 15 kHz. For best results, however, a threshold detector and demodulator should be added as shown in Fig. 4.

Note how this circuit uses one half of an MC1458 (other dual op amps can also be used) as a preamplifier and the second half as a comparator or threshold detector. Demodulation (actually, integration) of the transmitted intelligence is provided by *R5* and *C2*. The recovered audio is then amplified by the LM386 power amplifier. Potentiometer *R6* controls the signal level at the input of the LM386 and is therefore used as a gain control.

Potentiometer *R4* permits adjustment of the comparator. Standard dual op amps such as the MC1458, however, may have insufficient bandwidth for the threshold circuit to work properly. In such cases, the narrow incoming pulses from the transmitter are stretched by the preamp and detector stages until they merge to form an amplitude- rather than pulse-modulated signal. You can obtain true threshold detection by using selected 1458's or using the LF353 or another of the better-quality dual op amps. For optimum results, you might prefer to use individual op amps rather than a dual IC to prevent inadvertent triggering of the comparator at very high preamp-gain levels.

Once the receiver is working, you can reduce the gain of the preamp by increasing the value of *R2*. And you can change the values of *R5* and *C2* to alter the tone response.

Use care when tinkering with the receiver, because inadvertently touching a lead might produce an ear-splitting squeal from the speaker. To protect your ears, you can insert a few hundred ohms of series resistance between *C5* and the speaker, at least until the receiver is ready to be buttoned up and there is no chance of inadvertently touching off a spasm of shrieks and whoops.

Analog Data Transmission System. In the October and November 1979 installments of this column, we experimented with voltage-to-frequency converters and an analog light-wave data-transmission system designed around a pair of 9400 V/F chips. The LM331 V/F converter can also be used in this application. Since those columns appeared, V/F chips have become more widely available. Nevertheless, I have long wanted to design an analog light-

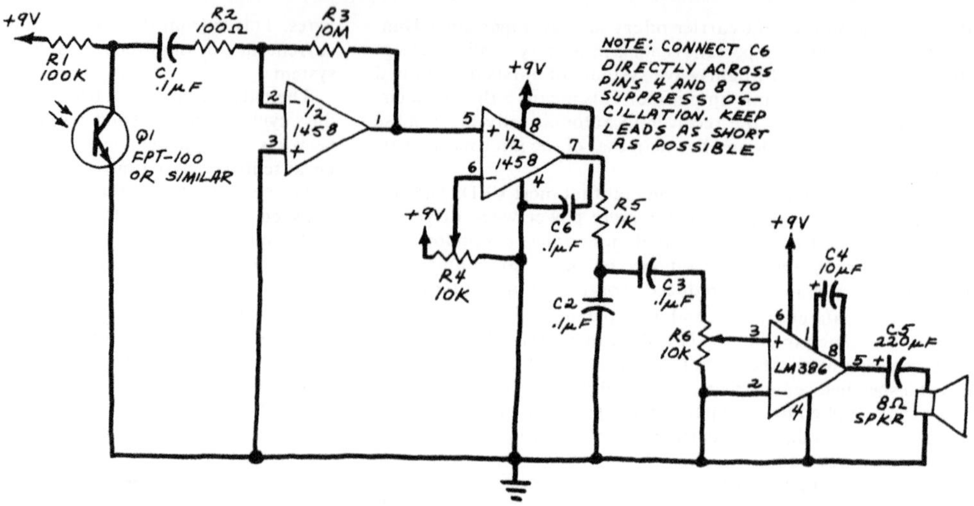

Fig. 4. A PFM receiver with a threshold detector and demodulator added.

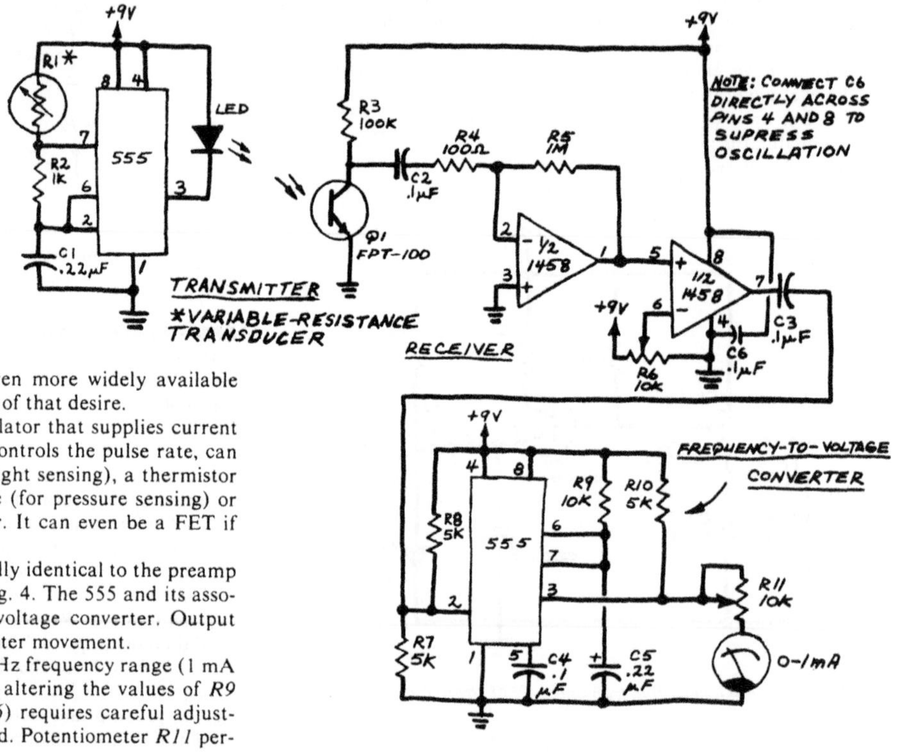

Fig. 5. An analog light-wave transmission system using a 555 astable oscillator transmitter with a receiver similar to that shown in Fig. 4.

wave transmission system around the even more widely available 555 timer chip. Figure 5 is the realization of that desire.

The transmitter is a basic astable oscillator that supplies current pulses to a LED. Resistance *R1*, which controls the pulse rate, can be a cadmium-sulfide photoresistor (for light sensing), a thermistor (for temperature sensing), a strain gauge (for pressure sensing) or some other variable-resistance transducer. It can even be a FET if remote monitoring of a voltage is desired.

The front end of the receiver is essentially identical to the preamp and threshold detector stages shown in Fig. 4. The 555 and its associated components form a frequency-to-voltage converter. Output monitoring is provided by a 0-to-1-mA meter movement.

This circuit works best over a 0-to-360-Hz frequency range (1 mA = 360 Hz), but this can be extended by altering the values of *R9* and *C5*. The threshold potentiometer (*R6*) requires careful adjustment, particularly if slow op amps are used. Potentiometer *R11* permits calibration of the output meter.

The receiver's front-end phototransistor must not be exposed to ambient light if the system is to operate properly. LEDs and PIN photodiodes, which can also be used as detectors, are less susceptible to the deleterious effects of ambient light.

Going Further. For more information about optical-fiber communications, see W.S. Boyle's excellent article "Light-Wave Communications" in the August 1977 issue of *Scientific American*. A good general introductory book is *Light-Beam Communications* (F. Mims, Howard W. Sams, 1975), and a more technical text is *Fundamentals of Optical Fiber Communications* (edited by M.K. Barnoski, Academic Press, 1976). You can keep abreast of the latest developments in this field and get the names and addresses of fiber manufacturers by reading such trade magazines as *Laser Focus* and *Electro-Optical Systems Design*. All these and many other publications on light-wave communications are available at well-stocked libraries. ■

5. Light-Wave Voice Communicator

THIS MONTH'S project is an amplitude-modulated light-wave voice communicator that you can assemble from inexpensive, readily available components. You can use the communicator to send and receive high-quality voice signals over distances of hundreds of feet through the atmosphere or through an optical fiber "waveguide."

The Transmitter. The transmitter, which is shown schematically in Fig. 1, employs a 741 op amp as a high-gain audio amplifier which is driven by a microphone. The output of the 741 is coupled to *Q1*, which

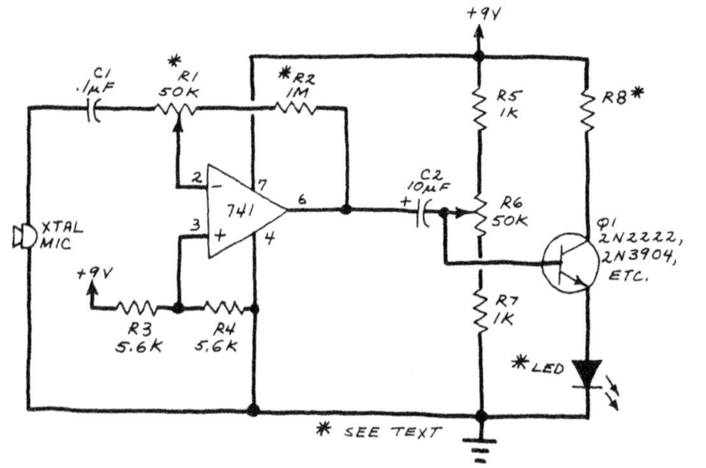

Fig. 1. Schematic of a light-wave voice transmitter.

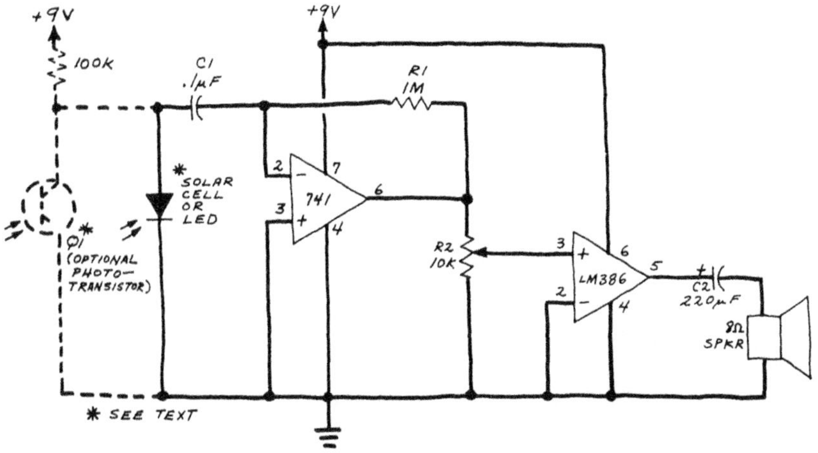

Fig. 2. A light-wave receiver to go with the transmitter.

serves as the driver for a LED. Potentiometer *R1* is the amplifier's gain control. Miniature trimmer resistor *R6* permits adjustment of the base bias of *Q1* for best transmitter performance.

Gain control *R1* can be eliminated if *C1* and *R2* are connected directly to pin 2 of the 741. For maximum sensitivity, increase the value of *R2* from one to ten megohms and use a crystal microphone with a large diaphragm such as the Radio Shack Model 270-095. The miniature crystal microphones sold by many parts suppliers will also work, but they generate less output.

If you prefer, fixed resistors *R5* and *R7* and potentiometer *R6* can be replaced with two fixed resistors after *R6* has been adjusted for best transmitted voice quality. Disconnect *R5* from +9 volts and *R7* from ground, measure the resistance between the wiper of *R6* and the disconnected ends of *R5* and *R7*, and substitute fixed resistors having similar values.

The transmitter works best with near-infrared emitting GaAs, GaAlAs and GaAs:Si LEDs. GaAsP red LEDs can also be used, but they emit considerably less optical power and therefore are best suited for optical fiber links.

Whichever LED you select, it is important to limit its forward current to a safe operating level. A reasonable range of quiescent current is from 10 to 40 milliamperes. High-level audio inputs will raise the current substantially. Resistor *R8* determines the quiescent current, and its resistance should be 100 or more ohms. In my prototype, 330 ohms gave a standby current of 22 milliamperes.

For best results, insert a milliammeter between the emitter of *Q1* and the LED's anode and substitute a 1000-ohm potentiometer for *R8*. Adjust the potentiometer until the desired current level is achieved. Then remove the pot, measure its resistance, and replace it with a fixed resistor.

The Receiver. The light-wave receiver, which is shown in Fig. 2, consists of a 741 operated as a preamplifier and an LM386 power amplifier. Potentiometer *R2* is the gain control.

You can use various kinds of detectors as the front end of the receiver. Phototransistors are very sensitive, but they do not work well in the presence of too much ambient light. Note that a 100,000-ohm series resistor is required if you use a phototransistor. Solar cells and photodiodes work well. So do LEDs of the same semiconductor as the transmitter.

An interesting aspect of using LEDs as detectors is that, although they are not as sensitive as phototransistors, they are much less sensitive to the adverse swamping effects of ambient light. Using a LED as a detector also means you can switch the LED's anode between the input of the receiver and the output of the transmitter to form a light-wave voice transceiver capable of bidirectional communications through a single optical fiber. Of course, you'll need two complete transceivers to fully use this operating mode.

Going Further. This transmitter and receiver system will send voice across a room without the need for external optics. For ranges of hundreds of feet, you must use a lens to collimate the light from the LED. You must also use a lens to collect and focus light on the receiver's detector. For more information on the use of lenses and related subjects, see *Light-Beam Communications* (F. Mims, Howard W. Sams & Co., 1976). ■

6. Experimenting with a Light Pen, Part 1

AMONG the most interesting data-entry devices for computers and remote terminals are those that are sensitive to light. The most sophisticated optical data-entry devices are solid-state television cameras. When such a camera is used with a computer having a large complement of RAM storage capacity, complex operations such as pattern recognition, equipment monitoring, and area surveillance can be performed easily.

Television cameras provide perhaps the ultimate in optical data entry, but their cost (as well as that of the necessary interface circuit) varies from high to exorbitant. Two much more common—and cheaper—optical data-entry devices are *light wands* and *pens*. Television cameras contain many hundreds or thousands of resolution elements, but most light wands and pens incorporate a single-element light detector such as a photodiode or a phototransistor.

It's important at the outset to understand the differences between light wands and pens. Light wands are designed to detect the presence or absence of contrasting marks such as bars of ink printed on paper or plastic. Therefore, light wands usually include built-in light sources to illuminate the marks. They also include precisely focused optics that assist the detector and the light source in their work.

Light pens, on the other hand, are designed to detect a point of light on the screen of a video display such as a cathode ray tube. Simple light pens do not include an internal light source because most video displays are light emitters. However, some pens designed for use with high-resolution displays include a pinpoint light source so that the operator will know precisely where the pen is pointed.

Applications. The light wand is a one-way data entry device. You've probably seen sleek-looking wands attached by flexible cables to some late-model cash registers. A sales clerk can record a purchase merely by sweeping the wand past the bar code printed on the label or package of many different products. Some wands can even read the printed information on a price tag!

Because light wands can be used by unskilled operators and provide faster and more reliable data entry than keyboards, their use is rapidly expanding. They are currently being used in portable inventory monitoring systems in some department stores and supermarkets. They are also used in some libraries to read information from bar-coded identity cards and books.

Figure 1 is a photograph of a light wand made by Hewlett-Packard. The wand's cable plugs into the company's HP-41C programmable calculator and allows bar-coded programs to be quickly loaded into the calculator. If you've ever spent ten or more tedious minutes loading a long program into a calculator, you can readily appreciate the convenience and speed provided by such a wand.

Fig. 1. Hewlett-Packard's bar code-reading Optical Wand plug into an HP-41C calculator.

Light-Pen Applications. Light pens are simpler and therefore physically slimmer than light wands.

How the light pen allows information to be "drawn" on the screen of a CRT is not immediately obvious—at least it wasn't to me when, as a high school student, I viewed a film which showed computer operators using light pens!

Actually, the light pen's principle of operation is remarkably simple. In a typical CRT/light-pen system, for example, the entire screen is repeatedly scanned by a tightly focused electron beam. This produces a fast-moving dot of light too dim to be seen by the human eye but easily detectable by a phototransistor or photodiode.

The computer knows the precise location of the moving dot at any given instant. Therefore, if a light pen is connected to an input port, the computer knows exactly where the light pen is pointed. Depending upon the computer's software, this permits the operator to select specific data to be displayed on a CRT for any desired purpose, and to "write" information, including complex graphics, onto the screen and into the computer's memory.

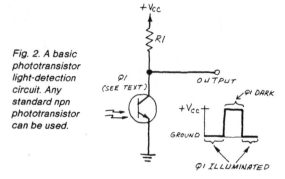

Fig. 2. A basic phototransistor light-detection circuit. Any standard npn phototransistor can be used.

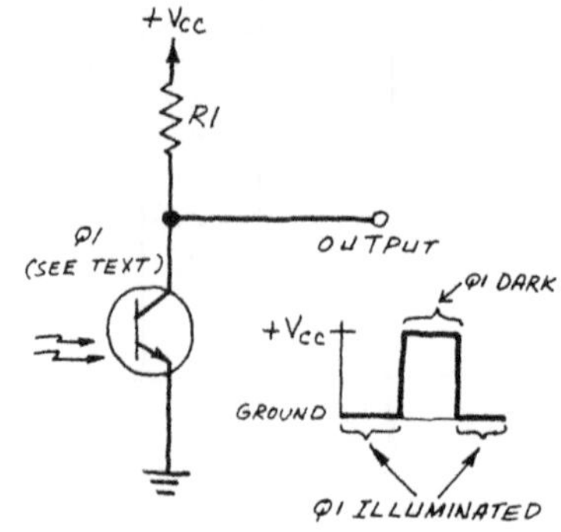

A Homemade Light Pen. A light pen is very easy to make. Both photodiodes and phototransistors make suitable sensors. The former are faster but the latter are more sensitive.

The basic phototransistor light-detection circuit in Fig. 2 illustrates how a straightforward detector responds to a light pulse. Any standard npn phototransistor such as the FPT-100 can be used for *Q1*. When *Q1* is dark, its collector-to-emitter resistance is much higher than *R1*. The output voltage of the circuit therefore rises very close to $+V_{CC}$. When photons strike the device's light-sensitive region, *Q1* becomes forward-biased and its collector-to-emitter resistance falls far below that of *R1*. The circuit's output voltage thereupon approaches ground potential. Summing up, the output of the circuit is normally a high voltage. When light strikes phototransistor *Q1*, the output voltage is low.

This basic circuit can be used in some light-pen applications. A much better circuit, however, is shown in Fig. 3. An operational amplifier is used without a feedback resistor to provide the highest possible gain. The gain is so high that the op amp functions as a comparator whose output switches from +5 volts to ground when the voltage applied to its noninverting input falls below the reference voltage provided by *R2*. This occurs when *Q1* is illuminated.

When *Q1* is dark, the voltage at the noninverting input of the op amp rises above the reference voltage. The comparator output then swings from ground potential to +5 volts. Potentiometer *R2* can be adjusted to alter the light level at which the comparator switches. Those readers who have experience with op amps are probably wondering about the function of potentiometer *R3*. In a working version of this circuit lacking *R3*, the output voltage when *Q1* is illuminated can be greater

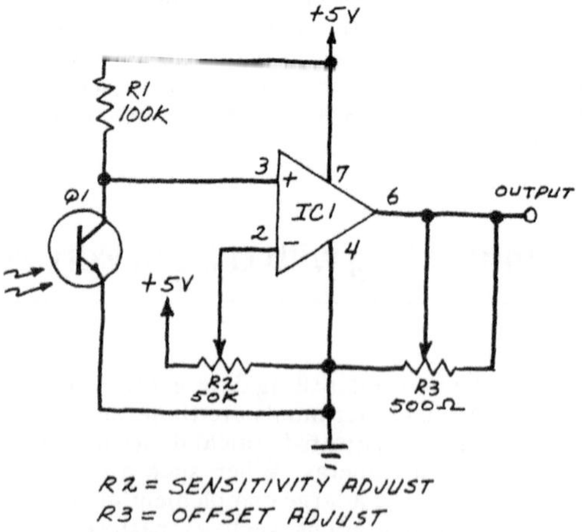

Fig. 3. An expansion of the circuit in Fig. 2. using an operational amplifier for higher gain.

than 1 volt. This exceeds the maximum allowable TTL logic 0 level of about 0.85 volt. Therefore, if TTL logic is to be controlled by the circuit shown in Fig. 4, it is necessary to adjust *R3* to pull the output down a few tenths of a volt.

Incidentally, the basic phototransistor circuit shown in Fig. 2 will directly drive TTL logic without the need for a pull-down resistor. However, it is less sensitive than the circuit that appears in Fig. 3.

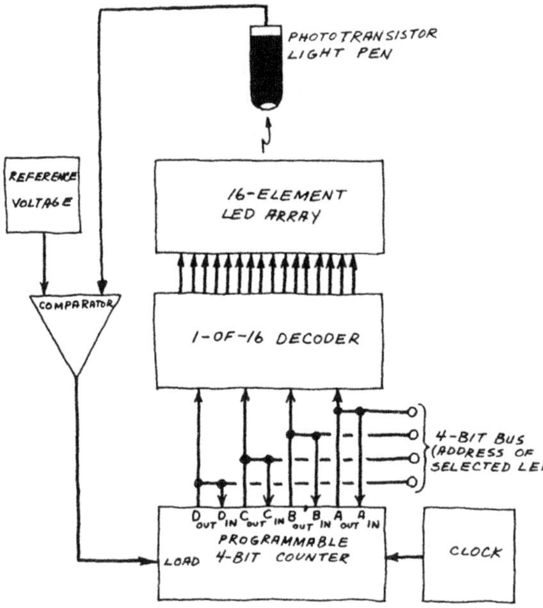

Fig. 4. Block diagram of a 16-position data-entry circuit controlled by a homemade light pen.

Light-Pen Data-Entry Circuit. It's relativitely easy to design a data-entry circuit controlled by a homemade light pen. Figure 4 is the block diagram of one such circuit I've designed.

The operation of the circuit is straightforward. The clock supplies a stream of pulses to a programmable 4-bit counter. The counter's binary output is decoded by a 1-of-16 decoder which sequentially illuminates each of sixteen LEDs.

When the light pen is dark, the LEDs are scanned at a rate determined by the clock frequency. When the light pen is brought near any of the LEDs, nothing happens until that LED glows during the scan sequence. The output from the comparator then changes state and causes the counter to be loaded with whatever data is present at its data inputs. Because these data inputs are connected to their respective outputs, the current count is loaded into the counter. This freezes the counter even though the clock continues to supply pulses to it. The address of the selected LED then appears on the 4-bit bus.

Incidentally, the usual way to block clock pulses is to insert a gate between the clock output and the counter input. The method employed here eliminates the need for such a gate.

Figure 5 is the schematic diagram of my prototype data-entry circuit. A 555 timer operating in the astable mode (*IC3*) serves as the circuit's clock. The clock frequency, and hence the LED scan rate, can be adjusted by means of potentiometer *R5*. You can also increase the value of *C1* to slow the scan rate.

Counter *IC4* is a 74193 programmable up-down counter. Note how the programming data inputs are tied to their respective outputs. The 1-of-16 decoder (*IC2*) is a 74154. The anodes of the 16 LEDs connected to the decoder outputs are all tied to a single current-limiting resistor because only one LED is illuminated at any given instant.

The light pen circuit appears above the LED array. Note that the output of operational amplifier *IC1* is connected to the LOAD input of counter *IC4*.

You can assemble a working version of this circuit on a solderless breadboard in less than an hour. The selection of devices for use as *IC1* and *Q1* is not critical. Any general-purpose op amp such as a μA741 is suitable, and any standard npn phototransistor such as the FPT-100 can be used. The phototransistor should be connected to the circuit by means of clip leads. Power can be provided by a +5-volt supply or you can use a 6-volt battery if you first connect the cathode of a 1N4001 diode to those points in the circuit marked +5 volts and the anode to the battery's positive terminal.

Fig. 5. Schematic diagram of a complete data-entry circuit. The circuit for the light pen appears above the LED array.

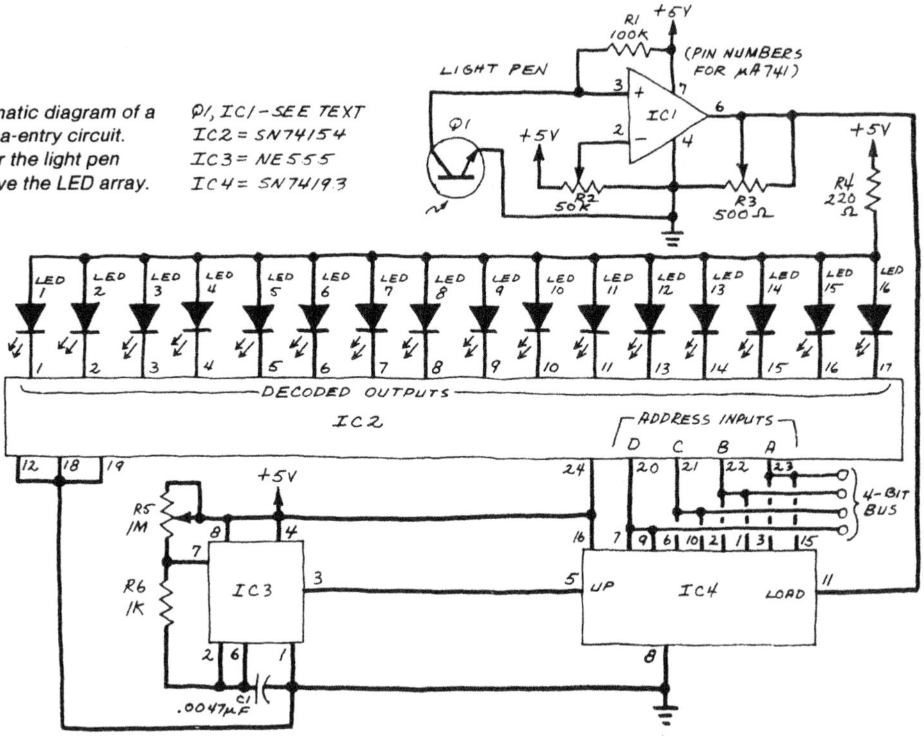

When you apply power to the circuit, the LEDs will either flash off and on in rapid sequence or all the LEDs will appear to glow dimly. If the latter occurs, the clock frequency is so great that the LEDs switch on and off faster than your eyes can respond. For initial tests, adjust *R5* to achieve this latter condition.

Before attempting to use the circuit, you must trim the light-pen circuit. A trial-and-error approach will eventually produce useful results, but a much better approach is to temporarily disconnect the grounded lug of potentiometer *R3* from ground and connect a voltmeter between the output of the op amp and ground. Illuminate *Q1* with a flashlight and adjust *R2* until the output voltage of the op amp falls to its lowest value, which was approximately 1.2 volts in the prototype circuit. Don't turn the rotor of *R2* beyond this point once you have found it.

When you remove light from *Q1*, the output voltage of the op amp should immediately increase several volts. (It reached 3.4 volts in the prototype.) The light pen is now adjusted for maximum sensitivity. Indeed, it is probably so sensitive that ordinary room lighting will be able to switch the comparator. Therefore, you should wrap a cylinder of black electrical tape one-half inch in diameter around *Q1* to block ambient light. Heat-shrinkable tubing can also be used for this purpose.

Next, reconnect the lug of potentiometer *R3* to ground and again illuminate *Q1* with a flashlight. Adjust the rotor of *R3* until the LEDs stop scanning and only a single LED remains on. The light-pen circuit is now trimmed and ready for use.

Test the circuit by bringing the aperture of *Q1* close to any of the LEDs in the array. Depending upon the scan frequency, the selected LED should immediately or very quickly glow brightly and all the remaining LEDs will darken. The binary address of the selected LED will then appear on the 4-bit bus between *IC2* and *IC4*.

It's interesting to move *Q1* back and forth along the row of LEDs and watch them appear to track its movements. For best results, the scan rate should be adjusted so that all the LEDs glow dimly when none has been selected. ■

Experimenting with a Light Pen, Part 2

Adding a Bus Register. The data terminal which was described in Part I can be made more compatible with external circuits by adding a register to its 4-bit bus. The register will ignore any logic signals on the bus until a WRITE switch issues a command to load the register with whatever is on the bus.

Figure 1 shows one simple way to add such a register. The register, a 74175 quad D flip-flop, follows the bus data when its LOAD input is brought to logic one by means of the WRITE switch. The data remains in the register until the WRITE switch is again toggled from HOLD. The contents of the register can be cleared to 0000 by toggling the RESET switch from HOLD to CLEAR.

You'll need to insert three-state transmission gates between the register outputs and the bus. Alternatively, you can use a 74173 4-bit D register with self-contained three-state outputs. This approach is shown schematically in Fig. 2. Note that the 74173 has more control inputs than the 74175. The system clock loads into the 74173 the data present at its inputs when both DATA ENABLE inputs (pins 9 and 10) are grounded by means of *S1*. This means that you must be sure to depress *S1* for at least one clock cycle—which can be a significant interval when the clock rate is very slow.

The outputs of the 74173 and therefore the output bus reflect the data stored in the chip when both OUTPUT CONTROL inputs (pins 1 and 2) are grounded. Should either OUTPUT CONTROL input go to $+V_{cc}$, the outputs enter the high-impedance state and, for practical purposes, the 74173 disconnects itself from the output bus.

A Light-Pen-Controlled LED Display. The basic light-pen data-entry terminal forms the nucleus of a 16-element LED display that can be illuminated in a pattern selected by the light pen.

If, for example, the LEDs are arranged in five rows, four having three LEDs and one having four, then each of the ten decimal numerals plus a decimal point can be formed. Figure 3 shows one possible way to form each decimal numeral or any of the letters of the alphabet on such a display. As you can see, in spite of the limited number of display elements, the legibility and appearance of the characters generated are surprisingly good. Arranging the LEDs in a 4 x 4 array makes possible the display of many graphic symbols and some upper- and some lower-case letters. Figure 4 shows some examples.

Hopefully, you are by now as interested as I've long been in experimenting with a circuit having such capabilities. Assuming that you are already familiar with the basic light-pen data-entry terminal described in Part I, we can now begin modifying that circuit for video-graphics applications.

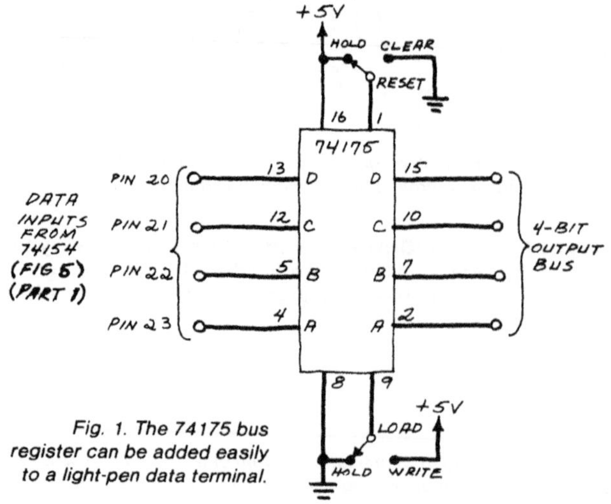

Fig. 1. The 74175 bus register can be added easily to a light-pen data terminal.

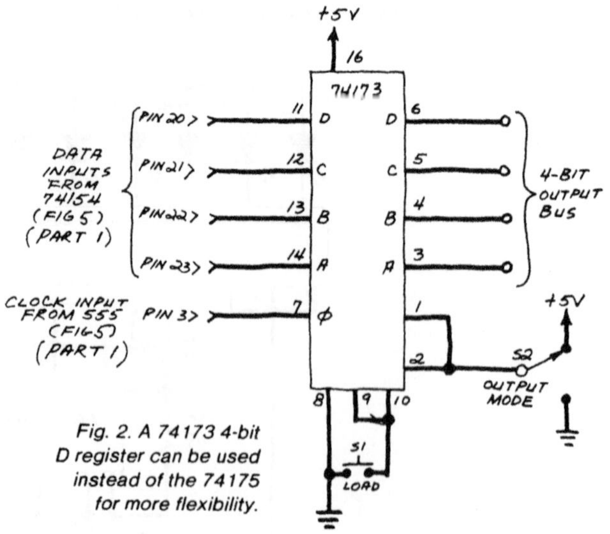

Fig. 2. A 74173 4-bit D register can be used instead of the 74175 for more flexibility.

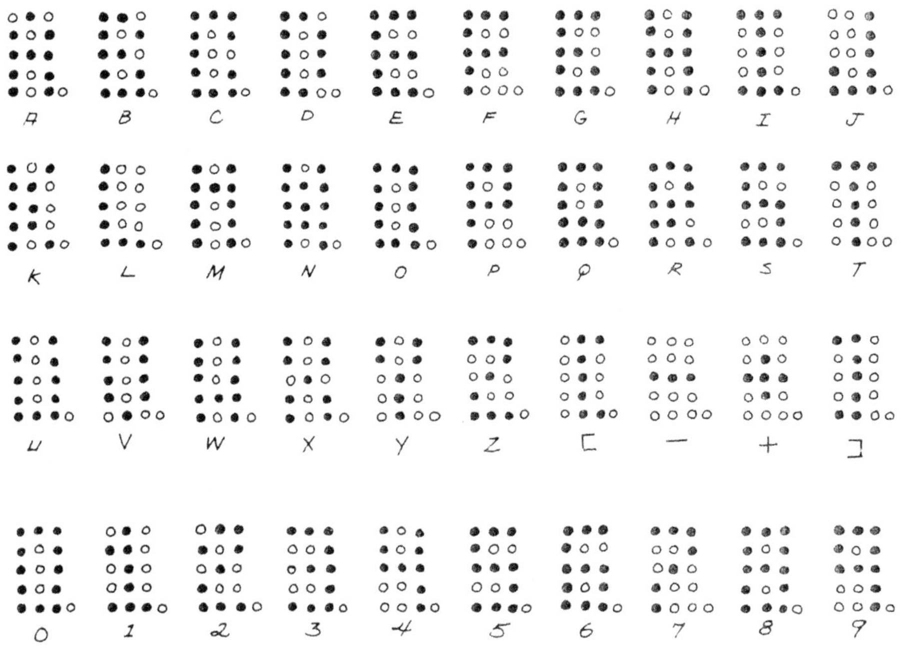

Fig. 3. Letters, symbols, and digits created by 5-by-3 dot matrix.

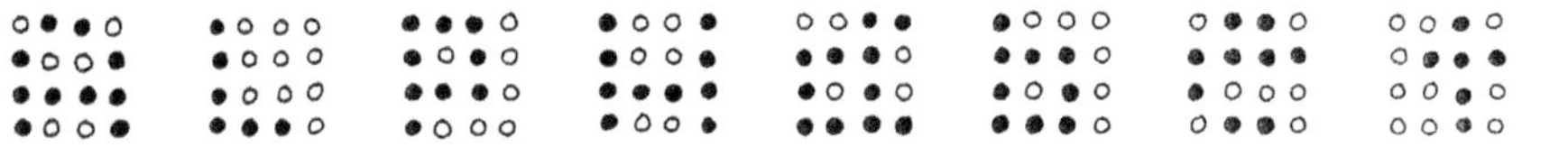

Fig. 4. Upper- and lower-case letters of the alphabet on a 4-by-4 array.

Two principal requirements must be satisfied. First, the circuit must be able to remember each LED location selected by the light pen. Second, the selected LEDs must be substantially brighter than the unselected LEDs.

The solution to the first problem is simply to add a RAM. The second requirement is trickier. Ideally, only the selected LEDs should glow. That's impossible, however, because all of the LEDs must be sequentially strobed to make them eligible for future selection by the light pen. There are several solutions to this apparent contradiction. You can better understand the one that I chose by referring to the block diagram of the complete light-pen-controlled display in Fig. 5.

If you compare Fig. 5 of this part with the block diagram of the light-pen data-entry terminal (Fig. 5 of Part I), you'll immediately notice several important similarities. For example, the configurations of the clock, counter, decoder, bus and light pen are identical in both circuits.

You will also notice some important additions to the circuit. One major addition is a 1 x 16-bit RAM whose address lines are connected to the 4-bit output bus. Also, a LOAD switch connected to the RAM has been incorporated into the light pen.

When a particular LED has been selected by the light pen, closing the LOAD switch records its new status in the RAM. The RAM is able to keep track of the selected LED because

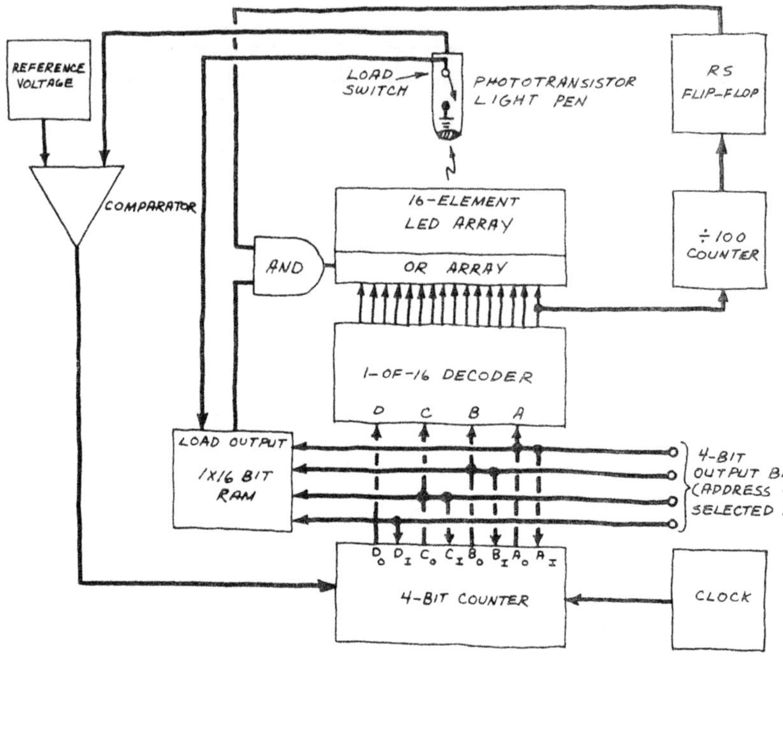

Fig. 5. Block diagram of a light pen controlling a 16-bit LED display.

the count supplied to the decoder is equivalent to the address furnished to the RAM.

Notice the AND gate that has been connected to the OR array between the decoder and the LEDs. This gate network permits the LED array to be strobed and therefore to display selected LEDs. Without the assistance of the additional logic, the multiplexing action would be divided equally betweeen the LEDs being sequentially strobed by the scanning circuit and the LEDs selected by the light pen. This means that both the scanned and the selected LEDs would appear equally bright.

This problem is solved by means of a divide-by-100 counter and a set-reset flip-flop. Connecting the flip-flop to the counter results in a combined circuit that is a modified divide-by-10 counter. The output of the modified counter is at logic 0 for ten clock pulses. It then goes to logic 1 for 90 clock pulses. The cycle then repeats.

The outputs of the RAM and the counter are ANDed and the result ORed with the decoder outputs. This is done so that the LEDs selected by the light pen are strobed 90 times during an interval of 100 clock pulses. All of the LEDs are then strobed 10 times during the remainder of the 100-pulse clock interval. The net effect is that the LEDs selected by the light pen are substantially brighter than the remaining LEDs.

It's even possible to extingui[illegible]elected LEDs for intervals of one second or mor[illegible]own the clock. The selected LEDs will appear [illegible]ously and the other LEDs will blink on ever[illegible] during the scan period.

Figure 6 is the [illegible] practical circuit that corresponds to the [illegible] of Fig. 5. Although the circuit appears muc[illegible]lex than the relatively simple light-pen data-in[illegible] described in Part I, that circuit forms the core of [illegible] You can verify this for yourself by noting the almost identical connections of the 555 timer, 74193 counter, 74154 decoder and the light pen comprising *Q1* and operational amplifier *IC6*.

The lowest-order bits in each nibble of a 7489 4-by-16-bit RAM (*IC9*) provide a 1-by-16-bit RAM for the circuit. Although three-fourths of this RAM is not utilized, it's always available should you wish to expand the display.

Two series-connected 4017 CMOS decade counters (*IC10* and *IC11*) and an RS flip-flop made from two gates in a 7400 (*IC5C* and *IC5D*) comprise the circuit's modified divide-by-10 counter. The remaining two gates are configured as a gate that ANDs the outputs of the RAM and the divide-by-10 counter. The AND output is ORed with each of the 16 outputs of the 74154. The LEDs are illuminated sequentially.

Modifying the Circuit. There are several modifications that can be made to the circuit that was just presented. You can eliminate counter *IC11* by connecting pin 10 of *IC5* to pin 11 of *IC10*. This will provide divide-by-10 operation, but the unselected LEDs will be strobed once for every nine times that the selected LEDs are strobed.

There are several ways to substitute other memories in place of the 7489 (*IC9*). You can, for instance, use a MOS or CMOS 256-by-1-bit RAM if you prefer. While only the first 16 bits will be used, the remaining 240 will be available for future expansion of the circuit. You can eliminate RAMs entirely by using an array of flip-flops. The resulting circuit, however, will employ more ICs than the RAM version.

A Long-Range Light Pen. During my experiments with the light-pen circuits that have been described in this two-part series, it often occurred to me how convenient it would be to have a long-range light pen. This would not be possible with red LEDs, however, because the optical power typically radiated by red emitters is measured in tens of microwatts. Also, their spectral emission peaks at approximately 670 nm, halfway down the response curve of most phototransistors.

An infrared LED made from gallium arsenide is much more powerful than a red LED. When, for example, a forward current of 20 mA is flowing, an infrared LED might emit more than one milliwatt of optical power. Also, its near-infrared emission corresponds closely to the wavelengths at which a silicon phototransistor exhibits peak response.

On the assumption that an infrared LED should increase the detection range of the light pen, I connected a General Electric 1N6266 near-infrared emitter in series with one of the red LEDs in the display in Fig. 6. The detection range increased from a fraction of an inch to several inches. I then removed the red LED and connected the infrared emitter in its place. This increased the current through the infrared emitter and resulted in a further increase in the detection range.

Of course, visible emitters must be used in light-pen-controlled displays. Therefore, I tried a GE SSL-3 LED which

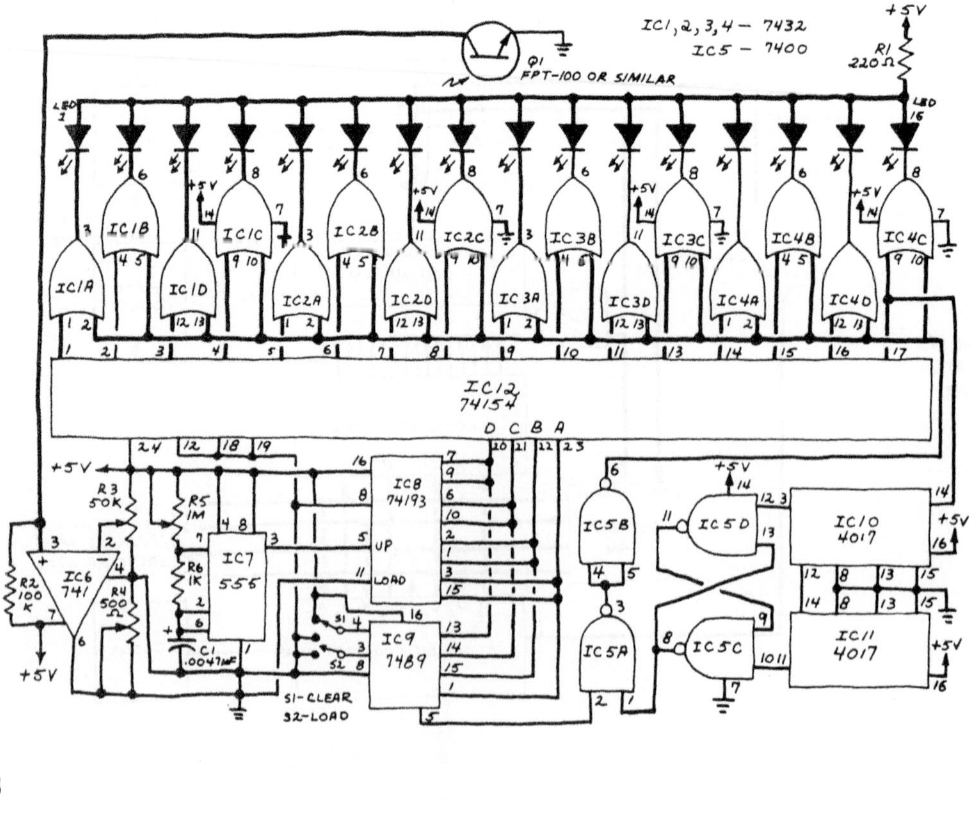

Fig. 6. Complete circuit of a 4-by-4 LED display controlled by a light pen.

emits both near-infrared *and* visible green light. This LED, which is no longer manufactured, consists of a gallium arsenide chip coated with an infrared-sensitive phosphor. When forward biased, the chip emits infrared light. This stimulates the phosphor into emitting green light. The result is a visible green beam superimposed upon an invisible infrared beam.

Using the SSL-3 resulted in a light-pen detection range of several inches. But it proved impractical to use an array of such LEDs in the display because the low duty cycle resulted in a barely visible green glow.

My final attempt to extend the light pen's detection range used a simple pnp driver delivering several hundred milliamperes to the LED during each strobe pulse. This resulted in the radiation of enough infrared power to trigger the light pen at a distance of up to ten inches.

The range resulting from the use of infrared LEDs can be further extended by adding a collection lens to the light pen. Theoretically, the detection range will be doubled each time the collection area is doubled. Adding lenses to the infrared emitters is not advisable. This would restrict the detection region of the light pen to sixteen narrow cones of invisible light. The use of only the self-contained lenses of typical infrared LEDs results in a much broader detection region. ■

7. A Light-Sensitive Tone Generator

IF YOU like circuits which respond to light, you'll love the light-sensitive tone generator shown in Fig. 1. Most light-sensitive tone generators have a single light-sensitive component. The circuit in Fig. 1 has *two*. This provides an unusual up-down audible tone.

In operation, the 741 is connected as an oscillator which produces a clipped sine wave at its output. The output waveform therefore resembles a square wave with slow rise and fall times.

Photocells *PC1* and *PC2* control the frequency of oscillation of the 741. For best results, use cadmium sulfide photocells having a very low resistance when illuminated and a high dark resistance. Most of the photocells available from electronics part suppliers that cater to hobbyists and experimenters fall in this category.

When both *PC1* and *PC2* are dark, the frequency of the 741 oscillator will fall to about 700 Hz. If *PC1* is illuminated with a small flashlight while *PC2* remains dark, the frequency will suddenly fall to a few Hz and then rise quickly to 1 kHz or more.

If *PC2* is illuminated with the flashlight while *PC1* remains dark, the frequency will suddenly rise to 3 kHz or more. When both *PC1* and *PC2* are illuminated, the frequency will level off at about 1.2 kHz.

I made these measurements when the wiper of balance potentiometer *R3* was at its center position. The setting of *R3* can be altered to produce a wider range of tone frequencies.

The signal from the 741 is converted into an audible tone by an LM386 audio amplifier. Potentiometer *R4* serves as a gain control. When you first apply power to the circuit, make sure *R4's* wiper is rotated toward the ground connection. Otherwise the tone from the speaker may be uncomfortably loud. **Caution**: Don't substitute an earphone for the speaker! The sound level may be too high.

This circuit has more than its obvious novelty value. You can, for example, use it as a light-controlled sound effects generator. For this application both photocells should be placed at one end of individual plastic or cardboard tubes which have been coated on their inside with black paper or paint. The circuit can then be "played" by flashing light down the tubes or by blocking a continuous light source with your finger tips.

The circuit has educational value as well. It nicely demonstrates the feedback which makes possible the conversion of an amplifier into an oscillator. And it demonstrates the "memory effect" of cadmium sulfide photocells. You'll quickly notice this phenomenon while experimenting with the circuit. A cadmium sulfide photocell requires several seconds or even minutes to resume equilibrium following the removal of light, and this gives rise to a gradually changing tone even when the light source is removed.

Finally, you can replace the two photocells with two potentiometers in a joystick to obtain full manual control of the circuit. Pushing the stick back and forth will sweep the output frequency across its full spectrum. Moving the stick in a circular fashion will produce a sound like a siren. ■

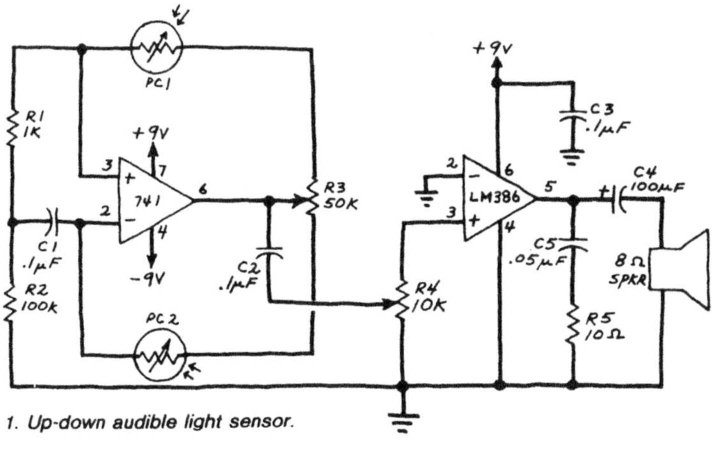

Fig. 1. Up-down audible light sensor.

Three
LED Circuits

1. A High-Resolution LED Display

A thin, high-resolution, two-dimensional display with X-Y addressability would have many applications in such fields as television, oscilloscopes, electronic games, micro-computer displays and alphanumeric and graphic data displays for pocket calculators and data loggers.

Experimental flat-screen displays have been made based on gaseous discharge, electrofluorescence, light-emitting diodes and liquid crystal technologies. The liquid crystal method appears to offer the greatest economy and certainly the thinnest configuration, but this display medium cannot yet change states fast enough for television applications.

The technology needed to build large-area LED displays has been available for a decade, but the high cost of the LEDs themselves and the addressing circuits they require has thus far restricted their use to military and laboratory applications. Now that low-cost visible LEDs are available, you can assemble a 16 × 10, 160-element LED display for less than $20—assuming you can procure the LEDs for less than 10¢ each.

Figure A is the circuit diagram of the array. The ten 330-ohm resistors limit current to the LEDs, providing about 10 mA to each LED if a 5-volt power supply is used.

The exact construction method employed in the assembly of the display depends on the lead arrangement of the LEDs. Figure B is a photograph of the 160-element display assembled on a perforated board with 0.1-inch hole centers and a copper solder pad at most holes (Radio Shack 276-1551 or similar).

I used yellow LEDs, but you can use red or green LEDs if you prefer. I also painted the LED side of the board black *before* installing the LEDs to enhance the display's contrast. The current-limiting resistors can be seen near the lower left of the display.

Although the electrical circuit of the array is very simple, its construction requires a good deal of patience. First, all the LEDs must be soldered to the board. That alone requires 320 separate solder connections. Then all the anodes in each horizontal row and all the cathodes in each vertical column must be connected together with bus wires. This requires 320 additional solder connections.

The resistors and output connections to the board's copper fingers require another 72 solder connections, resulting in a total of 712 solder connections! Don't be discouraged though. I was able to complete the board shown in Figure B in less than four hours—and that included plenty of short breaks to relieve eye strain.

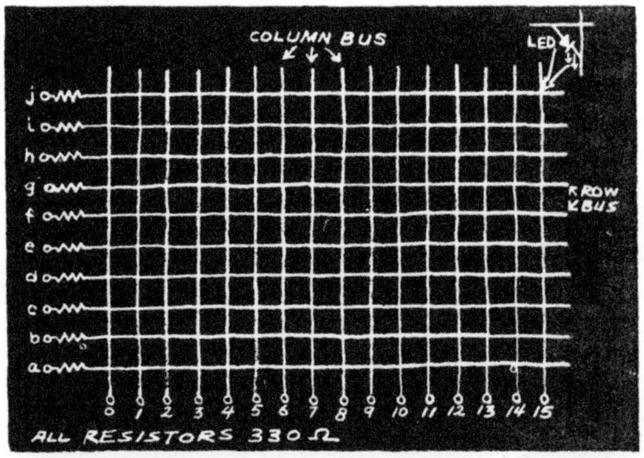

Fig. A. 10 x 16 160-element LED display.

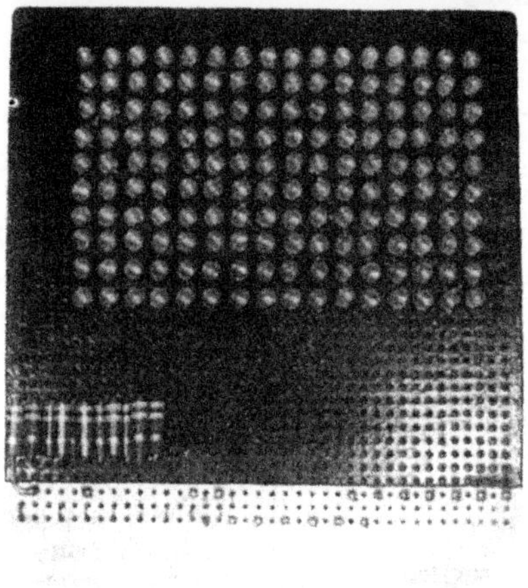

Fig. B. Prototype 160-element flat-screen LED array.

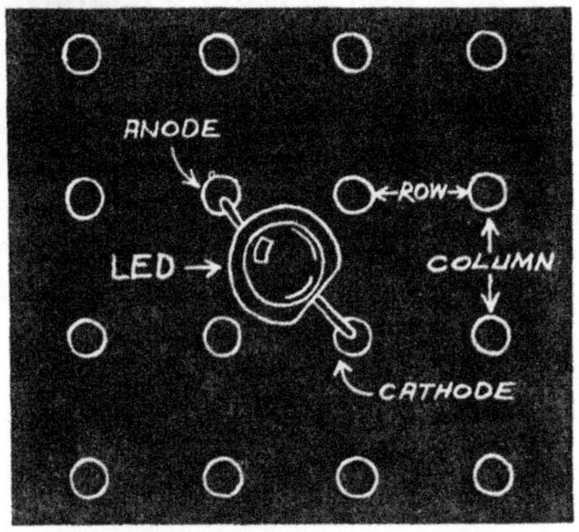

Fig. C. Orientation of individual LED in array.

The following tips will simplify the assembly of the display board:

● Select LEDs with leads parallel to their viewing axis and make sure the leads fit the holes in the board you select.

● Test *each* LED *before* it is soldered in place. You can make a temporary test jig from a 6-volt battery, 330-ohm resistor and some clip leads.

● Install one column of ten LEDs at a time. Make sure the diodes are oriented as shown in Figure C. The cathode lead is usually indicated by a notch or flat area in the epoxy encapsulant.

● Bend the leads of each LED outward slightly on the back side of the board. Turn the board over and place it on two supports so the LEDs hang from their leads.

● Use a low-wattage iron and small-diameter solder to solder *one* lead of each LED to its copper foil pad. Turn the board over and make sure the LEDs are aligned properly. Then solder the remaining leads.

Follow these steps to solder all sixteen columns. Be sure to keep the LEDs perfectly aligned for best results. If you have trouble keeping the columns straight, tape a pencil to the board adjacent to each column while the diodes are being soldered.

Use tinned, small-diameter (e.g. No. 26) wire or stripped Wire-Wrap wire for the column and row buses. Solder the row buses first. The easiest way is to lay a wire along the anode leads in one row so the wire touches each solder pad. Then use a very small amount of solder to tack the wire to each pad. The column buses must be soldered *above* the row buses since the bus wires are uninsulated. Figure D is a small portion of the completed board.

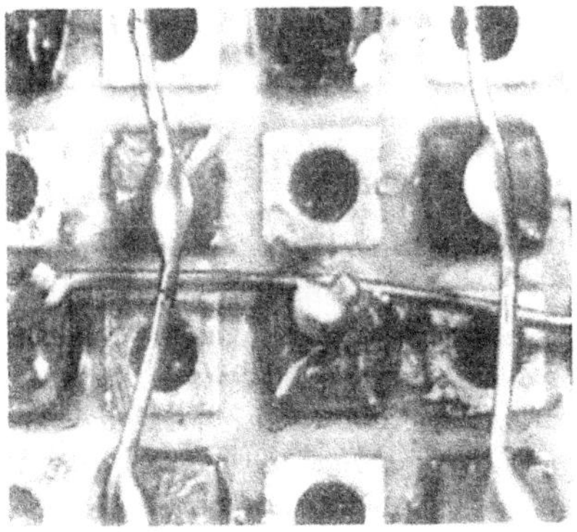

Fig. D. Row and column buses are soldered to LED leads on back of board.

The display board is completed by soldering lengths of Wire-Wrap wire from points a through f and 0 through 15 in Fig. A to the copper fingers on the board. Try to select an orderly pin-connection arrangement to simplify the interface to a driving circuit. The connection pattern can be considered a bus, and the bus connections for the prototype board that I built are listed according to the pin designations of a 44-terminal edge connector socket (the pin designations are marked on the socket):

LED Array Connection	S-44 Socket	LED Array Connection	S-44 Socket
0	7	13	20
1	8	14	21
2	9	15	22
3	10	a	A
4	11	b	B
5	12	c	C
6	13	d	D
7	14	e	E
8	15	f	F
9	16	g	H
10	17	h	J
11	18	i	K
12	19	j	L

2. LED Bargraph Display Chips

THE HEART of many LED bargraph circuits is the quad comparator, a chip that contains four independent comparators. Connecting two or three such chips to a voltage divider comprising a string of series-connected resistors (Fig. 1) results in a straightforward but complex bargraph readout.

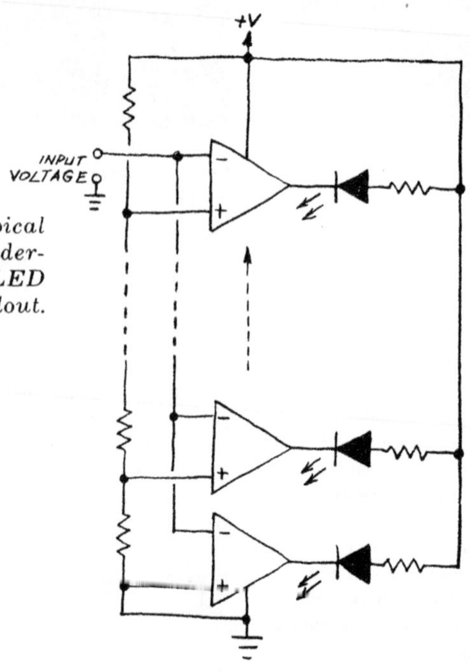

Fig. 1 Typical voltage divider-comparator LED bargraph readout.

Recently, however, semiconductor manufacturers in the United States, England, and Japan have announced new ICs that combine on a single chip the voltage divider and comparators required for a multiple-level bargraph LED display. The new chips have many fascinating applications and are very easy to use. This month, we'll take a close look at three of these chips: Texas Instruments' TL490C/TL491C and National Semiconductor's LM3914.

TL490C/TL491C Bargraph ICs. With the exception of their outputs, these two 10-step analog level detectors are functionally identical. Each contains a resistor voltage divider and ten comparators. They will light a 10-element row of LEDs in adjustable increments of 50 to 200 millivolts per LED.

Both chips incorporate output transistors that allow direct drive of the LEDs. The TL490C has open-collector outputs capable of sinking as much as 40 mA at 32 volts max. The TL491C, on the other hand, has open-emitter outputs capable of sourcing a maximum of 25 mA at up to 55 volts. Figure 2 shows how LEDs are connected to both a current source and a current sink.

These new devices are very easy to use. Figure 3, for example, is a simple 10-element readout. I assembled it on a solderless breadboard using a Texas Instruments data sheet as a guide. Potentiometer *R1* provides a variable voltage to the circuit for demonstration purposes. Varying the setting of *R1* lengthens or shortens the bar of glowing LEDs as the input voltage increases or decreases.

Note that the circuit requires a supply of 10 to 18 volts for proper operation. The chip *can* be powered by a 9-volt battery; but, if that is done, the highest-order LED will fail to glow. A pair of 9-volt batteries connected in series makes an excellent power supply for portable operation. Be sure to use alkaline batteries for best results.

Both the TL490C and TL491C incorporate a THRESHOLD input that allows the sensitivity of the bargraph readout to be

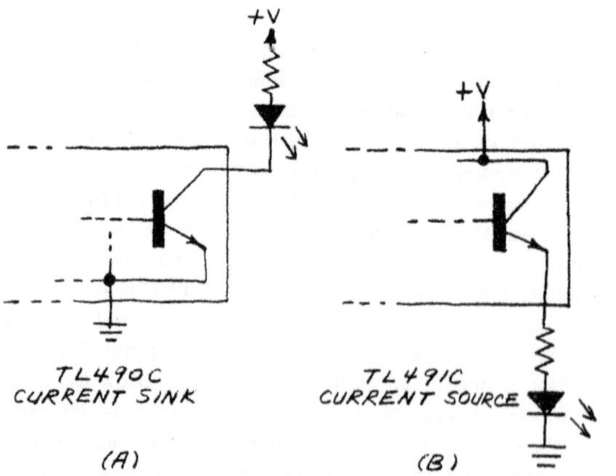

Fig. 2. LEDs connected to current sink (A) and current source (B).

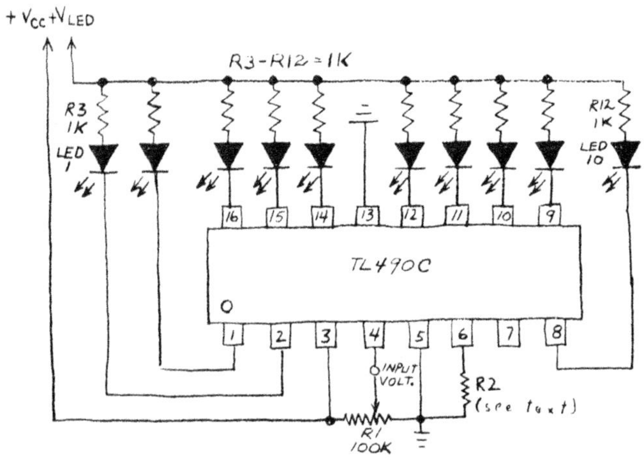

Fig. 3. Basic 10-element bargraph readout using TL490C.

varied from 200 millivolts per LED to 50 mV/LED. This is accomplished by connecting pin 6 to ground via a series resistor. TI provides an elaborate formula for calculating the input voltage required to activate the first LED: $0.84/V_{IN} = 1 + (R2 + 700)\ (2240)/(700R2)$ where V_{IN} is the threshold voltage and *R2* is the resistance between pin 6 and ground.

If this formula seems a little cumbersome, connect a 1000-ohm potentiometer between pin 6 and ground and adjust *R1* so that the first LED just begins to glow. The input threshold voltage can then be measured by placing the probes of a multimeter between the wiper of *R1* and ground. Of course, if you prefer to work with figures, you can algebraically manipulate and simplify the given equation, solving it for V_{IN}.

Considering the number of comparators within each IC package, the total current drain of one of these chips is moderate when no LEDs are on. However, with even a fairly high-value current-limiting resistor connected in series with each LED, current consumption is substantial when all the LEDs are glowing. (Each output pin can sink up to 60 mA of current.) Here are representative values I measured when a TL490C was connected as shown in Fig. 3.

Total Current Demand

V_{CC}	+10 V	+12 V
No LEDs on	11 mA	11 mA
5 LEDs on	54 mA	64 mA
10 LEDs on	93 mA	114 mA

The basic circuit in Fig. 3 has many interesting applications. With *R1* removed and a CdS photocell connected from the positive supply to the input of the IC (pin 4), the circuit functions as a light meter. As the light level at the sensitive surface of the photocell is increased, more LEDs will glow. The photocell may respond to light from the LEDs so be sure to point its sensitive surface away from the rest of the circuit.

You can even measure resistance with the circuit by connecting a resistor between V_{CC} and pin 4. Moisten your index fingers and touch these two points if you want to see the LEDs respond to your body resistance.

For practical, "ballpark" resistance measurements, you'll need to calibrate the circuit with some resistors of known values. If my preliminary results are valid, the circuit will not necessarily respond in a linear fashion to resistance changes.

Connecting a capacitor between pin 4 and ground provides an interesting demonstration of the effects of capacitance. Assuming the capacitor is discharged initially, all the LEDs will glow when the capacitor is first connected to the circuit. They will then wink off in sequence as the capacitor charges. For best results, use a component with a large amount of capacitance (at least 1000 microfarads). Smaller capacitors charge too quickly to allow you to follow the flashing LEDs.

Both the TL490C and TL491C include a CASCADE input that permits the user to cascade up to ten chips to form a 100-element bargraph. Figure 4 shows two TL490C chips connected in cascade. Note how a two-resistor voltage divider provides a 2-volt bias for the CASCADE input of the second of two TL490Cs. The second chip subtracts this reference voltage from the input voltage at pin 4 to automatically arrive at the correct threshold.

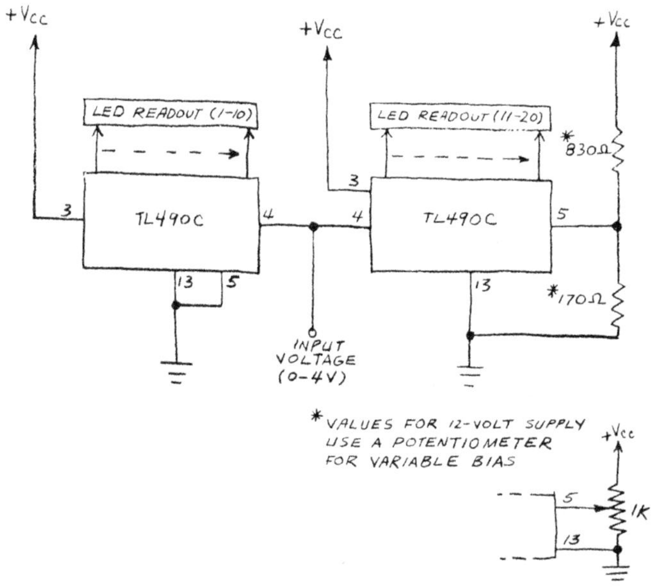

Fig. 4. How to cascade two TL490C chips. A voltage divider provides bias for cascade input of second chip.

LM3914 Dot/Bar Display Driver. This new National IC does everything the TI chips do and more! Like them, it has a self-contained voltage divider and ten comparators, the nucleus of a 10-element bargraph readout. Of more importance, however, is its self-contained decoding network that converts the chip from a straightforward bargraph driver into a more sophisticated moving-dot driver. A single MODE CONTROL input (pin 9) allows easy selection of either mode.

The National Semiconductor data sheet doesn't explain how the LM3914 achieves moving-dot operation. The moving-dot circuits I've described in previous columns required a fair amount of logic to convert a bargraph output into a moving-dot readout. It would be interesting to discover which approach National has selected.

Why is the moving-dot mode so important? One application that comes immediately to mind is a simplified solid-state os-

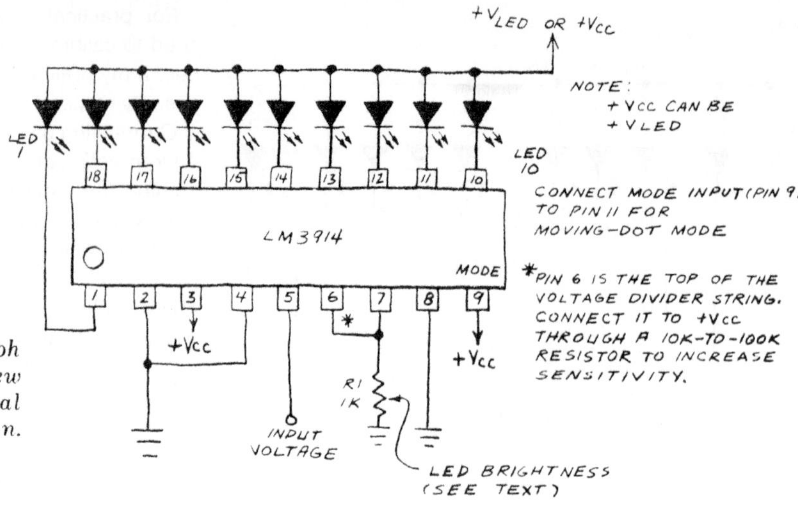

Fig. 5. A 10-LED bargraph readout employing the new LM3914 IC from the National Semiconductor Corporation.

cilloscope with a LED screen. More about this later. Another advantage of the moving-dot readout is that one of ten outputs can be selected by a variable voltage. Think of the possibilities! You can connect one or more outputs to relays, drive transistors, optoisolators, or SCRs. In this way, you can make motors, alarms, and many other devices responsive to such variables as changing temperature, humidity, wind speed, weight, pressure, light, or any other analog function that can be converted into a continuously variable voltage by a suitable low-cost transducer.

Figure 5 shows how to use the LM3914 as a basic bargraph driver. Compare this circuit with the TL490C version in Fig. 3. You'll note the circuits are very similar. One major difference, however, is the use of a fixed resistor (*R1*) to control the brightness of the LEDs. This single resistor effectively programs the current available to each LED, thereby eliminating the need for individual current-limiting resistors.

The operation of *R1* as the LED-brightness control is dependent upon an internal reference voltage available at pin 7. The current passing from pin 7 through *R1* to ground is approximately one-tenth of the current passing through each illuminated LED. Since the voltage reference output is typically 1.3 volts, the LEDs will receive 13 mA of current when the resistance of *R1* is 1000 ohms. (Why? Ohm's law says that the current flowing through a resistor equals the voltage across it divided by its resistance. In this case, the current is 1.3/1000 or 0.0013 A. The LED current is ten times greater or 13 mA.)

To use the LM3914 in the moving-dot mode, mode control pin 9 should be disconnected from the positive supply voltage and connected to pin 11. This modification is easily made to the circuit in Fig. 5.

The LM3914 can be cascaded to form a moving-dot readout having 200 or more elements by connecting pin 9 of the first chip in the series to pin 1 of the next higher chip. This connection pattern is continued for each chip except the last. Pins 9 and 11 of the last IC are tied together. The only other requirement is a 20,000-ohm resistor in parallel with *LED9* of each chip (between $+V_{CC}$ or $+V_{LED}$ and pin 11) except for the first one. For details, see the National data sheet.

Figure 6 shows an interesting circuit from the National data sheet that causes the bargraph readout to flash. This circuit can be programmed to flash when the input voltage reaches a specified level by connecting the junction of *R1* and *C1* to any of the ten LED outputs. (The LEDs flash when the input voltage is sufficient to activate the selected LED output.) This flashing mode is very noticeable and makes for an eye-catching warning indicator. ■

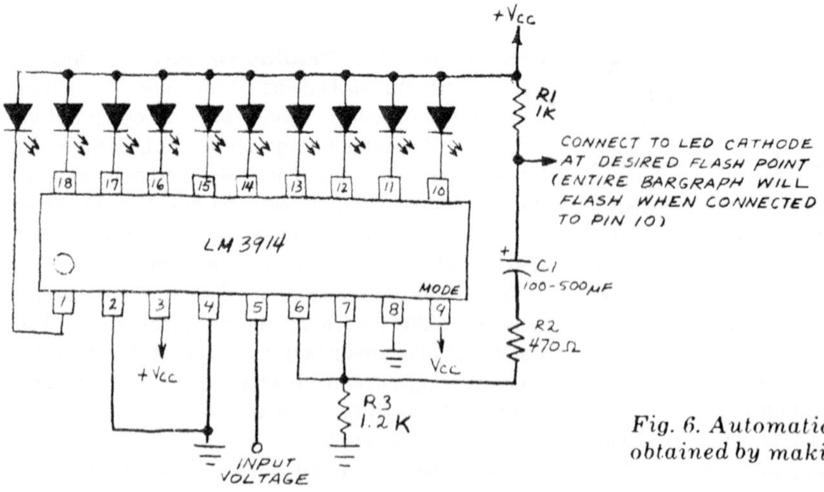

Fig. 6. Automatic bargraph flashing can be obtained by making connection suggested.

3. High-Current LED Pulser

INFRARED LEDs make ideal optical sources for remote controls, intrusion alarms, reflective and break-beam object sensors, signaling devices and TV commercial killers. However, unless an efficient heat sink is employed, most infrared LEDs are restricted to a maximum continuous forward current of no more than 100 milliamperes. At this current, a high-quality GaAs:Si LED will deliver from 6 to 10 milliwatts of optical power. This is roughly equivalent to the *visible* radiation emitted by a small one- or two-cell penlight with a prefocused lamp.

Rapidly pulsing a LED at very high current levels makes it possible to obtain much higher power outputs. For example, a G.E. 1N6264 LED that emits 6 mW at 100 mA of forward current will emit 60 mW when driven by 1-ampere pulses a few microseconds wide.

Figure 1 shows a simple circuit that can deliver high current pulses to an LED. This pulser is considerably more powerful than the LED transmitter module that was the Project of the Month for February 1979. With the parts values shown, it will apply hefty 2.7-ampere pulses at a rate of about 100 Hz to a LED. The pulses are about 17 microseconds wide. They can be readily detected by a simple phototransistor receiver such as the Project of the Month for January 1979. Current drain from a small TR175, 7-volt mercury battery is 5 mA.

Many different LEDs can be used with the pulser. For most LEDs, the peak current exceeds by a factor of three the component's maximum continuous rating. Applying even larger pulses will not necessarily destroy a LED, but might shorten its useful life. For best results, use infrared emitters made from GaAs:Si rather than GaAs diodes. Good choices include the TIL-32 (Texas Instruments), 1N6264 (General Electric), OP-190 and OP-195 (Optron) and 276-142 (Radio Shack).

You might have difficulty finding the transistors specified in Fig. 1. If so, you can substitute a common npn silicon device such as the 2N3904 or 2N2222 for *Q1*. The choice of *Q2* is more critical, however. If maximum current is to be delivered to the LED, *Q2* must be a *germanium* transistor. A germanium pn junction has a smaller forward voltage drop than a silicon pn junction, and this causes a germanium transistor to have a lower effective "on" resistance. The LED therefore receives more current if a germanium device is used.

The 2N1132 works better than any other germanium transistor I've tried. The 2N1305 is easier to find and will deliver about 2 amperes to the LED. If you can't find a suitable germanium transistor you can substitute a common pnp silicon switching transistor such as the 2N3906 or 2N2907. Less current will be delivered to the LED, but the optical output will still be adequate for many applications.

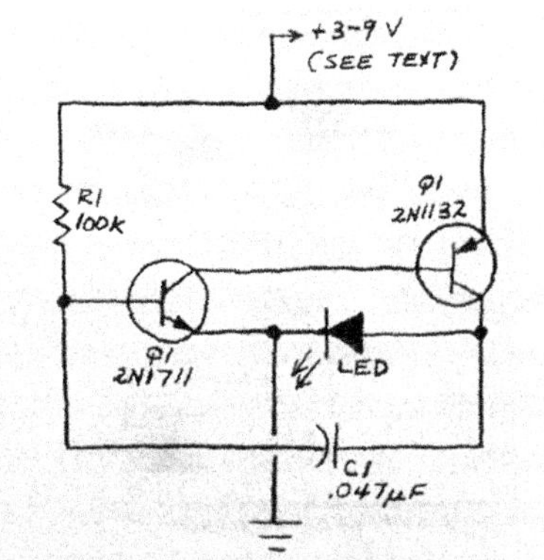

Fig. 1. High-current LED pulser.

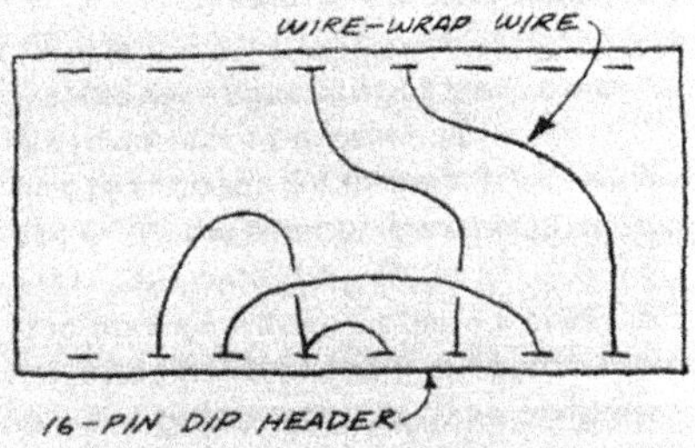

Fig. 2. Connections for LED pulser.

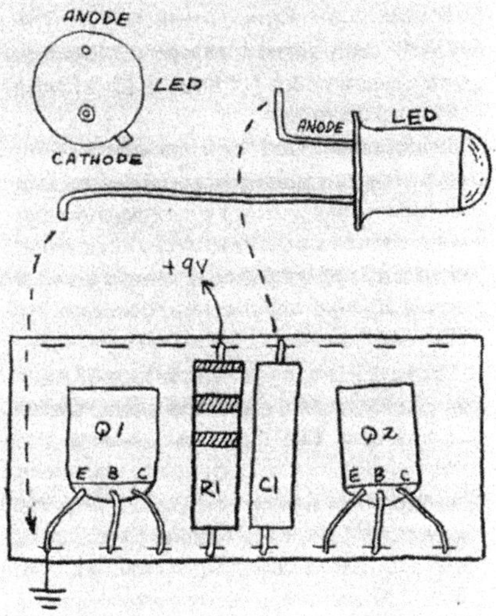

Fig. 3. Pulser component placement.

For example, if *Q1* is a 2N3904, *Q2* is a 2N3906 and the circuit is powered by a standard 9-volt battery, 1.1-ampere pulses will be delivered to a LED. Because of the different characteristics of the silicon transistors, the repetition rate will jump to 1400 and the current demand will increase to about 100 mA. That's enough to quickly deplete even an alkaline battery, so for best results the resistance of *R1* should be increased to reduce the pulse-repetition rate and the operating current. For example, if the value of *R1* is changed to 1 megohm, the repetition rate will decrease to 120 Hz and the current drain to a much more reasonable 8 mA.

Once you've made a final selection of component types and values, you can assemble a permanent version of the LED pulser on a DIP header or postage-stamp-sized perforated board. I took the latter approach for my germanium-transistor unit because the transistors are packaged in TO-5 cans. It was still possible to install the pulser, TR-175 battery, switch and adjustable lens in a brass tube measuring 0.5" × 3.25" (1.3 cm × 8.3 cm).

Figure 2 shows how to assemble the pulser on a DIP header if silicon transistors in plastic packages are used. Interconnect the pins on the header with Wire-Wrap leads, but don't solder them in place yet. Use lengths of wire that are longer than necessary, securing them in place by wrapping their free ends under the header.

Figure 3 shows where the components go. To make things as compact as possible, use a miniature tubular capacitor for *C1* instead of a ceramic disc. Any capacitance from 0.01 μF to 0.05 μF is satisfactory, but the smaller values will increase the pulse-repetition rate and reduce the current to the LED somewhat. If you must use a disc for *C1*, try bending it over the top of the header so that it will present a lower profile and leave room for the LED.

If you use a miniature tubular capacitor for *C1*, the completed circuit will use only half the space in the DIP header's cover. Instead of installing the cover, I clipped all the pins from the header and mounted it on a snap terminal salvaged from a discarded 9-volt battery. The conductive strips at each terminal were trimmed to size and folded over each end of the header to secure it in place. Taking care to observe the polarity, I soldered short connection wires from the header to the two metal strips. The result is a tiny but powerful LED transmitter that snaps directly onto the terminals of a 9-volt battery.

Whether you use germanium or silicon transistors, with a little care you can install the complete pulser in a pen-light, lipstick tube, pill bottle or other small container. Although the germanium unit is more powerful, even the silicon pulser projects a beam that can be received at 1000 feet or more at night using a simple phototransistor receiver—provided you use a 2- or 3-inch lens at each end of the link. ■

4. Tri-State LED Demonstrator

THE TRI-STATE LED is one of the most interesting optoelectronic components available to the experimenter. The most common version incorporates separate red and green LED chips mounted very close to one another in a clear or milky-white epoxy package. The two chips are connected as shown in Fig. 1 in what is called an inverse parallel configuration. This ensures that one of the two diodes is forward-biased regardless of the polarity of the applied voltage.

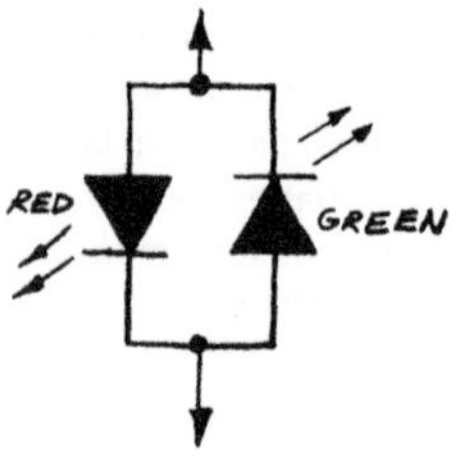

Fig. 1. Schematic symbol for a tri-state LED.

The three states of a tri-state LED usually are defined as red, green and off. Actually, a total of seven optical states is available: off, steady, or flashing red, green, or yellow. Yellow radiation is obtained by rapidly switching the polarity of the applied voltage. The pulsed red and green radiation from the two chips visually merge. Although the color the eye perceives is not a true yellow, it is distinctly recognizable as being neither red nor green.

The schematic diagram of a circuit that has been adapted from one given in the data sheet of Monsanto's MV5491 tri-state LED is shown in Fig. 2. The circuit incorporates two series resistors to provide an optimized current to each LED to balance their brightness. The 1N914 diode (*D1*) bypasses *R2* when the green LED is selected. This compensates for the green LED's higher barrier potential so that the same forward current flows through each diode.

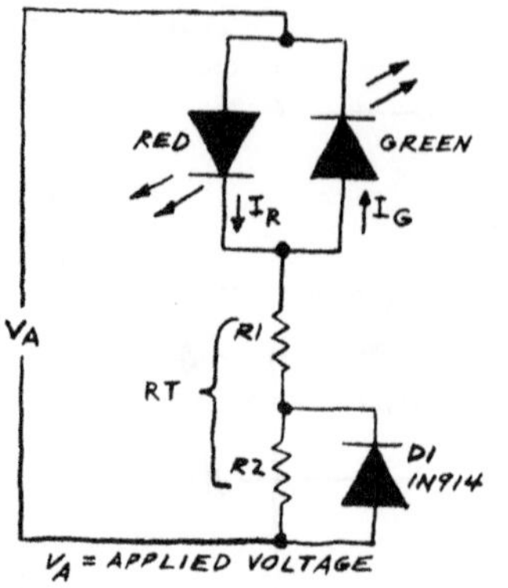

Fig. 2. Circuit used to calculate needed resistances.

The formulas employed to calculate the values of *R1* and *R2* for specific red and green LED forward currents are: $R1 = (V_A - 3.3)/I_G$; $R_T = (V_A - 1.63)/I_R$; $R2 = R_T - R1$; where I_G and I_R are the forward current through the green and red LEDs, respectively, and V_A is the applied voltage. For example, to bias both diodes at 20 mA when V_A is 5 volts, *R1* and *R2* should be 102 and 68 ohms, respectively. The MV5491 data sheet includes a table that gives resistance for *R1* and *R2* for a range of forward currents.

Incidentally, don't be concerned if the exact resistor values the equations dictate are unavailable. Just try to obtain the closest standard value. If you're not concerned with matching brightnesses, simply insert a single 270-ohm resistor in series with the LED when powering it from a 5-volt supply.

Figure 3 is a simple astable multivibrator that demonstrates six of the seven states of a tri-state LED. You can assemble the entire circuit on a miniature solderless breadboard in several minutes. When the wiper of *R1* is at the midpoint of its travel, the LED will alternately flash red and green. The effect is visually striking, particularly if you are used to viewing monochromatic (single-color) LEDs.

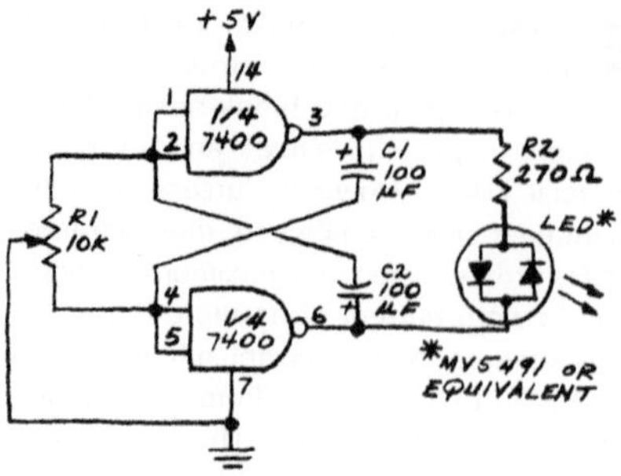

Fig. 3. Tri-state demonstration circuit.

Rotating the wiper of *R1* will increase or decrease the red-green flash rate. At one extreme, the red and green flashes will merge into a washed-out orange or yellow color. Both diodes are still flashing, but the flash rate is faster than the flicker response of the eye. (You can hear the flash rate as a series of clicks by connecting the input of a small audio amplifier to ground and through a 0.1-microfarad capacitor to either pin 3 or 6 of the 7400.) At the other extreme, the LED will stop flashing and glow a steady red or green depending on the direction it is connected.

So far, we've accounted for five of the seven states. The sixth state occurs when the circuit is turned off and the LED is extinguished. The seventh state, which this circuit does not provide, is flashing yellow. It can be obtained by gating the pulse train applied to the LED with a low-frequency pulse train at the cost of somewhat increased circuit complexity.

I've seen only a few commercial applications for tri-state LEDs. One is the indicator lamp on the power switch of the Realistic STA-2100 AM/FM stereo receiver. The LED glows red when the switch is pressed. After a few seconds, it glows green as the unit begins operation. ■

Four

Test & Measurement Circuits

1. Experimental Solid-State Oscilloscope

Figure A is the schematic diagram of the experimental scope. In operation, a signal applied to the noninverting input of the 741 op amp is amplified and routed to a flash A/D converter made up of LM339 quad comparators and a 74147 priority encoder. A detailed description of this A/D converter was given in the September and October 1978 Experimenter's Corner columns.

The digital output of the 74147 is a 4-bit BCD nibble. This nibble, after being decoded by a 74145 1-of-10 decoder, activates one of the ten horizontal rows of LEDs. One of the sixteen vertical columns of LEDs is activated simultaneously by a horizontal scanning circuit consisting of a 555 timer, 74193 4-bit counter and 74154 1-of-16 decoder. The 555 serves as a time base whose sweep frequency is controlled by a 1-megohm potentiometer. The 74193 and 74154 form a 0-to-15 sequence generator that sweeps the sixteen columns of LEDs one at a time.

Because only one row and one column of LEDs are activated at any time, only one LED in the array glows at any single instant. When the sweep rate is faster than 20 or 30 complete scans per second, the individual LEDs merge into a broken line that provides a rough pictorial representation of the positive half of the waveform appearing at the input of the 741 op amp.

Three of the gates in a 7400 quad NAND gate add a simple but very useful trigger feature. When pin 14 of the 74193 is grounded by placing the MODE switch in its FREE RUNNING position, the sweep circuit scans the LED columns continuously. A careful adjustment of the time base potentiometer will freeze the waveform being displayed. Any drift, however slight, in the incoming waveform will require a readjustment of the time base. Otherwise, the waveform being displayed will move across the screen from left to right or right to left, as it did before the potentiometer was adjusted.

When the MODE switch is placed in its TRIGGERED position, pin 14 of the 74193 is reset (or cleared) to 0000 when an input signal is *not* present. An input signal with enough amplitude to activate the lowest-order LM339 comparator causes pin 6 of the 7400 to go from high to low. This allows the 74193 to make a complete scan of the display columns.

If the input signal is still present when the scan is completed, another scan is immediately begun. Otherwise, the scope waits for the signal to recur before initiating a new scan. The result is that in the triggered mode the scope automatically locks onto a recurrent waveform, and displays it with its rising portion originating in the first column of LEDs. A LED connected to the trigger gate network indicates when triggering is occurring, which might not be known if the gain is set so high as to prevent observation of the waveform.

When the scope is in the triggered mode, the waveform being displayed can be expanded or compressed by changing the horizontal sweep rate. Similarly, the height of the waveform can be increased or decreased by adjusting the VERTICAL GAIN potentiometer. An overrange LED connected to the output of the highest-order LM339 comparator indicates when the gain is too high and the top of the waveform is therefore off the screen.

Although it is considerably easier to assemble the drive circuit than the LED display board described last month, care must be exercised when building it because wiring errors can be very difficult to find. I assembled the prototype driver circuit on the same kind of board used for the display (Radio Shack 276-152 or similar) using wrapped-wire, point-to-point construction. Figure B shows how the major components were arranged on the top side of the board. Note the miniature phone jack that serves as the vertical input for the scope. This jack should be installed from the front side of the board.

After the components on the board are

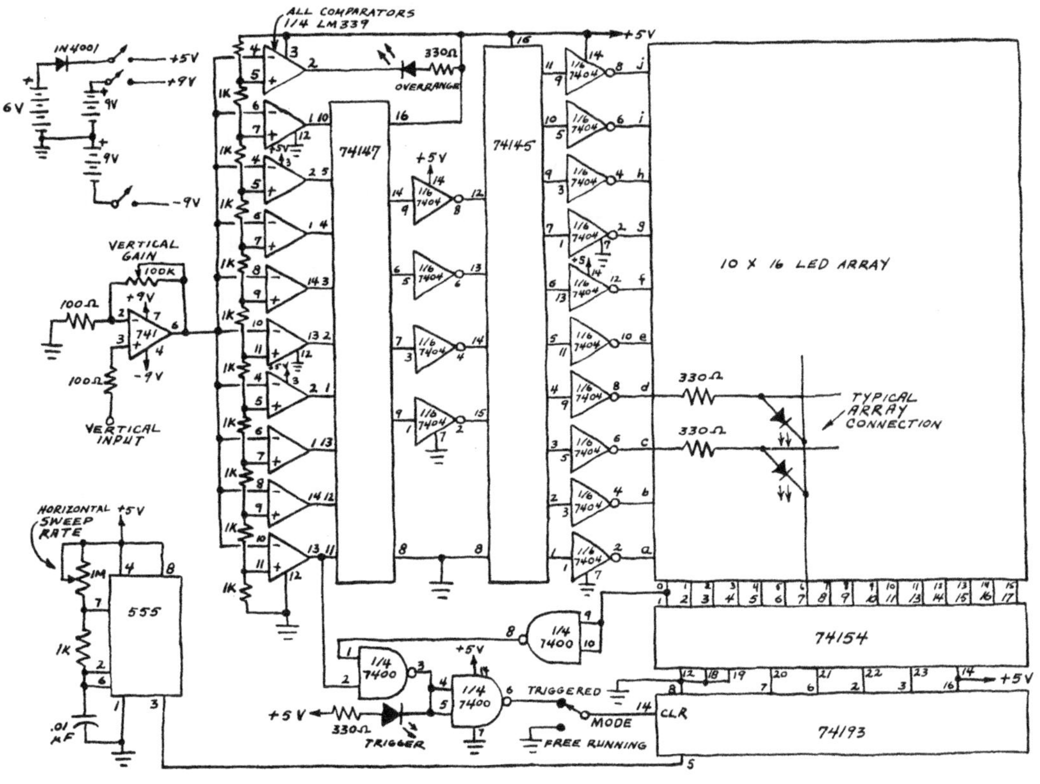

Fig. A. Schematic of a solid-state 10-by-16 LED array oscilloscope.

connected together, connections to the display board, the time base potentiometer, the vertical gain potentiometer and the power supply are provided by soldering wrapping wire to the appropriate copper fingers on the board.

Last month, the bus locations I used on the display board were given. The bus locations used for the remainder of the scope circuit are as follows:

Oscilloscope Circuit	***S-44 Socket***
+5 volts	1
Ground	2
Overrange LED Cathode	3
Vertical Gain Control	M
Vertical Gain Control	N
Horizontal Sweep Control	S
Horizontal Sweep Control	T
Mode Switch:	
Pin 6 7400	U
Ground	V
Pin 14 74193	W
Trigger LED Cathode	X
−9 volts	Y
+9 volts	Z

Don't feel compelled to use these bus locations. All of them can be changed so long as they don't interfere with the display bus. It's a good idea to reserve the *1, 2, Y* and *Z* locations for the power supply since these are closest to the two copper strips that traverse the perimeter of the circuit boards. You can use a small file to separate these into four separate strips.

After the scope circuit has been completed, saw off the upper portion of a second circuit board and install a power switch (3PST), trigger mode switch (SPDT), trigger LED and vertical gain and horizontal sweep rate potentiometer. Figure C shows the layout I used to make the prototype. This board becomes the control panel for the scope. Its components should be connected to the appropriate copper fingers in accordance with the bus given above or the one you select.

You'll need to make a card rack or mother board to hold the three boards that comprise the oscilloscope. If you can afford a commercial mother board, great. If not, do what I did and attach three 44-connector sockets with wire-wrap terminals to a couple of wood or plastic rails. Mount the sockets an inch apart from each other. Next, connect all the common connectors with wrapping wire. Then insert the scope card in the rearmost socket (ICs facing forward), display card in the center socket and control card in the foremost socket.

The prototype scope is powered by two 9-volt batteries and a 6-volt battery made from four AA alkaline cells in a plastic holder. The batteries are connected to the scope with three battery connector clips.

Before applying power to the circuit, be sure to take the time necessary to retrace all the wiring to make sure you've made no errors. The most common problem in a project using wrapped wiring is overlooked connections, particularly those to the power supply pins of the ICs.

Make a probe for the scope from a length of flexible shielded cable, a miniature phone plug, a test clip and an alligator clip. For initial tests, use a ramp or triangle wave having an amplitude of up to a few volts and a frequency of a few hundred hertz. Be sure to perform the tests in a darkened room, at least initially, since the LED array will not be as brightly illuminated as a conventional CRT screen.

If the scope fails to respond, make sure that power is being applied properly. Then try changing the gain and time base settings as well as the trigger mode. If the scope still does not work, you will have to troubleshoot the circuit. Hint: Proceed one step at a time. For example, troubleshoot the sweep circuit by beginning at the 555 (is it pulsing?). Then move to the 74193 (is it counting?), etc.

Once you have the scope working, you'll be able to display a variety of waveforms after a little practice. Spikes, triangles and ramps are the waveforms best reproduced. Sine waves are slightly distorted by the low resolution of the screen, and the rising and falling portions of square waves are usually very faint. Often several LEDs in a vertical column will appear to be on as shown in Figure D, but it's usually possible to resolve the general waveform being displayed.

Although the resolution of this experimental scope is poor and the design of the vertical section leaves much to be desired, it does demonstrate that a flat-screen, compact oscilloscope can be built. If you build the scope you'll want to spend some time calibrating the vertical and horizontal sections with the help of a voltmeter and frequency generator or a conventional oscilloscope. You might also want to modify or try to improve the design if you're experimentally inclined. The A/D converter of

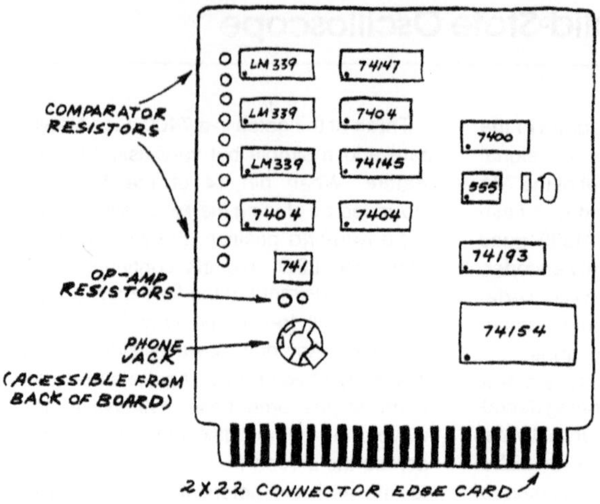

Fig. B. Layout of the major components of the experimental oscilloscope on a 2 x 22 connector edge board.

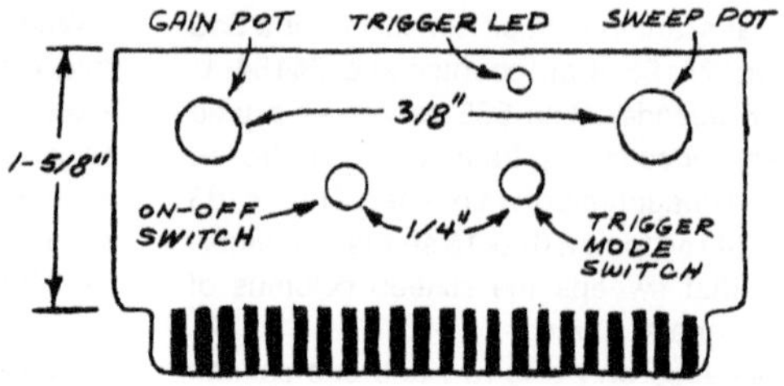

Fig. C. Prototype layout for the control board for the LED oscilloscope.

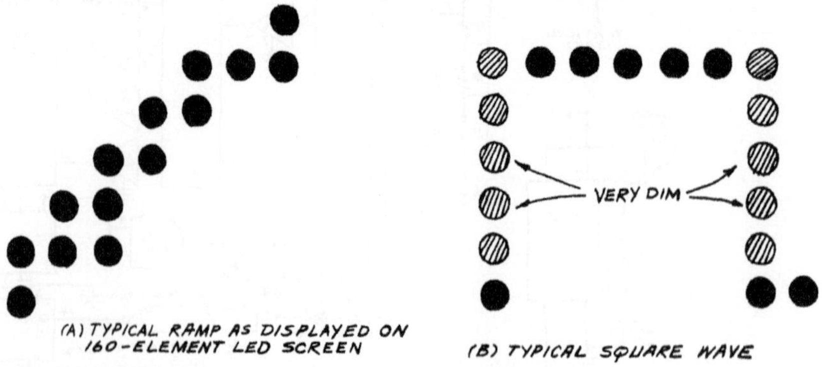

Fig. D. Portions of two waveforms as displayed on 160-element LED display.

the vertical section, for example, can be greatly simplified by using a single-chip A/D converter. The sensitivity of the vertical comparator string can be varied by replacing the 1-megohm resistor at the top of the voltage divider with either a fixed resistor having a different value or, better yet, a 1-megohm potentiometer. Similarly, a wide range of sweep rates can be obtained by replacing the single timing capacitor with a group of fixed capacitors of various values, one of which would be selected by a rotary switch. ■

2. Missing-Pulse Detectors

Missing-pulse detectors can be found in applications ranging from moderately sophisticated, break-beam intrusion detectors to adjustable-duration event timers. Figure 1 is the circuit for a simple but reliable missing-pulse detector made from a 555 timer.

The circuit, which was adapted from one given in the Signetics 555 applications note, is a modified monostable multivibrator. In operation, an input pulse applied to pin 2 triggers the one-shot. The output then goes high for a period determined by the values of timing components *R1* and *C1*.

A 555 monostable ordinarily ignores trigger pulses that arrive *during* the timing period. In this circuit, however, *Q1* fools the one-shot into accepting a trigger pulse during the timing cycle. Refer to the schematic and you'll see why. Normally, *Q1* is off, but a trigger pulse biases it into conduction. This discharges *C1*. Simultaneously, the trigger pulse initiates a new timing cycle.

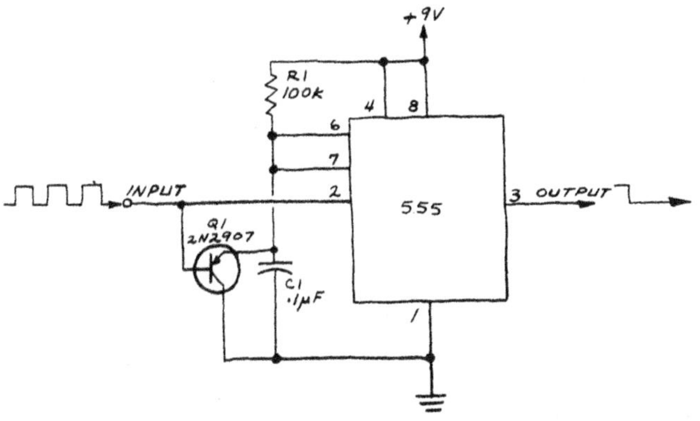

Fig. 1. Basic missing-pulse detector circuit.

If the interval between incoming pulses is *less* than the timing period, the output of the 555 will remain high. Should an incoming pulse not arrive until *after* the previous timing cycle has ended, the output will go low until the pulse arrives. By adjusting the time constant so the timing cycle is slightly longer than the interval between incoming pulses, the circuit will respond to missing pulses by switching low until a new pulse arrives. The circuit can also be adjusted to respond to a *decrease* in the frequency of incoming pulses.

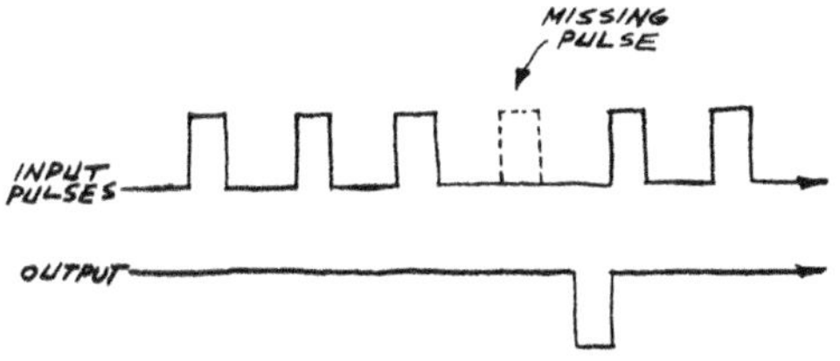

Fig. 2. Missing-pulse detector timing diagram.

If this explanation of how a missing pulse detector works seems complicated, the timing diagram in Figure 2 will help you understand what happens. Although the diagram illustrates a single missing pulse, a series of two or more missing pulses might also occur. Should this happen, the output will remain low until the pulse train is again received.

Simplified Missing-Pulse Detector. The circuit shown in Fig. 1 is commonly used in missing-pulse applications, but that shown in Fig. 3 is simpler. In this circuit, the reset pin is connected to the trigger input. A pull-up resistor connected to $+V_{cc}$ must be added, but the transistor across *C1* (*Q1* in Fig. 1) is no longer needed.

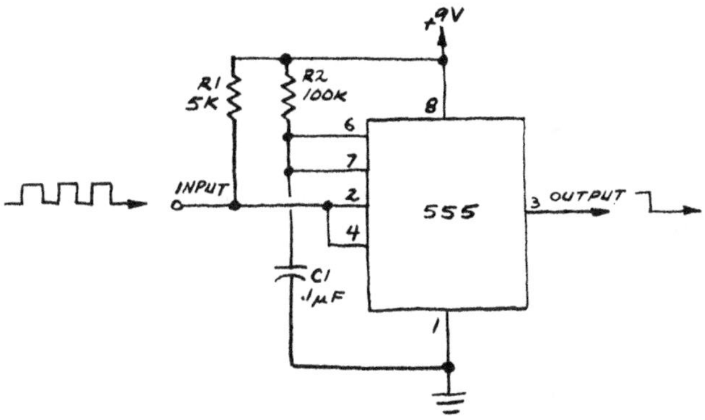

Fig. 3. Simplified circuit for a missing-pulse detector.

Break-Beam Object Detector. Figure 4 shows a simple but effective infrared, break-beam object-detection system comprising a pulsed LED transmitter optically coupled to a missing pulse detector. In operation, pulses from the transmitter are detected by phototransistor *Q3*, which is used to reset *and* trigger the one-shot before the timing cycle can be completed. Blocking the path between the transmitter LED *(LED1)* and *Q3* will cause the receiver LED *(LED2)*. to glow. The receiver LED will go off when the optical channel is reopened.

The sensitivity of the circuit is determined by *R2* and the phototransistor. The resistance of *R2* can be less than 33,000 ohms, but the receiver's sensitivity will be reduced. Sensor *Q3* can be a standard silicon phototransistor, but a Darlington phototransistor will provide higher sensitivity.

Timing components *R3* and *C2* determine the time constant of the one-shot. A fixed resistor can be used for *R3* if its value is such that the timing cycle is longer than the period between transmitter pulses. The time required for the circuit to respond

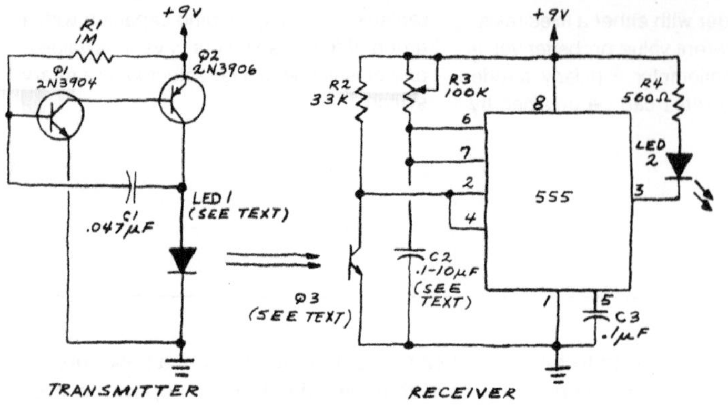

Fig. 4. Schematic for a break-beam object detector.

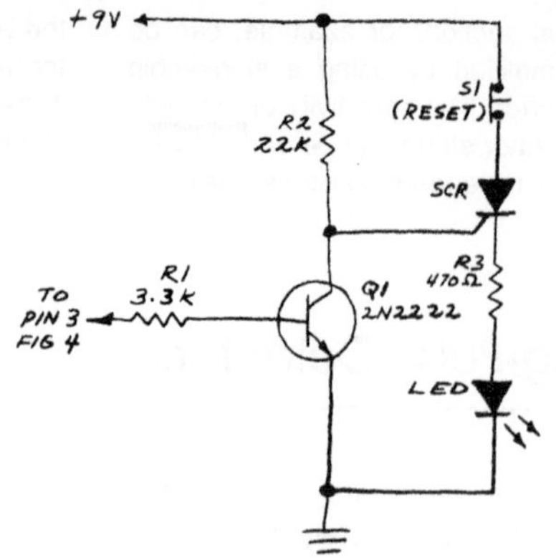

Fig. 5. SCR output circuit.

to a missing pulse is the difference between the transmitter-pulse interval and the receiver's time constant. Therefore, the circuit will appear to respond almost immediately to an obstruction placed in the optical path when the time constant is slightly longer than the pulse interval. On the other hand, the circuit will require as much as a few seconds to respond if the time constant is much longer than the pulse interval. Increasing *R3*, *C2* or both will increase the time constant.

Long time constants make possible such specialized applications as detecting slow-moving objects or long objects moving through the optical channel at the same velocity as short objects. A long time constant also provides a degree of false-alarm immunity when the system is used as an intrusion alarm because the detector can thus be adjusted to ignore falling leaves and other transient interruptions.

The range of the system is determined by the sensitivity of the receiver and the optical power radiated by the transmitter LED. For best results, use a photodarlington for *Q3* and stick to the relatively powerful transmitter circuit shown in Fig. 4. Be sure to use a GaAs:Si device for *LED1*. Suitable types include the Optron OP-190 or OP-195 and the G.E. 1N6264. Also, don't allow too much ambient light to strike *Q3* (although some dc illumination will provide base bias and increase *Q3*'s sensitivity).

With these components, the maximum detection range will be a few handbreadths. Adding lenses to both the transmitter and receiver will increase the operating range. Best results will be obtained with lenses having a focal length approximately equal to the diameter of the lens (which corresponds to an *f* number of 1). With 5-cm diameter, *f1* lenses, a range of a few meters or more can be achieved.

Adding an Output Latch. The output pin of the receiver (pin 3 of the 555) switches from a low to a high state when a missing pulse occurs and, after a timing interval, returns to its low state. In some applications, such as intrusion alarm systems, it's necessary to latch the output to a high state once a single missing pulse has been detected. Figure 5 shows one way the latching function can be achieved with the help of an SCR. This simple circuit is designed to be connected directly to pin 3 of the 555 in Figure 4.

An SCR is triggered by a positive gate voltage. Because the 555 output is normally high, *Q1* is required to invert the output. Resistor *R3* limits current flowing through the indicator LED. If the resistance of *R3* is too low, excessive current will flow through the LED and SCR. On the other hand, if the value of *R3* is too high, the current through the SCR will be less than its minimum *holding current*. This means the SCR will turn off and on, rather than latching on, when the 555 output changes states.

Reset switch *S1* is a normally closed pushbutton. If the 555 output is high (for example, when the transmitted signal is being received) and the SCR has been gated on by a previous missing pulse, pressing *S1* will turn off the SCR and prepare it to latch onto the next missing pulse.

Optically-Coupled Slot Switches. Slot switches are made by mounting a LED and phototransistor so they face one another across a narrow space in a plastic fixture. Applying a forward current to the LED switches the phototransistor. An opaque object (magnetic tape, paper card, etc.) inserted in the slot blocks the beam from the LED and turns the phototransistor off.

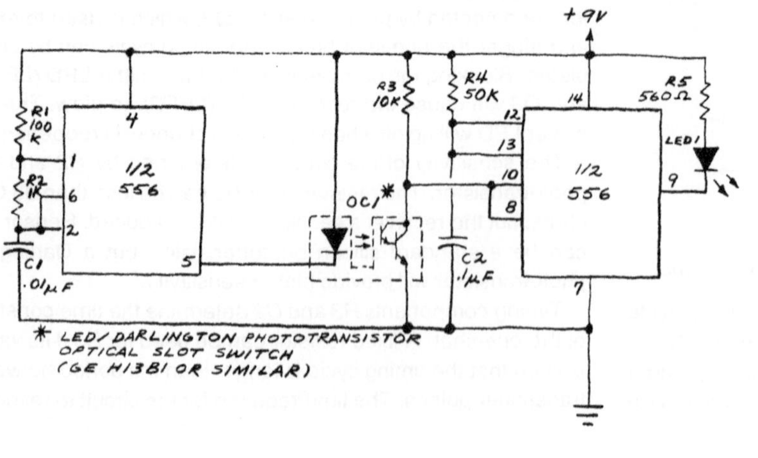

Fig. 6. In slot switch circuit, one half of 556 is a pulse generator and the other a missing-pulse detector. Blocking the slot between the LED and the photo transistor causes the detector to change states and energizes the light emitting diode.

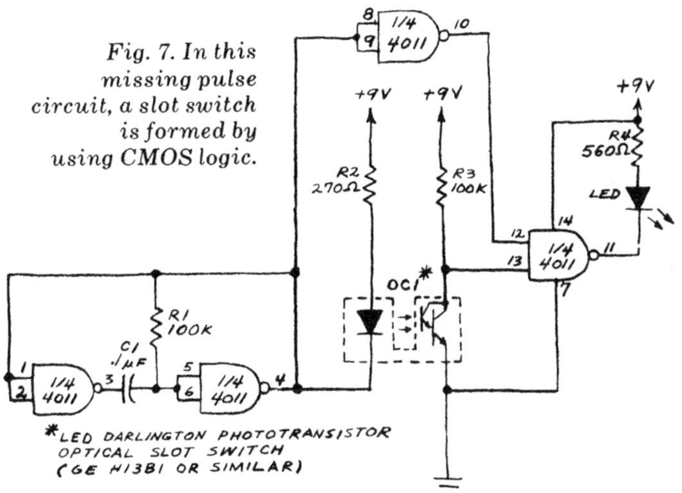

Fig. 7. In this missing pulse circuit, a slot switch is formed by using CMOS logic.

Many optoelectronics companies make various types of optical slot switches. If you can't find one, or if you don't like the prices of those you find, it's easy to improvise by mounting an infrared LED and photodarlington on a suitable jig. The gap between the two components should be a few millimeters.

Usually, a dc bias is applied to the LED in a slot switch. It's possible to achieve the same results—and at the same time save current—by pulsing the LED and connecting the phototransistor to a missing pulse detector. Here are two examples.

556 Slot Switch. In the circuit shown in Fig. 6, one half of a 556 dual timer serves as the pulse generator for a LED. The remaining half is connected as a missing pulse detector. Pulses from the transmitter continually reset and trigger the one-shot. Blocking the slot between the LED and phototransistor causes the missing pulse detector to change states and light the indicator LED.

The SCR latch in Fig. 5 can easily be added to this circuit. Also, you can experiment with *R4* and *C2* in the receiver portion of the circuit to alter its response time. For example, if the timing cycle of the receiver is 100 milliseconds longer than the period between pulses from the LED, the slot switch will ignore an interruption lasting less than 100 milliseconds.

CMOS Slot Switch. A single 4011 quad NAND gate can provide the bulk of the transmitter and receiver electronics for a pulsed break-beam slot switch based on the missing-pulse principle. Figure 7 is the schematic diagram of the slot switch.

In operation, the LED in the slot switch is pulse-modulated by the astable multivibrator formed by two of the gates in the 4011. Timing components *R1* and *C1* determine the pulse rate and *R2* limits the peak current through the LED. Pulses from the LED are detected by the Darlington phototransistor in the slot switch and presented to one input of a NAND gate. The inverted output from the multivibrator is presented to the second input of the NAND gate. When optical pulses are received by the phototransistor, its collector goes low, causing the output of the NAND gate to go high. When the slot is obstructed, both inputs to the NAND gate go high each time the slot switch LED is pulsed. This turns the indicator LED on.

Although the indicator LED appears to be glowing continuously when the slot is obstructed, it is actually flashing at the same rate at which the slot switch LED is pulsed. ■

3. A "Matchbox" LED Oscilloscope

Figure 1 is the schematic diagram of a compact LED scope that uses only three ICs and consumes only 15 mA. Operation of the circuit is fairly straightforward, especially if you're already familiar with solid-state scope basics.

The incoming waveform is applied directly to pin 5 of the LM3914, where its instantaneous amplitude is detected by a voltage divider/comparator chain. Decoding logic then drives one of the LM3914 outputs low.

Any LED in the row connected to the selected output is then eligible to glow. The remaining requirement is a positive voltage at the LED's anode. This is obtained from a horizontal sweep circuit made from a 4011 quad NAND gate and a 4017 Johnson counter.

The 4011 performs two important functions, one of which is to provide a stream of clock pulses. This is accomplished by two gates connected as a free-running or astable oscillator. The frequency of oscillation is determined by the values of *R4* and *C1*.

The 4017 counter is unique in that it includes a 1-of-10 decoder. This eliminates

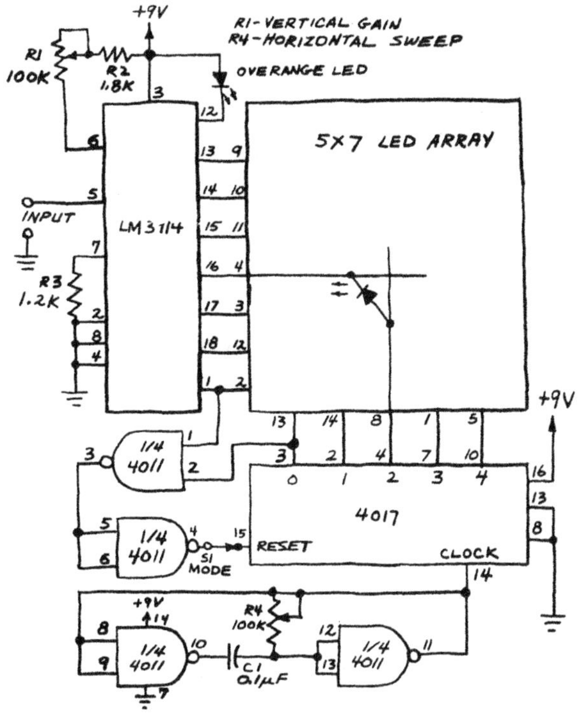

Fig. 1. Schematic diagram of matchbox LED oscilloscope.

the need for a separate decoder IC. Furthermore, because the activated output of the 4017 goes high when all other outputs remain low, the 4017 can be connected to the anode of an LED.

The remaining two gates in the 4011 form an AND gate that provides an automatic trigger. When MODE switch *S1* is closed, the gate resets (clears) the 4017 if the input voltage has sufficient amplitude to activate the lowest-order output of the LM3914 at the same time that the lowest-order counter output is high. This feature makes it relatively easy to freeze the waveform being displayed.

The "screen" of the scope is a single 5 × 7 dot-matrix LED display (Monsanto MAN-2A, Texas Instruments TIL305, Litronix DL-57 or equivalent). Although 35 LEDs provide very limited resolution at best, with experience it's possible to visualize square and triangle waves being displayed on the readout.

In case you're wondering where the current limiting resistors of the LED display are, they are not necessary! The LM3914 includes a novel feature that permits the current at the selected output to be externally programmed by a single resistor *R3* connected to pin 7. This pin provides a reference voltage of 1.2 to 1.3 volts, and the current through *R3* is 1/10 the LED current. According to Ohm's law, the current flowing through a resistor is the quotient of the voltage across the resistor divided by the resistance in ohms. The current through *R3* is therefore 1 mA, which means that the LED current is 10 mA.

Figure 2 is a photograph of a miniature, Wire-Wrapped prototype scope that I assembled on a perforated board measuring 1.2" × 1.9" (about 3 cm × 5 cm). Notice that pins 9 and 10 of the LM 3914 extend over the lower end of the socket. The small capacitor installed in the two unused pin positions of the 4011 socket is *C1*. The overrange LED is installed below the 5 × 7 LED array. Components *R1*, *R4* and *S1* are attached to the circuit board with cyanoacrylate adhesive. I used a miniature Micro Switch™ pushbutton switch as *S1* because I had one on hand, but any other, spst switch is suitable.

Some typical display patterns I have obtained are shown in Fig. 3. Often, the displayed pattern will bear little resemblance to the actual wave. Sometimes it's easier to visually integrate the approximate shape of a wave by switching off the automatic trigger and adjusting *R4* until the waveform slowly parades across the display.

For some interesting visual effects, try connecting a radio or audio amplifier to the input of the scope. Music and voice signals will stimulate a dynamic, miniaturized light show. For best results, leave the trigger switch off.

Finally, remember that it's relatively easy to expand the scope's display. You can add a second 5 × 7 display or make a 10 × 10 display from individual LEDs or 10-element LED bars. If you're really ambitious, you can add additional LM3914's and 4017's and make a scope having 20 × 20 or more LEDs. ■

Fig. 2. Photograph of the wrapped-wire prototype LED oscilloscope assembled on a small piece of perforated board about 1.2" × 1.2".

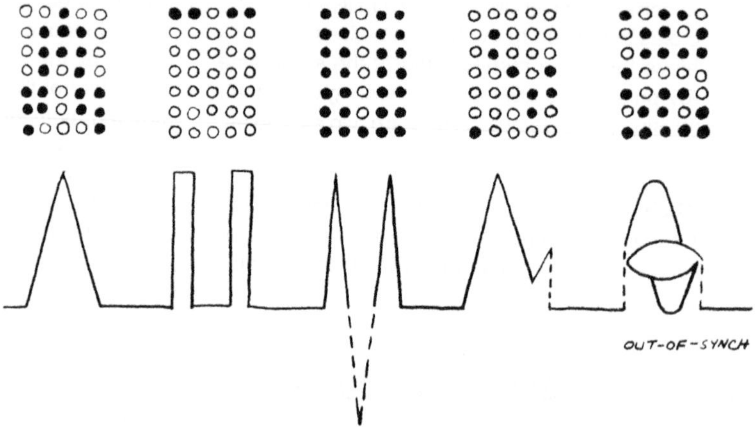

Fig. 3. Some typical display patterns obtained on a 35-element LED scope. Sometimes, the pattern bears little resemblance to the actual wave.

4. Solid-State Oscilloscope Wrap-Up

How They Work. Figure 1 is a block diagram of a basic solid-state scope. The horizontal sweep circuit is very straightforward and can be made up of standard TTL or CMOS ICs. An advantage of the CMOS logic family is its low power consumption and the flexibility of one of its members, the 4017, a chip that combines a counter and 1-of-10 decoder in a single package. The clock can be a simple 2-gate oscillator or an IC timer.

Most of the design variations in solid-state scopes occur in the vertical section's analog-to-digital converter. Early scopes required complicated voltage divider/comparator/decoder networks. Now, however, all of these functions are available in chips such as National Semiconductor's LM3914.

The operation of most solid-state scopes is straightforward. The display consists of an array of optical elements such as LEDs or liquid crystals. Only one vertical column in the display is enabled at any instant, and the instantaneous amplitude of the applied waveform determines which element within that column is activated. If the horizontal sweep is synchronized with the frequency of the input signals, the outline of the applied waveform will be displayed as a pattern of dots. Synchronization can be achieved either by manually adjusting the timebase or by means of an automatic trigger circuit that starts the sweep when the amplitude of the input signal exceeds a preselected value.

Early Solid-State Scopes. The development of visible LEDs in the early 1960's made possible compact graphic displays that could serve as the screens of solid-state scopes if teamed up with suitable driving and scanning circuitry. Back then, however, LEDs cost $10 *each* and the necessary drive circuits would have been very complex. Accordingly, most work in the area of LED displays was limited to military projects.

I began experimenting with solid-state scope designs when LED prices began to tumble in the early 1970's. For those readers who have requested additional information, here's a condensed history of the solid-state scopes that have appeared in POPULAR ELECTRONICS and *Electronics* magazines:

In February 1974, I assembled a crude scope with a display made from eight LEDs. Figure 2 is a simplified schematic of this scope. The vertical drive circuit was far too complex for convenient expansion, so I experimented with series-connected LEDs having slightly different turn-on voltages. A rising voltage applied to a string of such LEDs causes them to light up in sequence. In spite of its simplicity, this method proved impractical because the LEDs had to be selected for various turn-on voltages.

Another simple way to create a bar-graph display is to connect LEDs across each of the resistors in a voltage divider, as shown in Fig. 3. As the voltage applied across the divider is increased, the voltage drop across each resistor increases proportionally. If the resistance of each resistor in the chain is larger than the one above it, then the LEDs will light up in bar-graph fashion as the voltage applied across the entire divider network increases. For example, using the resistance values shown in Fig. 3, I constructed a circuit in which the LEDs began to glow when the applied voltage was increased to the values shown below.

Led	Voltage
1	8.1
2	9.0
3	10.0
4	10.7
5	11.7
6	12.7
7	14.0
8	15.0
9	16.0
10	17.5

This vertical-axis circuit is very simple, but it has some major drawbacks. The input voltage must attain a relatively high value before the first LED begins to glow. Also, only bar-graph (not moving-dot) operation is possible. Nevertheless, I've used this method to make a miniature solid-state scope with a 10 × 10 LED screen.

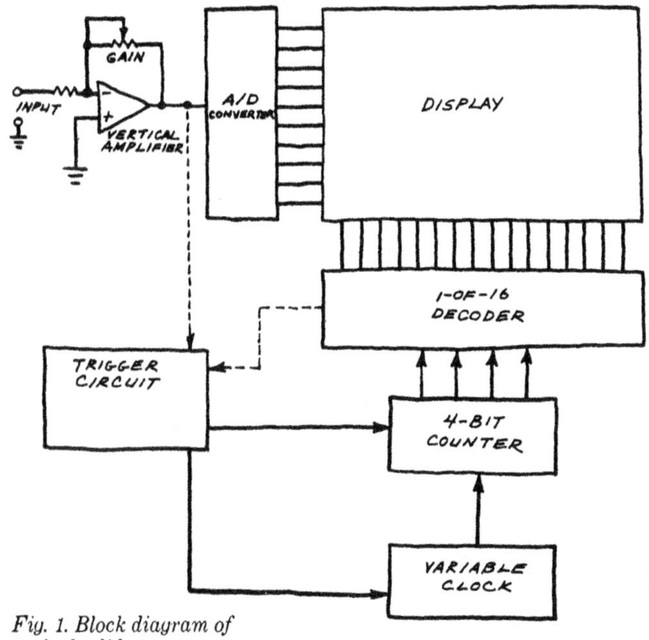

Fig. 1. Block diagram of typical solid-state scope.

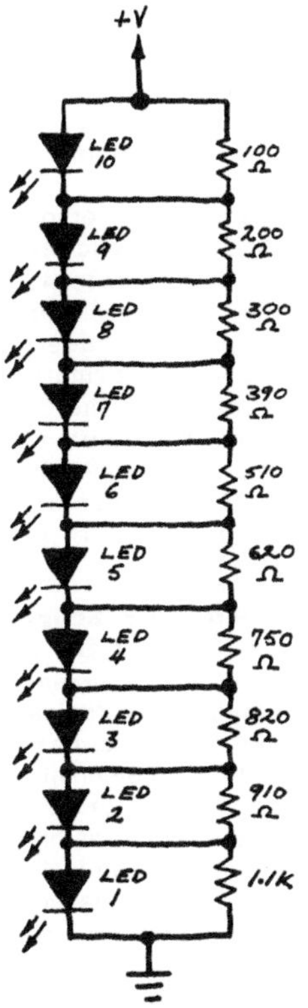

Fig. 3. Voltage-divider LED bargraph display.

Fig. 2. Simplified schematic of an original scope with display made from eight LEDs.

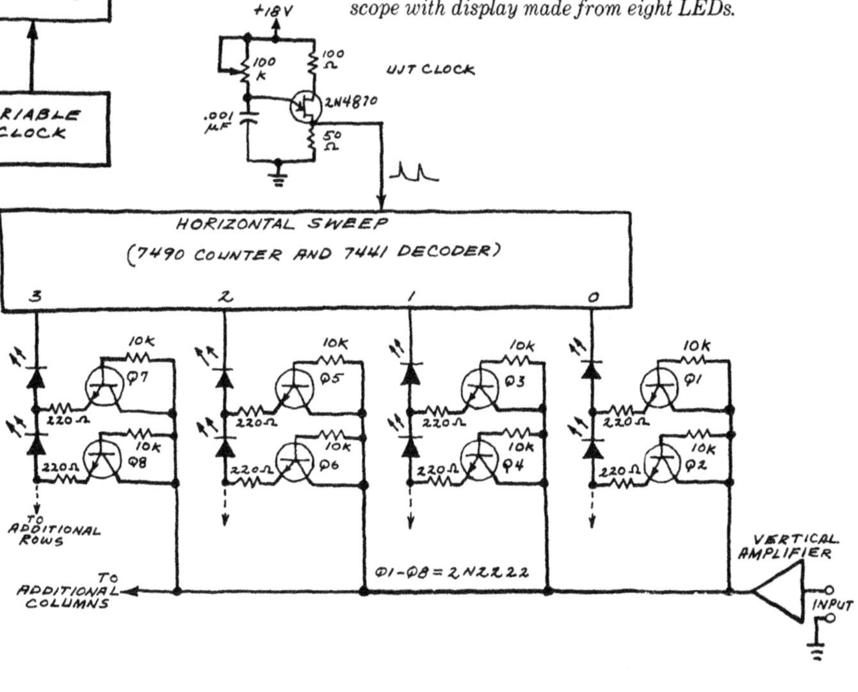

Figure 4 shows the circuit in simplified form. Excluding its power supply, this scope fits with room to spare in a pocket-calculator-size enclosure. Vertical sensitivity is adjustable from 0.01-volt per LED to 1.0 volt per LED, and horizontal sweep is adjustable from 20 microseconds per LED to 1.0 second per LED. Total power consumption of the display with all LEDs on is 308 milliwatts. The drive electronics consume another 54 mW. For more information about this scope, refer to "LEDs Replace CRT in Solid-State Scope" (*Electronics*, June 26, 1975, p. 110) and my book *LED Projects* (Howard W. Sams & Co., 1976, pp. 92-95).

A few years ago, Vernon Boyd described an improved solid-state scope in *Electronics* (November 24, 1977, pp. 128-130). His circuit employed a string of comparators and a decoder/driver network to generate a moving-dot readout. Almost one year later, the October 1978 installment of this column briefly described an improved scope with a 10 × 16 LED screen. Construction and operating details followed in the April 1979 "Project of the Month."

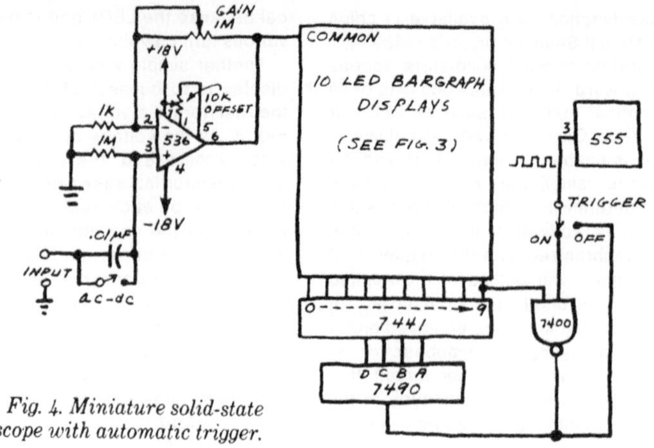

Fig. 4. Miniature solid-state scope with automatic trigger.

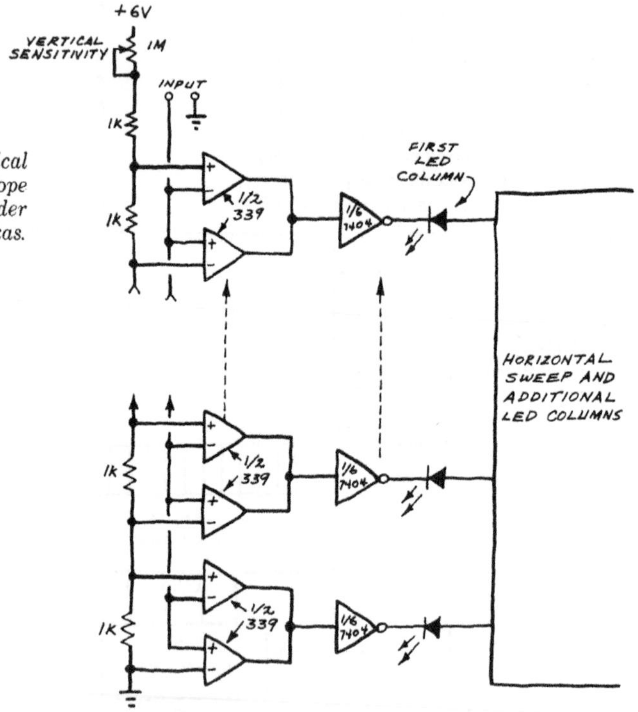

Fig. 5. Vertical section of scope designed by reader Bill Cikas.

Readers' Solid-State Scopes. Several readers have designed and built various solid-state scopes. For example, Bill Cikas, a self-taught electronics enthusiast living in Rockford, Illinois, has designed a scope with a 12 × 16 element LED display that fits on a 6" × 6" (15.3 × 15.3 cm) perforated board. A portion of the scope's vertical section is shown in Fig. 5, and a photo of the scope displaying a well defined sine wave appears in Fig. 6.

Steven Bolotin, a Chicago high school student, modified the scope described in the April 1979 installment of this column to permit both the negative and positive halves of a waveform to be displayed. The modification, which is shown in Fig. 7, is inserted between the vertical amplifier and analog-to-digital converter and causes the incoming wave to ride on an adjustable dc level.

Joe Sharp of Orange, Virginia, has worked on the resolution problem caused by the limited number of display elements in the screen of a solid-state scope. He has determined that at least thirty display columns are required for every ten display rows to minimize smearing of the trace. Joe's scope employs three cascaded LM3914s, thirty 10-element LED bar arrays, and provides automatic trigger, ac/dc operation and various other features.

Fig. 6. Photo below is of solid-state scope whose circuit is shown in Fig. 5.

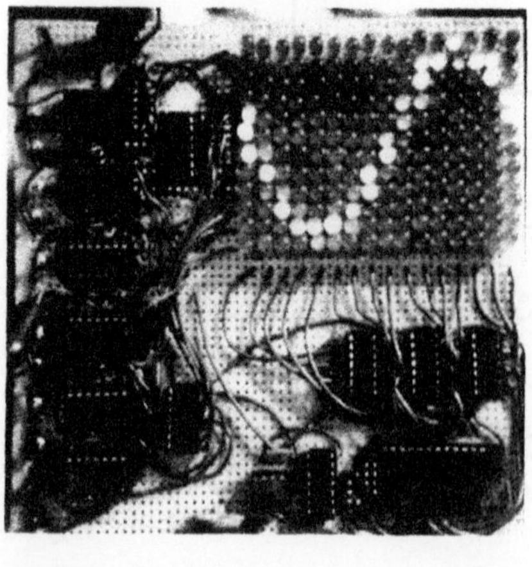

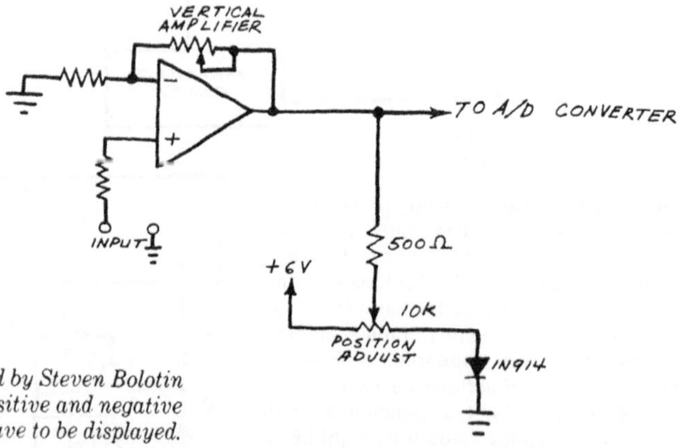

Fig. 7. Circuit designed by Steven Bolotin that permits positive and negative halves of a wave to be displayed.

Gregory Kovacs, a student at Eric Hamber Secondary School in Vancouver, British Columbia, has provided details of a sophisticated solid-state scope project he has undertaken. For his half-term electronics project, Greg designed and assembled a scope that can display the waveforms of signals with frequencies of several hundred kilohertz.

A noteworthy feature of Greg's scope is the use of a Siemens UAA170 IC dot generator. Like the more recently introduced LM3914, the UAA170 eliminates the complicated voltage divider/comparator/decoder network that would otherwise be required for the scope's vertical section. In a paper describing his project, Gary observes that a scope using a single UAA170 vertical display driver would be limited to a maximum input frequency of only 50 Hz or so. Therefore, he assembled ten separate vertical display boards, each with its own UAA170 and column of 16 LEDs, and scanned each board with a conventional counter/decoder circuit.

Each display board contains an op-amp sample-and-hold circuit controlled by a Siliconix DG181 analog switch. A horizontal sweep sequentially strobes each display board's sample-and-hold circuit and the sampled voltage level is held until the next strobe pulse arrives. The result is the capability of displaying input waveforms having frequencies of up to 1 MHz.

An important consequence of Greg's decision to use individual sample-and-hold circuits is that his circuit can function as an analog *storage scope*. It can sample and display on its screen for several minutes any waveform before degradation caused by leakage in the sample-and-hold capacitors occurs. Figure 8 is a photo that shows the high-quality trace the circuit provides.

Looking Ahead. Solid-state scopes have a very bright future. Thanks to the LM3914 and similar moving-dot display drivers, experimenters and hobbyists can easily design their own scopes. By cascading vertical and horizontal driver ICs, oscilloscopes with displays containing hundreds of LEDs are possible. Although building such a scope would be tedious (the display board in one of my scopes has more than 650 solder connections), cost is no longer the limiting factor it was before LEDs could be purchased in volume for less than 10¢ each.

Homebrew scopes will become even simpler when manufacturers produce LED dot/bar displays with integral solid-state decoder/drivers. A complete scope could then be assembled in building-block fashion simply by connecting display columns to a standard horizontal sweep circuit. Individual, discrete LEDs are fine for low-resolution scopes, but are for a few reasons unattractive if high resolution is desired. Several alternative display technologies are available, the least costly being liquid crystals. Unfortunately, liquid-crystal displays are too slow for conventional multiplexing techniques.

Recently, however, Ian A. Shanks of the Royal Signals and Radar Establishment in Malvern, England, solved the liquid-crystal addressing problem with a design that continuously applies signals to all elements of the display. Shanks has assembled a prototype storage oscilloscope with a 100 × 100 element display measuring 2.5″ × 2.5″ (6.45 cm × 6.45 cm). The waveform is displayed as a black line on a light background. It can be projected onto a screen by removing the display's reflective back and using it like a transparency in a slide projector. Shanks is building a new scope with a 128 × 256 element display and believes that a 1,000 × 1,000 element display can be made.

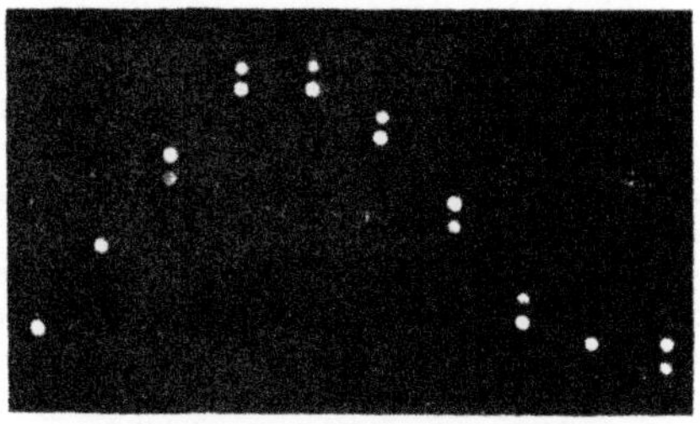

Fig. 8. Photo of waveform displayed on scope designed by Gregory Kovacs.

In Conclusion. The oscilloscope is the most important and useful piece of test equipment available. Hopefully, these solid-state scope developments will lead to pocket-size, high-resolution, full-feature scopes affordable by most experimenters and hobbyists. In the meantime, those described here show how *you* can make contributions to this new technology. ■

5. Do-It-Yourself Counters

THE MC14553B is a 3-decade BCD CMOS counter that you can use to make various event and frequency counters. Each of the three counter stages in this chip is teamed to a set of latches, which are registers comprising four flip-flops permitting the count at any given instant to be sampled and stored. The displayed count is thus periodically updated and "frozen" on the associated digital readout while the counters continue counting. This flexibility and the other attributes (modest power consumption, relatively high counting capabilities in a single IC) make the MC14553B ideally suited for use in many experimenter projects. Let's take a closer look at this chip.

MC14553B Operation. Figure 1 is the pin outline of the MC14553B. Here's an explanation of the pin functions:

- *CLOCK* (pin 12)—Counter input.
- *LE* (pin 10)—Latch Enable. When *LE* is at logic 1, the latch is loaded with the current count.
- *DIS* (pin 11)—Disable. Must be at logic 0 for counting to occur. Inhibits the input (blocks incoming clock pulses) when at logic 1.
- *MR* (pin 13)—Master Reset. Must be at logic 0 for normal operation. Resets all four BCD outputs to logic 0 when brought to logic 1. Keep LE at logic 1 during reset operations to preserve the latest count in the latch.
- *A, B, C, D* (pins 9, 7, 6, 5)—BCD outputs (TTL compatible).
- *DS1, DS2, DS3* (pins 2, 1, 15)—Digit select outputs (TTL compatible).
- *C1A, C1B* (pins 4 and 3)—Connection points for the external capacitor that controls the speed of the on-chip digit-select multiplex oscillator.
- *OF* (pin 14)—Overflow Output. Normally at logic 0, this pin goes to logic 1 when count exceeds 999.

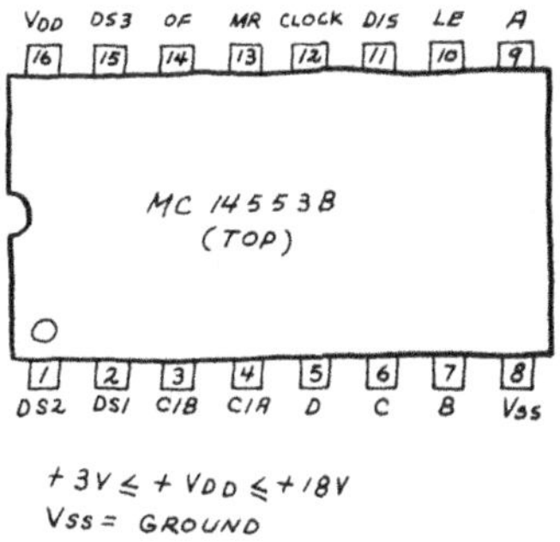

Fig. 1. Pin arrangement of the MC14553B counter.

Having acquainted ourselves with the basic layout of the MC14553B, let's now examine some circuits designed around this versatile IC.

Three-Decade Event Counter. A very simple application for the MC14553B is the three-decade event counter shown in Fig. 2. This circuit will count pulses arriving at pin 12 of the MC14553B when both the *Master Reset* (pin 13) and *Disable* (pin 11) inputs are low. The maximum count rate is dependent upon power supply voltage $+V_{DD}$, and is typically 1.5 MHz at +5 volts, 5.0 MHz at +10 volts and 7.0 MHz at +15 volts.

The BCD output of the MC14553B is decoded by an MC14543B BCD-to-seven-segment latch/decoder/driver. This chip was designed specifically to drive liquid-crystal displays. It can safely drive

LED displays, however, if the current to each LED segment does not exceed 10 mA or if buffer transistors are used. Incidentally, both the MC14553B and MC14543B are 16-pin DIPs. Because their part numbers differ by only one digit, use care to avoid interchanging the two chips inadvertently when assembling this circuit! Also, both are CMOS ICs, so be sure to follow the appropriate handling procedures for such devices.

The common-anode LED display shown in Fig. 2 is a multiplexed unit containing three or more digits. It can be purchased new or surplus, or can even be salvaged from a defective pocket calculator. I've not included pin numbers because many different types of displays are available. You can even make your own multiplexed display by connecting together the segments of three individual common-anode readouts. The common anodes of each display are connected to driver transistors *Q1, Q2* and *Q3*. These transistors are switched on and off in rapid sequence by *digit select* (*DS*) outputs 1, 2 and 3 of the MC14553B at a multiplex frequency determined by the value of *C1*.

If you prefer, you can use common-cathode readouts. The circuit described in the next section and shown schematically in Fig. 3 incorporates the appropriate modifications.

Current through the LED segments is limited by resistors *R1* through *R7*. It's important to restrict the amplitude of this current to a *maximum* of 10 mA. A convenient formula for determining the values of *R1* through *R7* for a segment current of 10 mA is: $R = 100 \times (+V_{DD} - 2)$. If $+V_{DD}$ is supplied by a 9-volt battery, for example, then *R1* through *R7* should be 700 ohms each. A more convenient value, 1000 ohms, will limit the forward current to 7 mA per segment, enough to provide ample display brightness for most applications.

Six-Decade Event Counter. Two or more MC14553Bs can readily be cascaded to provide additional decades of counting. Figure 3 illustrates how easy it is to double the number of decades of the basic event counter that we just discussed. As you can see by comparing the two circuits, the 6-decade counter is actually two 3-decade counters in series. Two convenient simplifications are that only one multiplex oscillator capacitor is required and that the digit-select transistors for one stage also control the displays of the second stage.

Note that the circuit in Figure 3 is designed to drive *common-cathode* displays, not common-anode displays as used in Fig. 2. You can use common-anode readouts by following the display configuration shown in Fig. 2. There are three significant differences between the two circuits resulting from the use of different displays. Pin 6 of the MC14543B is connected to ground instead of $+V_{DD}$ when common cathode displays are used. Also, npn digit-select transistors are employed, not pnp devices. Finally, the emitters of these digit-select transistors must be connected to ground, not $+V_{DD}$.

Frequency Counter. Figure 4 shows a network that can be connected to the 6-decade event counter to convert it into a frequency counter. This network consists of a crystal-controlled timebase employing an MM5369 IC oscillator/divider and four NAND gates. The MM5369 generates a 60-Hz pulse train when it is connected to a readily available 3.579545-MHz color-television crystal and a few passive components. A 4017 CMOS decade counter connected as a divide-by-six counter divides the output of the MM5369 into a 10-Hz signal. An additional 4017 connected as a divide-by-ten counter divides the 10-Hz output of the first counter into a 1-Hz pulse train. The frequency of a counter's timebase determines the resolution of that counter. For example, a 10-Hz timebase samples the accumulated count and updates the display ten times each second, but this means that the frequency shown on the display is only one tenth of the actual frequency. A 1-Hz timebase causes the readout to be updated only once each second, and the displayed frequency is the actual frequency of the input signal.

Both the timebase and the input signal whose frequency is to be counted are applied to a control circuit made up of all four gates in a 4011 CMOS quad NAND gate. One gate allows the input signal to reach the counter during each timebase cycle. A second gate is connected as a half-monostable which activates the *Latch Enable* input of the counter. This results in the storage of the total count accumulated during one timebase period. The remaining two gates strobe the *Master Reset* input of the counter after each counting interval to clear the counters prior to the next count cycle.

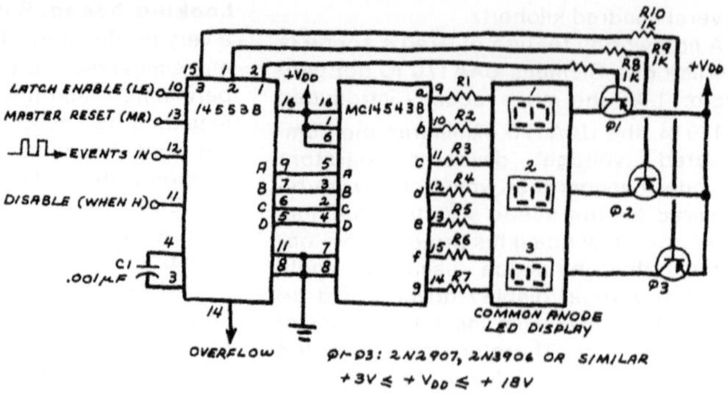

Fig. 2. Schematic diagram of a three-digit event counter using common-anode LED display.

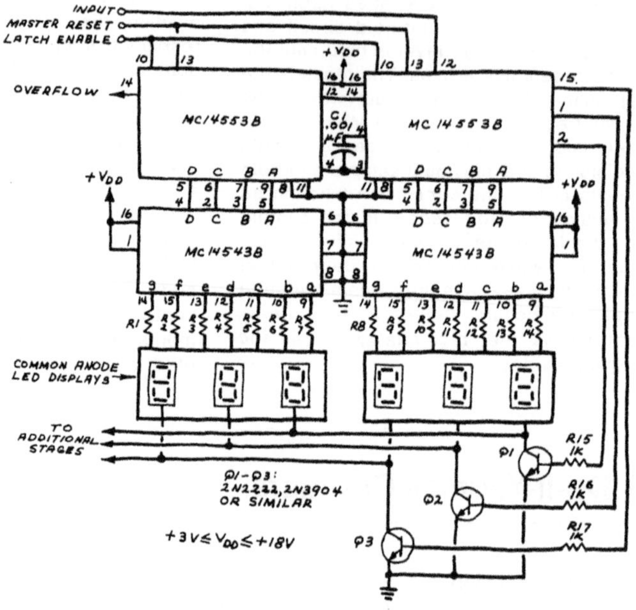

Fig. 3. Circuit for a six-digit event counter made from cascaded MC14553Bs.

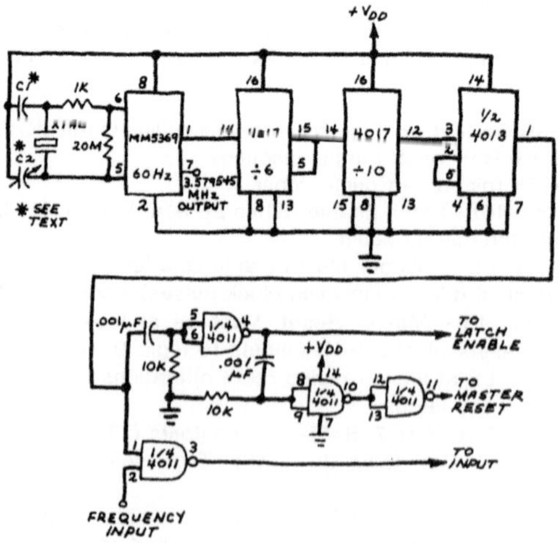

Fig. 4. A time-base network added to a six-decade event counter to convert it into frequency counter.

The accuracy of the frequency counter is of course dependent upon the accuracy of the timebase oscillator. National Semiconductor, the manufacturer of the MM5369 suggests capacitance values of 30 pF for *C1* and 6.36 pF for *C2*. For best results, use a small trimmer capacitor (0-to-30-pF or similar) for *C2*. Carefully adjust it until the output frequency at pin 7 of the MM5369 is 3.579545 MHz.

You'll need a frequency counter of known accuracy for this calibration procedure. If you don't have one, a polite request directed to a college electrical engineering department or an electronics repair shop might result in permission to use a counter for the few minutes it takes to adjust *C2*. Incidentally, you can use a "gimmick" capacitor consisting of two lengths of twisted wrapping wire for *C2*. Start with a 2-inch length and carefully trim short bits from the free end of the twisted pair until the correct frequency is obtained.

There's an alternative procedure you can follow to adjust the counter's timebase. It requires a 100-kHz or 1-MHz crystal-controlled oscillator whose output is rich in harmonics (such as a crystal calibrator designed for communications applications) and also requires a shortwave receiver capable of receiving the National Bureau of Standards radio station WWV or WWVH. During an interval when no audio tones modulate the station's carrier, zero beat the oscillator against the carrier. Then couple a portion of the oscillator's output to the input of the frequency counter, verifying that zero beat is maintained. (A signal-conditioning input circuit might be needed to square up the output of the oscillator to CMOS-compatible levels.) Finally, adjust *C2* so that the counter displays the nominal output frequency of the oscillator. Make sure that the oscillator remains in zero beat with WWV's carrier while you adjust *C2* for the proper readout on the circuit's LED display. ■

6. X100 Frequency Multiplier

LOW-FREQUENCY signals are more difficult to measure than you might think. First of all, you must have a frequency counter whose amplifier has a response that goes far enough down toward dc. Many counters are rated to 50 or 100 Hz, and will display meaningless information when driven by lower-frequency signals. Secondly, no degree of precision can be obtained unless your frequency counter has a gate time of 10 or more seconds. Even a 10-second gate time gives an accuracy of only one decimal place (0.1 Hz). To make matters worse, a 10-second gate time means that the display is updated only six times a minute. This can make critical circuit adjustments tedious.

Figure 1 shows schematically a simple but very useful ×100 frequency multiplier or *prescaler*. This device dramatically simplifies low-frequency measurements by inserting a divide-by-100 counter between the vco and the wideband phase comparator of a CMOS 4046 phase-locked loop. With the component values shown, the circuit will multiply signals at frequencies ranging from less than one to a few hundred hertz up to a range that even counters with relatively restricted low-frequency responses can accommodate.

In operation, properly conditioned pulses are applied to the input of the 4046. The loop generates an error voltage which is filtered and applied to the input of the vco, thus beginning the capture process. Because a divide-by-100 counter is placed between the output of the vco and the phase comparator's second input, the vco frequency will be exactly 100 times that of the input signal when the loop is phase-locked. The signal frequency can then be measured to an accuracy of one decimal place (0.1-Hz resolution) with a standard frequency counter having a one-second gate time.

The capture time of this phase-locked loop prescaler can easily exceed several seconds, especially when very low frequencies are applied to its input. If you intend to use the circuit to measure frequencies consistently higher than 50 Hz, you can speed up the loop's capture time by reducing the value of *C2* to 0.22 µF.

If the signals to be multiplied are noisy or have slow rise and fall times, it will be necessary to condition them before they are applied to the input of the 4046. Figure 2 shows a simple snap-action comparator made from a BiMOS CA3130 op amp that can be used for this purpose. This circuit can process signals with frequencies as high as 50 kHz. Other op amp types can be used in place of the CA3130.

In Fig. 1, a 4518 is used as a divide-by-100 counter. This IC was selected principally because it includes two decade counters in a single DIP, but other counter arrangements can also be used. A pair of 4017 decade counters, for example, can be cascaded to form a suitable divide-by-100 counter.

Applications for this month's project are plentiful because the circuit can be used to multiply virtually any low-frequency signal. Of particular interest is the measurement of such low-frequency biological signals as brain waves and cardiac-pulse, respiration and blink rates. Other possibilities include measuring the revolution rates of wind turbines, bicycle wheels and low-speed motors. ◇

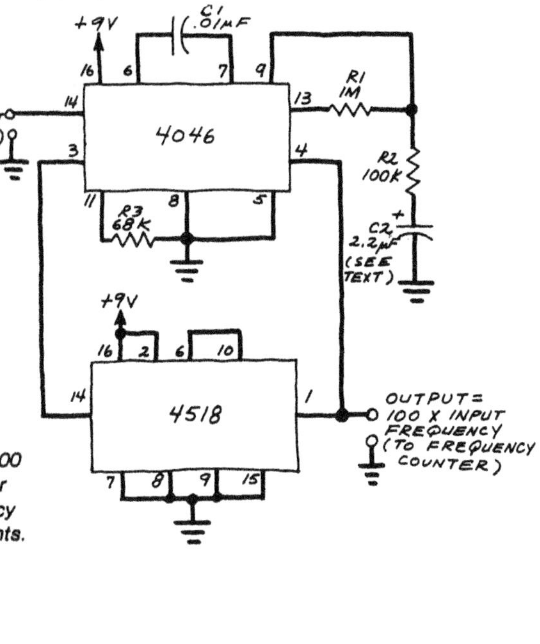

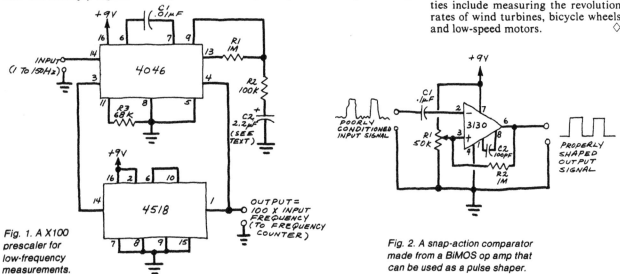

Fig. 1. A X100 prescaler for low-frequency measurements.

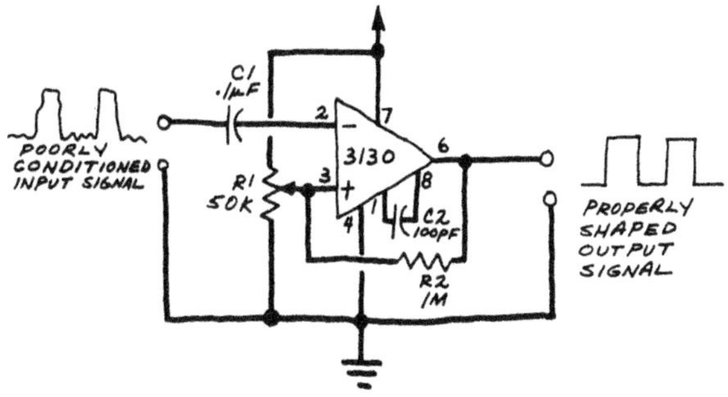

Fig. 2. A snap-action comparator made from a BiMOS op amp that can be used as a pulse shaper.

7. Audible Pulse Indicator

HOW MANY times have you wondered if the clock section of a circuit was functioning properly? Finding out can sometimes be a difficult job, particularly if you don't have access to an oscilloscope.

An excellent way to detect pulses when a scope isn't available is to use a logic probe. But, as with a scope, you must keep an eye on the test instrument to determine whether or not pulses are present.

Shown here is a circuit that provides both visual *and* audible indication of the presence of pulses. The circuit is designed around three timers, two of which are integrated onto a single chip.

Timers 1 and 2 are monostable multivibrators, each having a timing period of about 1/3 of a second. The pulse source is connected to the trigger input of Timer 1 through attenuator *R1*. If a pulse occurs, Timer 1's timing cycle is begun. Subsequent pulses which occur *during* the timing are ignored.

Ordinarily, after its timing cycle is complete, Timer 1 would be retriggered by the next incoming pulse. This is acceptable for slow-repetition rate signals. If the time between pulses is very brief, however, it would not always be possible to visually or audibly recognize the presence of pulses since one stretched pulse would be immediately followed by another. In other words, a train of closely spaced pulses would appear continous to the relatively slow eye or ear.

Timer 2 solves this problem by disabling Timer 1 by means of *Q1* for about 1/3 second immediately after each of Timer 1's timing cycles. Timer 1, therefore, responds to an incoming train of fast pulses by switching on and off at 1/3-second intervals.

Indicator *LED1* provides a visual response to the presence of incoming pulses. It stays on during Timer 1's timing cycle.

An astable audio-frequency oscillator provides the circuit's audible output. When Timer 1 has not been triggered, its output is low. Since Timer 1's output is connected to Timer 3's reset input through *R8*, Timer 3 is disabled when no pulse is present at Timer 1's input. When a pulse occurs, Timer 1 is triggered, which, in turn, enables the audio oscillator formed by Timer 3. Note that Timer 3, like Timer 1, is disabled for 1/3 second following the completion of Timer 1's timing cycle. Therefore, a very fast train of pulses is indicated by a slow series of tones spaced 1/3 second apart.

This circuit may need modification for some applications. For example, a high input impedance section can be added to prevent the circuit from loading down the clock being checked. Similarly, an input amplifier can be added to beef up weak pulses. The circuit can even be added to existing circuits so that it becomes an integral audible/visual pulse indicator.

In its present form, the circuit responds to pulses having an amplitude of from a few volts to V_{CC}. Though I used a 556 and a 555 for the three timers, you can use three 555's or a pair of 556's. If you choose the latter approach, you'll have an extra timer section for use in possible circuit modifications. ■

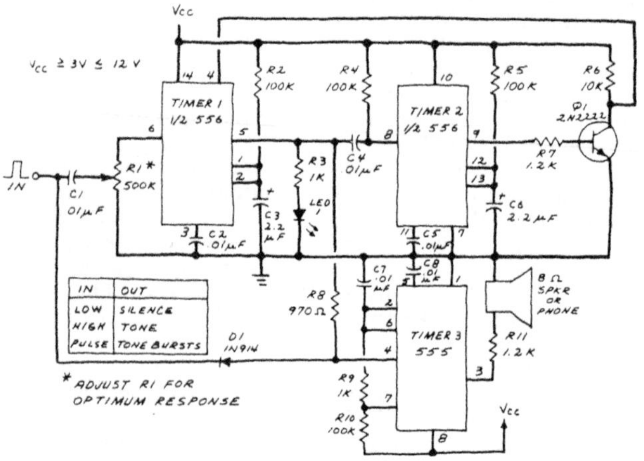

8. A Solid-State Panel Meter

ANYONE who has assembled an LED dot-bar driver from individual comparators, resistors and LEDs has a special appreciation for single-chip LED dot-bar drivers like National's LM3914/15/16 series. But even dot-bar driver chips require connections to each readout LED.

Last year National solved this problem with the introduction of its NSM3914/15/16 modules. These new modules can be used as low-cost, solid-state replacements for conventional panel meters. Each module consists of an LM3914/15/16 chip installed on a small printed circuit board measuring 1.99 x 0.850 inches. The chip is protected by an opaque plastic cover.

Also installed on the circuit board is

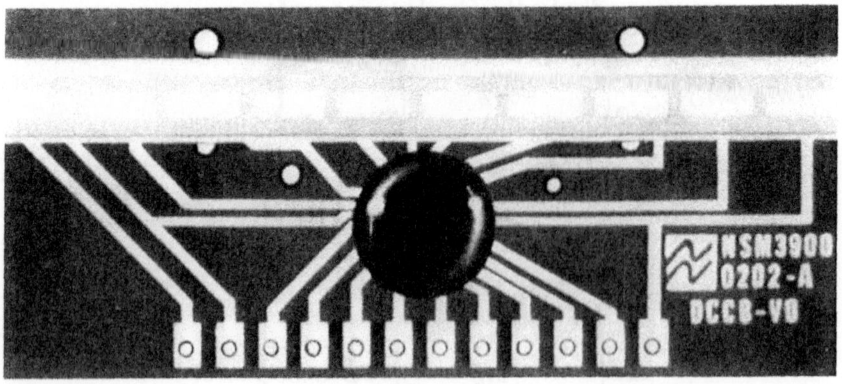

Fig. 1. Combining a dot-bar driver chip and a row of LEDs in one module provides a low-cost, solid-state meter.

a row of ten red LEDs. The LEDs are protected by a plastic bar having individual windows for each LED. Connections to the module are made via a row of edge terminals. The LED strip is oriented to make possible end-to-end stacking of multiple modules.

Figure 1 is a pictorial representation of the NSM3914/15/16 module. The terminals are numbered 1 through 12 beginning at the left. The connections to each terminal are as follows:

Pin	Electrical Connection
1	V_{LED}
2	LED 1
3	Ground
4	V+
5	R_{LO}
6	Signal In
7	R_{HI}
8	Reference Out
9	Reference Adjust
10	Mode (dot-bar)
11	LED 9
12	LED 10

For a more detailed explanation of these functions, see the data sheet for the LM3914, LM3915 or LM3916.

National's data sheets for the LM3914/15/16 provide plenty of application ideas for the modular displays which use these chips. Figure 2, for instance, shows how to connect the NSM3915 in accordance with one of the application circuits given in the data sheet for the LM3915.

This circuit provides a logarithmic display suitable for various audio applications. Each LED is activated at intervals of 3 dB. Power supply filtering is provided by *C1*; it may be omitted if the leads from the power supply to the module do not exceed six inches length.

The components in Fig. 2 can be soldered directly to the terminals on the NSM3915. Use care to avoid applying excessive stress to the circuit board since it is fragile. Also, do not overheat the terminals or the copper foil from which they are formed will peel away from the substrate.

The circuit in Fig. 2 is only an example of what can be done with one of the new National modules. I urge you to have a look at the data sheets for the various National LED dot-bar drivers before making a final circuit decision. ■

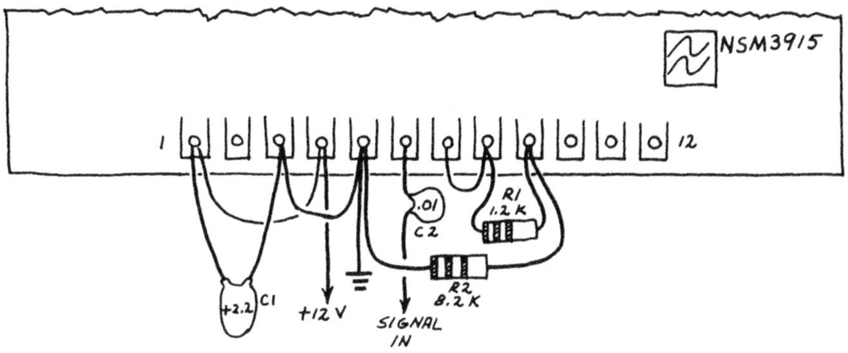

Fig. 2. How to connect the NSM3915 as a logarithmic readout. See the LM3915 data sheet for more information.

Five

Power Sources

1. Miniature DC-DC Converter

The LM3909 was originally designed as an LED flasher, but has many other applications. One that I've enjoyed experimenting with is a miniature power supply that allows a tiny watch battery to power a neon lamp or even a powerful semiconductor-laser pulser. Both these applications require 70 to 150 volts at relatively low current.

Figure A shows the circuit of the LM3909 dc-dc converter. In operation, the LM3909 rapidly switches *Q1* on and off at a rate determined by *C1*. The transistor can be considered a switch in series with choke *L1* and resistor *R2*. Each time *Q1* switches off, the magnetic field set up by the current flowing through *L1* collapses and induces a high voltage across the inductor. This voltage is rectified and stored in *C2*.

The LED is a bonus feature of the circuit. It glows to indicate when the circuit is operating. The neon lamp and 15,000-ohm series resistor shown in Figure A are optional. They provide a visual indication that the circuit is producing 70 or more volts. When powered by a 1.2-volt nickel-cadmium or 1.35-volt mercury "button" cell, the circuit produces enough voltage to flash the lamp when it is connected across capacitor *C2*.

If you don't like the orange glow of a neon lamp, try a *green* neon lamp (Radio Shack 272-1106 or equivalent). This lamp has a phosphor coating on its inside surface that glows green when illuminated by the radiation produced inside the lamp. In

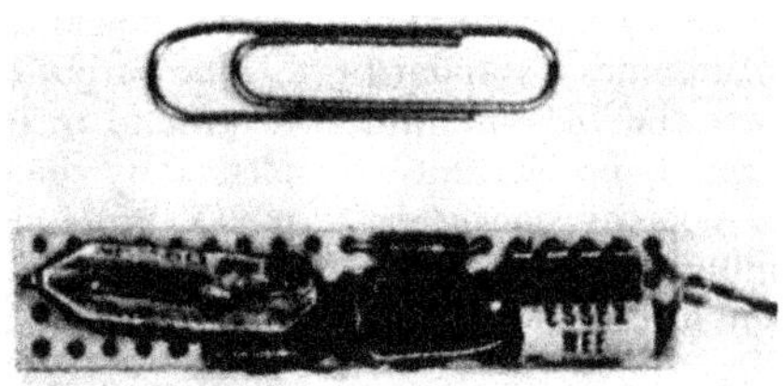

Fig. C. Photo of prototype version of the dc-dc converter.

any case, be sure to use a quality lamp because some of the surplus neon lamps I've tried do not work well.

The key components of the circuit are *L1* and *R2*. In the prototype circuit, I used a miniature Essex choke with an inductance of 1000-μH for *L1*. This choke is about the size of a ¼-watt carbon composition resistor. If this choke is used, the resistance of *R2* should be between 75 and 85 ohms.

If you can't find this choke, experiment with others until you find one that produces enough voltage to light a neon lamp. You'll find that many different chokes will produce a useful output. One version of the circuit that I built uses a miniature 33-mH choke (Aladdin) with excellent results. The 1979 Allied catalog (401 E. 8th St., Fort Worth, TX 76102) lists a number of subminiature r-f inductors on page 145 that should work fine.

If you don't use the 1000-μH choke specified, you'll need to experiment with the value of *R2*. As the inductance is increased, *R2*'s value can be decreased. Actually, *R2* is not even necessary beyond a few millihenries.

Assembly of this circuit should present no problems once you've selected a choke and determined the resistance of *R2* (if it is necessary). I used a piece of perforated board with copper solder pads at each hole. Figure B is a pictorial view of the assembled circuit.

Begin by inserting the components into the top side of the board and interconnecting their leads with wrapping wire. Then solder all the connections to their respective solder pads. Figure C is a photograph of the complete prototype. This circuit includes a neon lamp and series resistor to illustrate its operation as a dc-dc converter. Don't forget that the circuit has many other possibilities.

For example, most semiconductor lasers require current pulses of many amperes for proper operation. The circuit in Figure A can power a four-layer-diode laser pulser with ease, especially if *L1* has an inductance of 10 to 35 mH. Recently, I built a midget laser transmitter using the circuit in Fig. A as a power supply (*L1* = 33 mH, no *R2*). The circuit is completely self-contained and includes lens, mercury "button" cell and switch in a 0.5" × 3" (1.3 cm × 7.6 cm) brass tube.

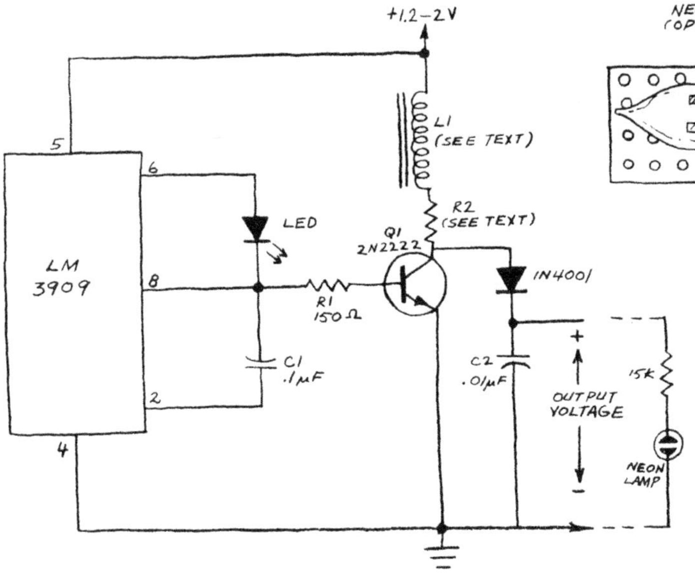

Fig. A. Miniature dc-dc converter circuit diagram.

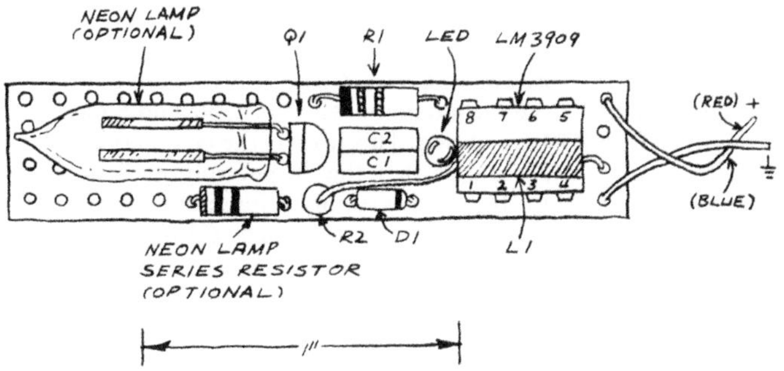

Fig. B. Arrangement of components on the board for the converter.

2. An Integrated Polarity Converter

EVERY few years, one of the semiconductor manufacturers introduces a new IC that quickly becomes a standard building block. Three classic examples are the 741 op amp, the 555 timer and the LM109 5-volt regulator. Recently, Intersil introduced its ICL7660 voltage polarity converter, which is sure to become a standard building block IC. The ICL7660 is a useful power-supply chip which generates a negative voltage from a positive input voltage. If you've ever wanted to employ a circuit that required a dual-polarity power supply in an application where only a single-ended supply was available, you can readily appreciate the practical implications of this new chip.

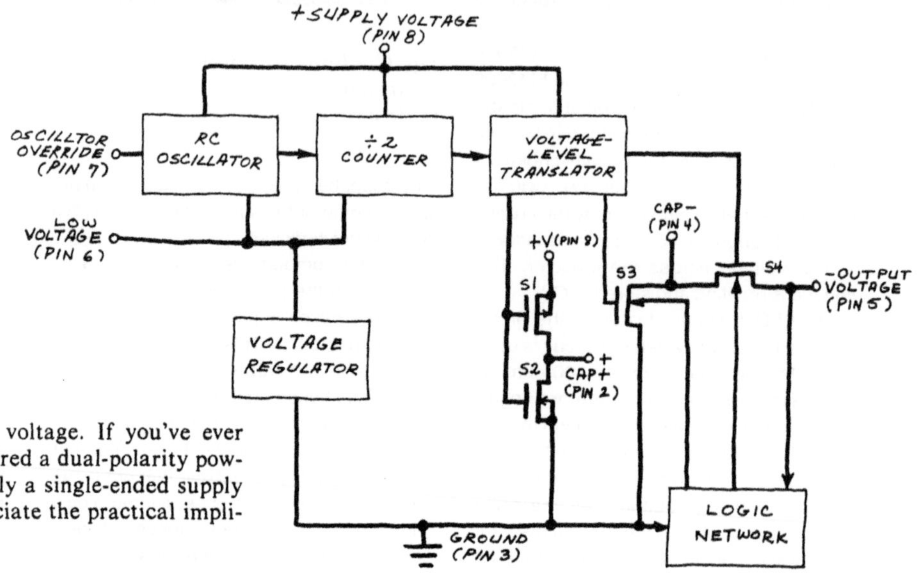

Fig. 1. Block diagram showing how the ICL7660 functions as a voltage polarity converter.

About the ICL7660. A block diagram of the ICL7660 appears in Fig. 1. It shows what might at first seem to be a relatively complex circuit. Actually, the operation of the ICL7660 is straightforward, and you can understand how the chip works by referring to the simplified diagram in Fig. 2. Switches *S1* through *S2* in Fig. 2 are the MOSFETs of Fig. 1. Capacitors *C1* and *C2* are external 10-μF electrolytics.

In operation, *S1* through *S4* are opened and closed by the switch logic at a rate determined by the oscillator. During the first half of the operating cycle, *S1* and *S3* are closed, and *S2* and *S4* are open. This allows *C1* to charge to the positive supply voltage. During the second half of the cycle, *S1* and *S3* are open, and *S2* and *S4* are closed. This inverts the connection of *C1* with respect to ground so that the output voltage is now equal in magnitude to the positive supply voltage but opposite in signal (polarity). The switches also connect *C2* in parallel with *C1*, thus causing *C2* to draw charge from *C1*. This enables *C2* to serve as a reservoir of charge during the first half of the cycle, when *C1* is being charged.

The MOS switches, which occupy most of the ICL7660 chip's real estate, required clever design techniques—although this might not be immediately obvious. The channels of *S3* and *S4* must remain nonconductive to prevent device latchup when power is first applied and when the output is short-circuited. The logic network (Fig. 1) monitors the output voltage, and the voltage-level translator applies bias when necessary to keep both switches (*S3* and *S4*) nonconductive. Without this precaution, Intersil reports that high power loss and device latchup (presumably resulting from overheating) would occur.

Using the ICL7660. The pinout diagram of the ICL7660 appears in Fig. 3. The chip can be powered by a supply voltage of +1.5 to +10 volts. It has a typical open-circuit voltage conversion efficiency of 99.9% and a power efficiency of 98%.

The output of the internal voltage regulator can be shorted to ground to improve low-voltage operation. This is accomplished by connecting pin 6, the LV (low-voltage) terminal, to pin 3. Note, however, that if the power-supply voltage is greater than +3.5 volts, the LV pin must *not* be grounded or device latchup might occur.

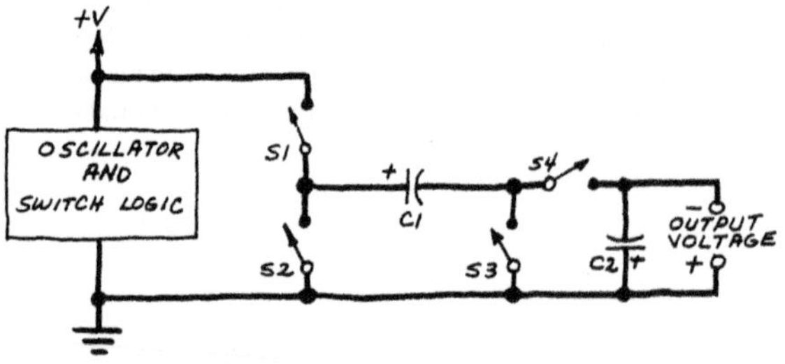

Fig. 2. Simplified version of Fig. 1.

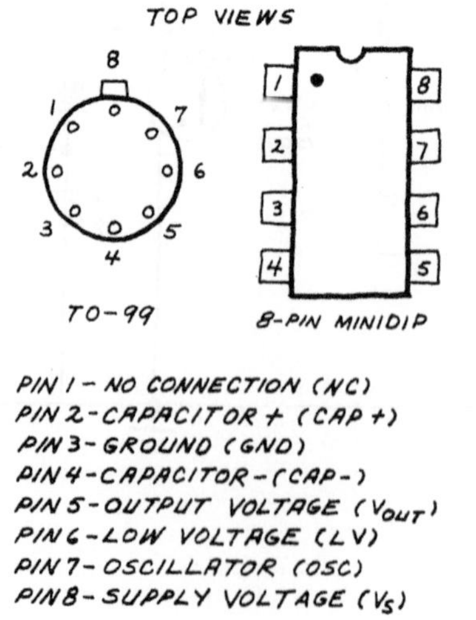

Fig. 3. Pin outlines of the ICL7660.

If the supply voltage exceeds +6.5 volts, a diode (*D1* in Fig. 4) should be connected in series between pin 5, the output terminal, and the load. Of course, this will reduce the voltage delivered to the load by an amount equal to the diode's voltage drop—about 0.6 volt for a silicon diode and 0.3 volt for a germanium diode. However, the addition of the diode permits the

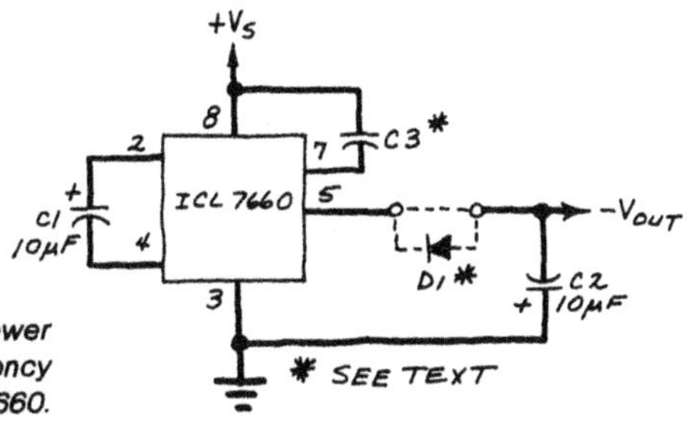

Fig. 4. How to lower the oscillator frequency of the ICL7660.

chip to operate over its entire rated temperature range at a supply voltage with a maximum of 10 volts without the possibility of device latchup.

The oscillator frequency of the ICL7660 is typically 10 kHz. If this frequency of oscillation causes interference to circuits powered by or close to the ICL7660, it can be changed. To *lower* the frequency of the oscillator, connect a capacitor (*C3*) between pins 7 and 8. The installation of a 100-pF capacitor will lower the frequency from 10 kHz to 1 kHz. This reduction in frequency must be accompanied by an increase in the capacitance of *C1* and *C2* by the same factor the frequency is reduced. Because 1 kHz is one-tenth of 10 kHz, *C1* and *C2* must be increased by a factor of ten(to 100 µF).

The frequency of oscillation can be *increased* by overdriving the internal oscillator with an external clock or oscillator. In this case, however, the values of *C1* and *C2* need not be reduced. Figure 5 shows how to connect CMOS or TTL circuits to the OSCILLATOR OVERRIDE input (pin 7) of the ICL7660.

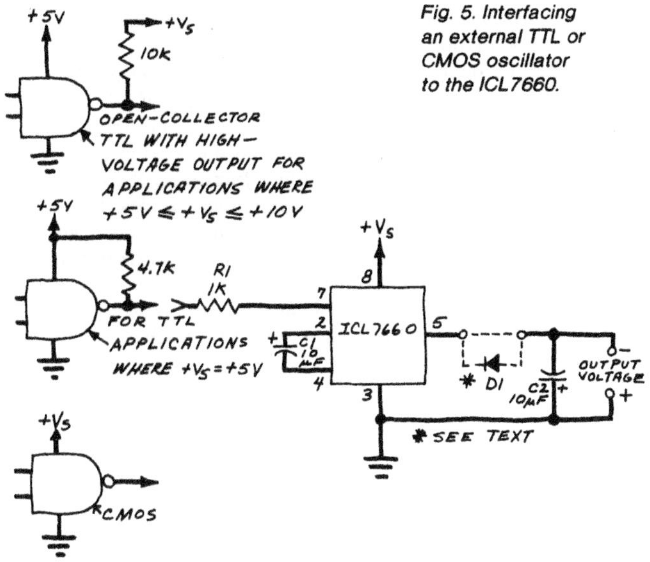

Fig. 5. Interfacing an external TTL or CMOS oscillator to the ICL7660.

Finally, keep in mind that the ICL7660 is a CMOS chip. Standard CMOS handling precautions should be observed. Don't use an ungrounded soldering iron. Also, be sure to protect the chip from static voltages when installing it.

The circuits which follow will give you some ideas about how to use the ICL7660 in specific applications. The chip is so easy to use that you should have few, if any, problems when you try the circuits on your own.

Basic Polarity Converter. Figure 6 shows a simple voltage-polarity converter employing a single ICL7660. The circuit has two design variations which are dependent upon the magnitude of the supply voltage. When $+V_S$ is less than 3.5 volts,

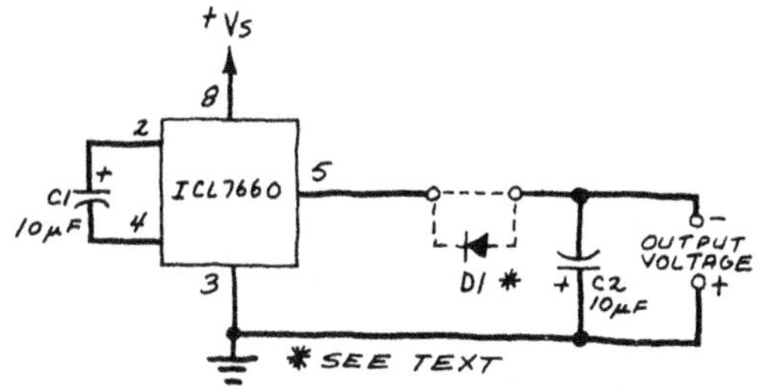

Fig. 6. Ultra-simple voltage polarity converter.

pin 6 should be grounded. When the supply voltage is greater than +6.5 volts, *D1* should be connected to pin 5 as shown. In no case should any pin be connected to a voltage higher than the supply voltage or below circuit ground. Otherwise, the chip might be destroyed.

Voltage Multiplication. Figure 7 shows how to cascade two ICL7660's to double the negative voltage available from a single chip. When the supply voltage exceeds +6.5 volts, *D1* and *D2* must be included. This will reduce the output voltage by the sum of the voltage drops across each diode. Thus, when a 10-volt supply is used and both *D1* and *D2* are silicon diodes, the voltage out will be −18.8 volts.

You can cascade up to ten ICL7660's using the cascade configuration shown in Fig. 7. Assuming that the $+V_S$ terminal of the first ICL7660 is connected to a +10-volt supply and that a silicon diode is connected in series with each IC's output terminal, the output voltage at the end of the multiplier string will be −94 volts!

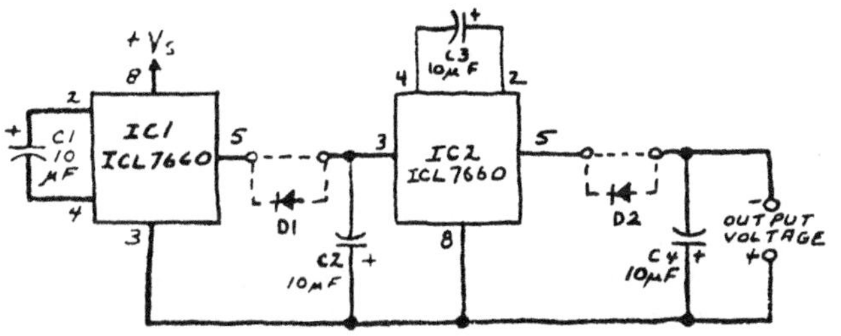

Fig. 7. Negative voltage multiplication by cascading ICL7660s.

Until now, all of the circuits we have described have had outputs whose polarity has been inverted with respect to the power supply (a positive supply voltage and a negative output voltage). The next circuit to be shown is a voltage multiplier whose output is of the *same* polarity as the supply. The circuit shown in Fig. 8 provides positive voltage multiplication using a single ICL7660. Capacitor *C1* charges to the supply voltage less the voltage drop across *D1*. The sum of the voltage across

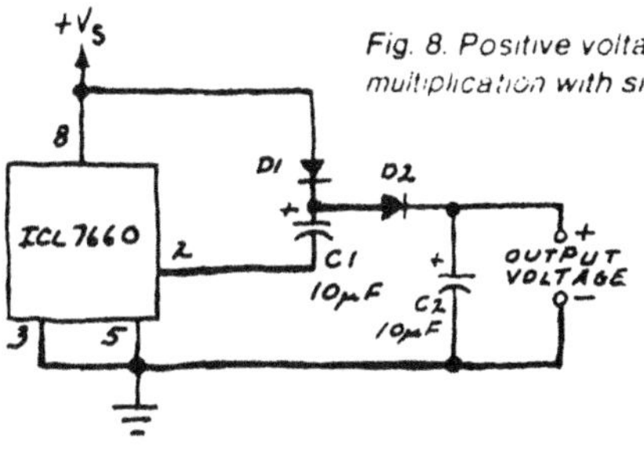

Fig. 8. Positive voltage multiplication with single ICL7660.

C1 and the supply voltage then charges *C2* through *D2*. The resulting output voltage is twice the supply voltage less the sum of voltage drops across the two diodes.

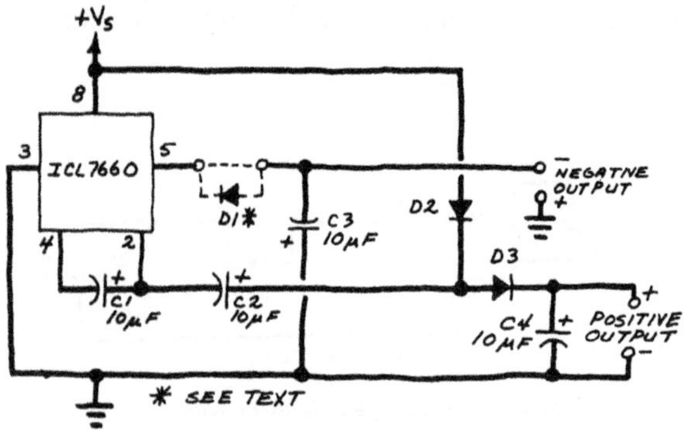

Fig. 9. Combined positive and negative voltage multiplication.

Dual-Output Supply. The circuit shown in Fig. 9 performs positive voltage multiplication and negative voltage conversion using a single ICL7660. This configuration can provide several different combinations of output voltage from a single positive supply. The voltage level appearing across the negative output is an inverted version of the positive supply voltage (less one diode drop if the supply voltage exceeds 6.5 volts, in which case *D1* must be included). The voltage level across the positive output is twice the positive supply voltage less the sum of two diode drops with *no* polarity inversion.

The high efficiency of the ICL7660 is apparent from an inspection of the following table. It is a list of output voltages that I measured when a sample chip was powered with a given sequence of positive supply voltages.

$+V_S$	Negative Output	Positive Output
+1.5	−1.5 V	+ 1.8 V
+3.0	−3.0 V	+ 4.2 V
+5.0	−5.0 V	+ 8.8 V
+6.0	−6.0 V	+10.8 V
+9.0	−8.4 V	+16.8 V
+10.0	−9.4 V	+18.8 V

Note that the voltage drops across *D2* and *D3* prevent the positive output level from approaching twice the positive supply voltage, especially at low supply voltages. Also, *D1* was included when $+V_S$ was raised above +6.5 volts, so the negative output levels measured when $+V_S$ was +9 and +10 volts reflect the forward voltage drop across the diode. This circuit can be made more versatile by adding one or more ICL7660s to multiply the negative output voltage level.

Square-Wave Oscillator. The ICL7660 is ideally suited for supplying the negative voltage to op-amp circuits that require a bipolar supply. Such a circuit is the square-wave oscillator appearing in Fig. 10. The circuit provides a true ac square wave that is *not* riding on a dc level. It can be powered by a single-ended 5-volt supply or a 9-volt battery.

The component values shown result in a square-wave frequency of approximately 900 Hz. Changing the values of *R1*, *C1* or both *R1* and *C1* will change the frequency of oscillation. More precisely, the frequency equals the reciprocal of the product of (2*R1C1*) and the natural logarithm of the quantity (2*R2*/*R3* + 1).

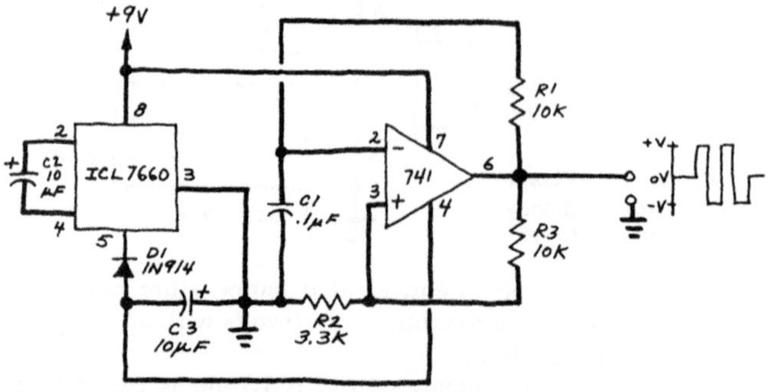

Fig. 10. Circuit for a square-wave oscillator.

The value of resistor *R2* should be approximately one-third that of *R1*. The resistance of *R3* should be from two to ten times that of *R2*. For those readers who lack access to scientific calculators or sophisticated slide rules, the component values specified in Fig. 10 cause the expression ln (2*R2*/*R3* + 1) to equal 0.5068. Therefore, the frequency of oscillation equals 1/(10,136) (*C1*).

Here are some results that I obtained using a breadboarded version of the circuit where f_O is the output frequency:

Nominal value of C1	Calculated f_O	Measured f_O
1 µF	98.65 Hz	95.8 Hz
0.1 µF	986.5 Hz	908 Hz
0.01 µF	9865 Hz	7768 Hz

As you can see, the calculated or predicted output frequency becomes less accurate as the value of *C1* decreases and the frequency of oscillation increases.

I originally built this circuit as part of an experimental high-voltage power supply. Connecting the secondary of a 6.3-volt filament transformer between pin 6 of the 741 and ground will cause 80 volts of ac to appear across the transformer's primary when the circuit is powered by a +6-volt supply. Increasing the supply voltage to +9 volts will boost the voltage induced across the primary of the transformer output to 160 volts ac. You can connect a diode-capacitor voltage multiplier to the output of the transformer to obtain even higher *dc* voltages.

Use caution if you add the transformer (and possibly a voltage multiplier) to the oscillator shown in Fig. 10. Such a combination can deliver a *hefty* shock! ■

3. The Polapulse® Wafer Battery

SOME of my relatives own Polaroid SX-70 Land cameras. One reason I look forward to family gatherings is the frequent use these cameras receive on such occasions. It's always nice to see the excellent photos the cameras provide, but perhaps my primary interest lies in the empty film packs, each of which contains a used Polapulse® 6-volt wafer battery. The battery powers the camera's electronics and drive motor.

The Polapulse battery has been deliberately overdesigned by Ray-O-Vac and Polaroid engineers to ensure that it will reliably power the SX-70 camera even under the most demanding circumstances. Probably the worst possible scenario is when all the available emulsions in the SX-70 film pack are exposed in rapid sequence under low-temperature conditions. Overdesigning the Polapulse battery provides plenty of reserve power for rapid-fire photo sessions. This also means that the battery still has plenty of life left in it even after its film-pack carrier has been tossed in the trash.

Recently, Polaroid began selling Polapulse batteries as

products in their own right. For $15, one can buy five Polapulse P100 batteries and a plastic holder that has connection wires.

This month, we will take a detailed look at the new Polapulse P100 battery. We will examine the battery's specifications as given in Polaroid's literature, and then look at the characteristics of Polapulse batteries that have been salvaged from discarded SX-70 film packs.

Polapulse Construction. The Polapulse battery utilizes the traditional LeClanche carbon-zinc chemistry in a physical configuration which differs radically from the cylindrical LeClanche cells with which we are all familiar.

The basic building block of the Polapulse has about the same physical dimensions as a playing card and is called a *duplex sheet*. As its name implies, the duplex sheet is a thin conductive card coated on one side with manganese dioxide and on the other with zinc. A single duplex sheet forms one-half of each of two adjacent 1.5-volt cells.

In the Polapulse, four duplex sheets are interspersed with separator layers made from synthetic fibers which have been impregnated with a gelled electrolyte. The conductive layer of the duplex sheet interconnects adjacent cells without the need for solder tabs, welded connections or wires. This provides a very simple, highly reliable, multicell battery.

The top side of the sandwich of cells is covered with an aluminum foil sheet coated on its lower side with zinc. This functions as the battery's negative terminal. The bottom side of the battery is covered with a second aluminum foil sheet coated on its upper surface with manganese dioxide. This becomes the positive terminal.

A paper insulating sheet, which has a square hole for electrical access, is bonded to the positive electrode. A protruding portion of the upper negative electrode is folded over the paper backing so that both electrodes are on the same side of the battery. The multilayer battery is then laminated into a sturdy package by adhesive borders around the perimeter of each separator sheet. A vent prevents excessive gas buildup.

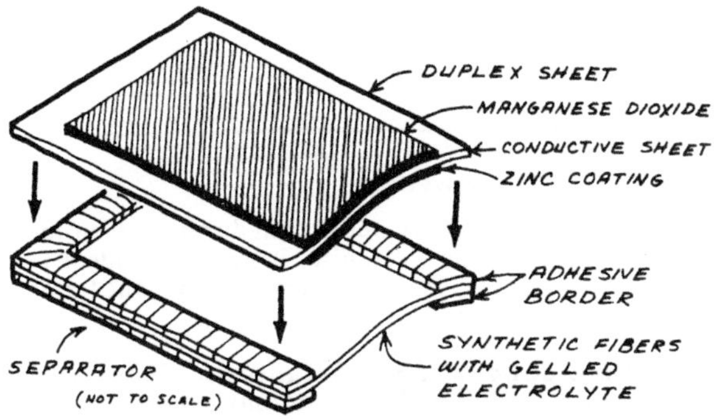

Fig. 1. Two layers in a Polapulse battery.

Figure 1 shows the construction of the duplex sheet and its orientation with respect to one adjacent separator. The overall dimensions of the Polapulse at this point are 3.73" x 3.04". The battery is a wafer-thin 0.18", and it weighs just under an ounce.

Polapulse batteries intended for use in SX-70 film packs are mounted on a card measuring 3.45" x 4.2" and are secured in place with laminated plastic film. Electrical access to the electrodes is provided by two circular apertures punched in the card.

The paper, plastic and foil construction of the Polapulse probably sounds very flimsy if you have not held one of these batteries in your hand. Actually, however, the battery is surprisingly sturdy. The adhesive bond around its perimeter provides a very rigid frame around the relatively fragile layers which form the battery.

More than 300,000,000 Polapulse batteries have been manufactured for SX-70 film packs. Thanks to the battery's adhesive bond and its protective outer layers, no instances of damage caused by leakage have been reported.

Electrical Specifications. The Polapulse delivers up to 6.8 volts, open circuit. Its useful temperature range is 20° F to 130° F (−7° C to 54° C). The most important feature of the battery is its surprisingly high capacity. For example, a fresh Polapulse can deliver a hefty 26 amperes on an instantaneous basis. After 30 seconds, the current decreases to a still-impressive 5 amperes. The current decreases to 2.5 amperes after 60 seconds.

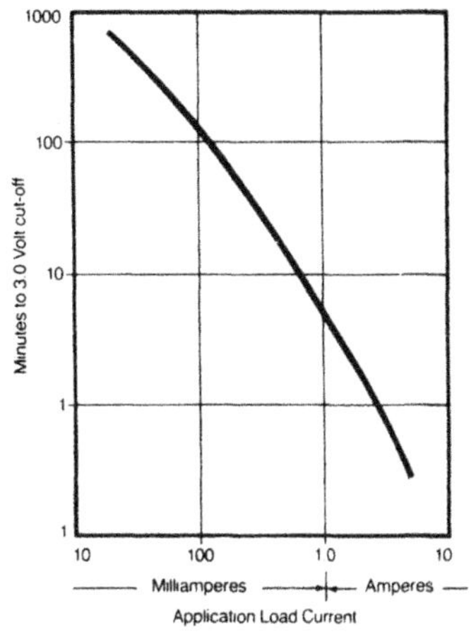

Fig. 2. Capacity (in minutes of time in use) of a Polapulse vs. load current.

Polaroid has published the following continuous discharge ratings for the Polapulse (3-volt cutoff).

Discharge Current (A)	Discharge Interval (Minutes)	mAH
0.02	690.0	230
0.05	234.0	195
0.10	102.0	170
0.50	14.7	123
1.00	5.7	95
5.00	0.4	33

These data are summarized in the graph of capacity versus load current in Fig. 2. A Polaroid brochure on the Polapulse P100 battery includes the family of constant-current drain curves in Fig. 3. How do these data compare with the performance of other types of batteries? Figure 4 shows plots of load current versus voltage for the Polapulse and several conventional batteries. The Polapulse provides superior performance under the conditions specified.

Experimenting with Polapulse. In preparing this column, I evaluated several Polapulse batteries salvaged from discarded film packs about a year ago. The results were fairly impressive, considering the batteries' age and prior use.

Figure 5, for example, is a graph of one battery's discharge voltage over a period of 30 minutes. The data from which the plot was derived were obtained by connecting a 100-ohm precision resistor across the terminals of the Polapulse. The voltage across the battery was then measured with a DVM at 30-second intervals for the first five minutes, and at 5-minute intervals for the remaining 25 minutes.

The discharge current during this test, discounting that which flowed during the initial 10-second surge, ranged from 47 mA when the battery developed 4.72 volts to 39 mA when battery voltage decreased to 3.9 volts. These discharge currents and output voltages mean that the Polapulse is more than adequate for powering many CMOS circuits. Intermittent operation will result in longer battery life and permit higher discharge rates.

I am particularly interested in emergency lighting. So I was curious to know if a used Polapulse battery would power a

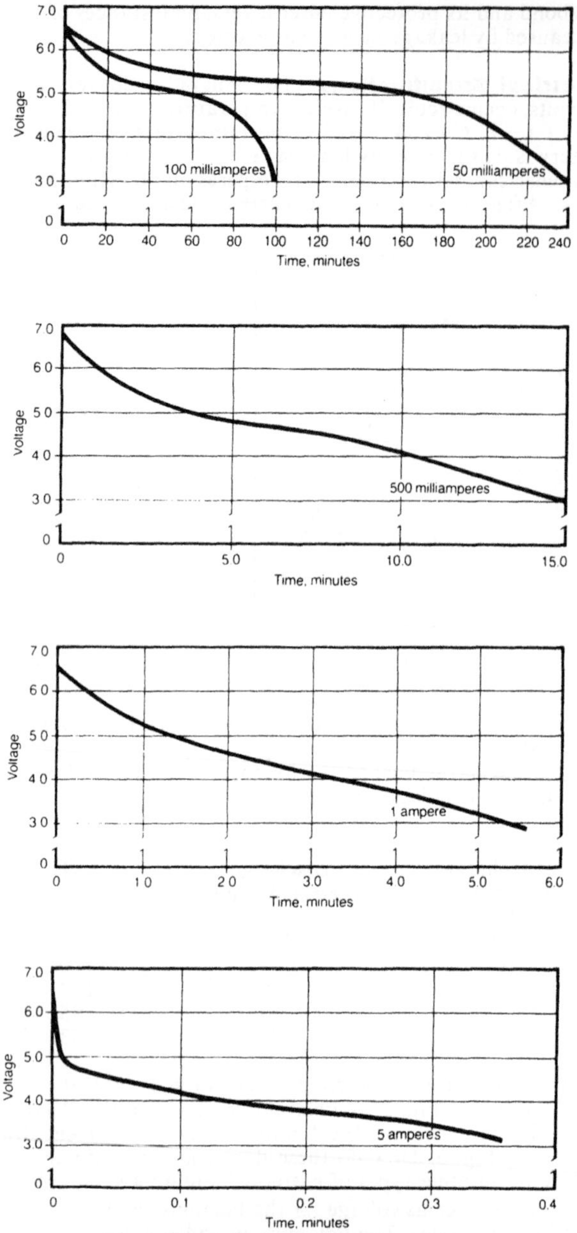

Fig. 3. Constant-current discharge curves for the Polapulse battery showing how the voltage drops off with time for various current drains. Top to bottom, currents are: 50 mA, 100 mA, 500 mA, 1A, and 5 A

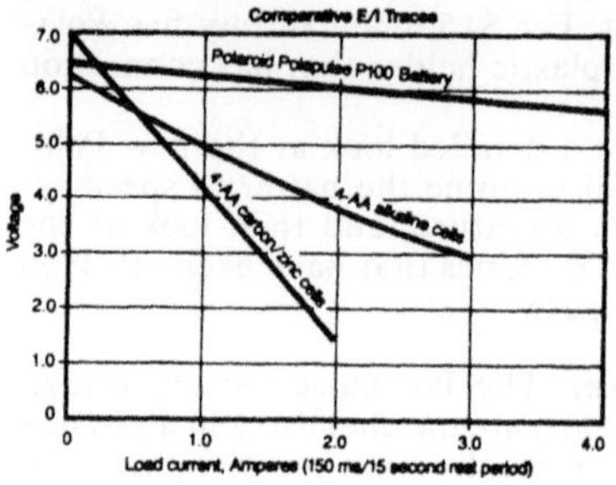

Fig. 4. Comparing the output of the Polapulse with other batteries.

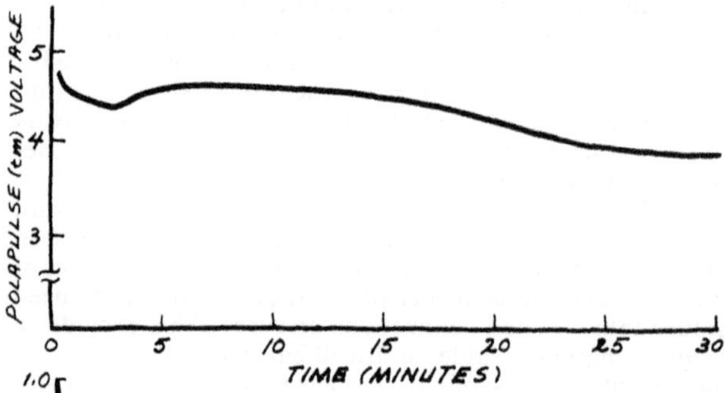

Fig. 5. Discharge curve of a salvaged Polapulse over 30 min.

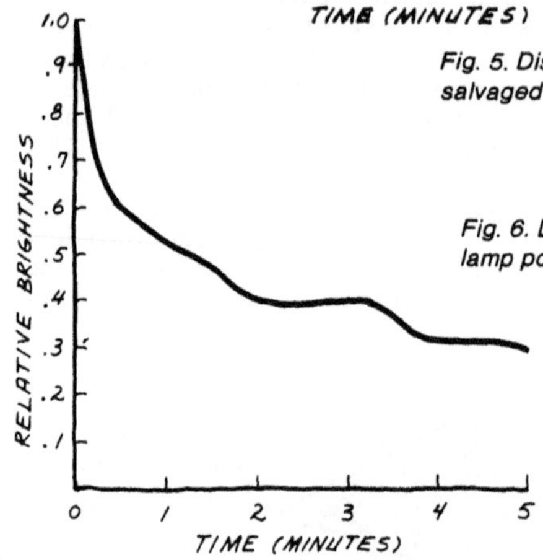

Fig. 6. Light output of a PR-13 lamp powered by a Polapulse.

PR-13 incandescent lamp. This lamp is often used in hand-held lights powered by 6-volt batteries. At 5.5 volts, they drain about 470 mA from the battery.

A quick check with clip leads showed that the Polapulse can easily power a PR-13 lamp at a high level of brightness. To monitor the lamp's output over a period of time, I placed the lamp adjacent to a solar cell whose output was monitored with a milliammeter. Figure 6 shows the results. When the light from the lamp became too dim for practical use, the meter gave a reading of 0.3 mA.

The salvaged Polapulse provided excellent light levels for approximately two minutes and usable light for the next two minutes. Four minutes isn't very long, but intermittent operation would result in even greater overall duration. Considering the compact size of the Polapulse, it makes an ideal power source for a miniature, short-term emergency light for camping trips and bike tours.

Making Connections to the Polapulse. The easiest way to solve the connection problem is to buy a Polapulse holder from a Polaroid dealer. This holder, however, will not accept Polapulse batteries salvaged from film packs. For these, you can use alligator clips to make temporary connections. Make sure the clips don't touch one another and don't penetrate the battery's protective covering. For best results, wrap a layer of vinyl electrical tape around one-half of each clip. The exposed half of each clip should be placed over the aluminum contact terminals of the battery.

Incidentally, you can reduce the size of salvaged Polapulse batteries by removing the outer protective envelope. Place the battery on a work surface with its printed side down. Then use a hobby knife to slit the clear plastic covering along the edge of the battery. This procedure will not harm the battery, but it will reduce its size.

For More Information. You can obtain more information about the Polapulse P100 wafer battery from the Polaroid Corporation's Commercial Battery Division (784 Memorial Dr., Cambridge, MA 02139). Ask for a specification sheet. ■

YOU can conduct a fascinating demonstration of electrical power generation by chemical means with a silver coin, a strip of magnesium and a piece of paper towel the size of a postage stamp. Dip the paper in lemon juice, place it over the coin and lay the magnesium strip on the paper. Then touch the cathode lead of a red LED to the magnesium. When the LED's anode lead is touched to the coin as in Fig. 1, the LED will glow brightly.

With the exception of the LED, this simple demonstration would have seemed fairly routine to pre-World War I experimenters. In those days many experimenters constructed their own primary and secondary (storage) cells. Commercial power cells were relatively expensive, and only a few kinds were available.

Today, literally hundreds of different kinds of batteries are available in a wide range of voltages and physical configurations. Nevertheless, for some special-purpose applications, a homemade battery may be more satisfactory than a commercial battery!

One example is powering a telemetry transmitter in an instrumented model rocket. The upward flight of such a rocket might last only a few seconds, yet a commercial battery could supply the necessary power continuously for days or even weeks. The penalty for this unnecessary capacity is excessive size and mass, both of which should be kept to a minimum.

This month, we will experiment with several electrochemical power cells that you can make from readily available materials. These cells are suitable for powering CMOS and other low-power circuits. Whether you assemble any of these cells or not, you may gain a better understanding of how conventional batteries work. If you do build cells of your own, you will certainly gain an appreciation for the convenience and drip-free operation of commercial power cells and batteries.

Some Definitions. Before proceeding any further, it is important to define a few basic terms:

Anode—The negative electrode of a cell.
Battery—Two or more electrically connected cells.
Cathode—The positive electrode of a cell.
Cell—A single two-electrode electrochemical generator.
Electrolyte—An ionized, electrically conductive paste, gel or liquid.
Primary Cell—A nonrechargeable cell.
Secondary Cell—A rechargeable cell.
Storage Cell—A secondary cell.

Electrochemical Generators. Alessandro Volta, an Italian physicist, invented the chemical generator. In March 1800, he demonstrated two of his generators before the Royal Society in London. One, called the Crown of Cups, consisted of a circular pattern of cups containing a solution of water and salt. One strip each of silver and zinc were immersed in each cup, and the zinc strip in one cup was connected to the silver strip in the adjacent cell. This arrangement formed a series connection of *wet cells*.

Volta's second generator was a stack of alternating disks of dissimilar metals separated by disks of paper soaked in brine. This device could produce more electromotive force in a smaller space than the clumsier Crown of Cups arrangement.

The electrochemical generators invented by Volta were used with little variation until about 1860, when other kinds of cells were developed. One such cell, patented by French scientist Georges LeClanche in 1868, was the predecessor of the modern zinc-carbon *dry cell*.

The basic design of the zinc-carbon dry cell (such as those used to power radios, flashlights and toys) has remained largely unchanged for more than sixty years. Each cell consists of a zinc cup or can (the anode) filled with a moist compound whose composition has changed through the years. One

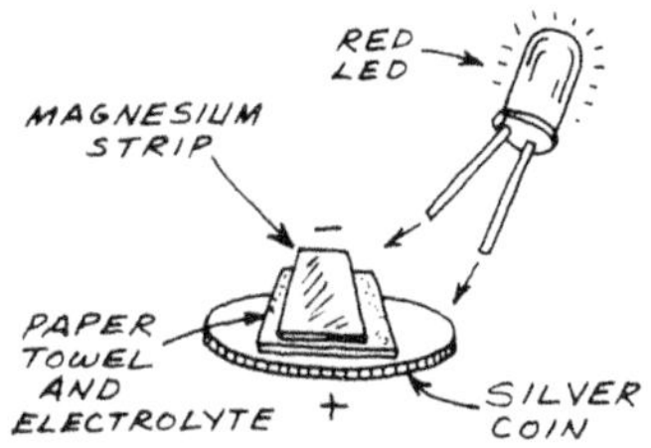

Fig. 1. An ultra-simple homemade power cell.

1924 recipe called for a mixture of one ounce each of zinc-chloride and ammonium-chloride, two ounces of water, and three ounces of plaster of paris, which served as a filler. Sawdust was also used as a filler material in some early cells. A carbon rod inserted into the compound serves as the positive electrode.

The moist compound of the 1924 recipe served as the cell's electrolyte. In today's cells, the electrolyte is a paper liner impregnated with ammonium- or zinc-chloride that slides inside the zinc can. The space between the liner and the cell's carbon rod is packed with a mixture of granulated carbon and manganese dioxide. The latter compound serves as the cell's cathode. It is considered a *depolarizer* because it prevents *polarization*, the formation of an insulating layer of hydrogen bubbles around a cell's positive electrode.

Figure 2 is a pictorial view of the inner construction of a typical zinc-carbon dry cell. Most such cells are well sealed to prevent leakage which might occur should the zinc become corroded. The zinc seal also keeps the electrolyte from drying out. Drying of the electrolyte and subtle chemical reactions at the electrodes over time eventually degrade a cell whether or not it is used.

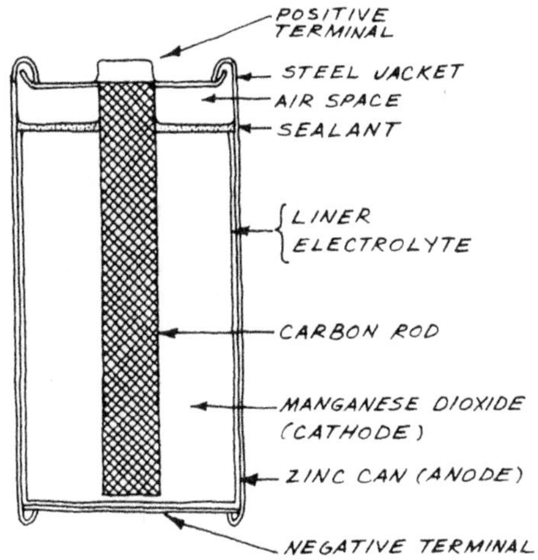

Fig. 2. Simplified internal view of a modern dry cell.

A Homemade Wet Cell. Figure 3 shows how you can make a simple wet cell from a plastic container such as a 35-mm film holder, a strip of copper and a strip of zinc. The metal strips are inserted through slits cut in the container's cap or lid. The container is then filled with an electrolyte such as salt water or lemon juice, and the cap and the electrodes are installed. This cell will produce about 0.7 volt.

If you connect a voltmeter to the cell's electrodes and pull the electrodes partially out of the electrolyte, the output volt-

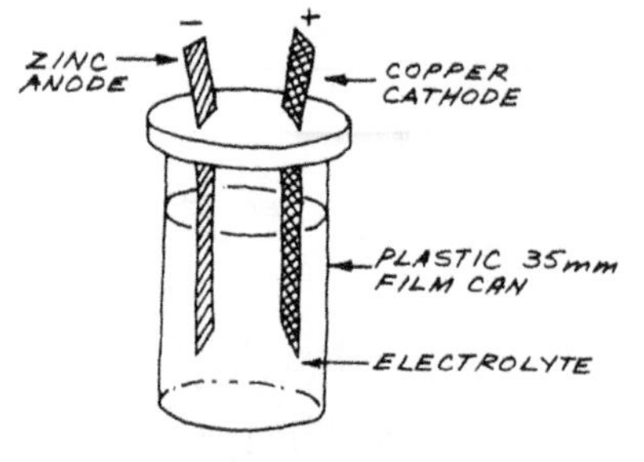

Fig. 3. Sketch of a homemade wet cell.

age will remain unchanged. Even when only a few millimeters of each electrode remain immersed in the electrolyte, the output voltage will remain unchanged. The cell's capacity to deliver current, however, is directly proportional to the area of the electrodes immersed in the electrolyte.

Electrode Materials. Any two dissimilar metals immersed in a suitable electrolyte will generate a voltage. Here are the voltages I measured for all possible pairs selected from the following group: a copper penny, a nickel, a silver dime, a magnesium strip, a zinc strip, and aluminum foil.

Cathode (+)	Anode (−)	Open-Circuit Voltage
Nickel	Copper	0.04
Magnesium	Zinc	0.05
Aluminum	Zinc	0.15
Silver	Nickel	0.19
Silver	Copper	0.20
Copper	Aluminum	0.70
Nickel	Aluminum	0.70
Copper	Zinc	0.72
Aluminum	Magnesium	0.78
Nickel	Zinc	0.81
Silver	Aluminum	0.84
Silver	Zinc	1.01
Nickel	Magnesium	1.44
Copper	Magnesium	1.45
Silver	Magnesium	1.65

For these measurements, I used as a combined electrolyte and separator several layers of paper towel soaked in salt water. The values you measure may differ slightly from mine, particularly if you use an acid electrolyte, in which case the values will be higher.

Note that the highest voltages are produced by pairing magnesium with nickel, copper or silver. Nickel and copper can be found in pocket change, but silver coins have not been minted in the U.S. since the mid-Sixties and are rarely found in everyday circulation. You can purchase magnesium strips at toy and hobby shops that sell Perfect brand chemicals.

Magnesium is highly reactive. I tried a magnesium strip in a lemon-juice wet cell and found that although the cell could easily power a LED, the magnesium was soon covered by a frothy layer of hydrogen bubbles. The cell functioned well in spite of the bubbles until the reaction with the citric acid in the lemon juice coated the magnesium with a black film.

Zinc is the next-best substitute for magnesium. You can get free zinc by cutting open a discarded zinc-carbon flashlight cell. If the cell is covered by an outer steel jacket, use pliers or diagonal cutters to peel it off. Be careful! The edges of the metal envelope will be very sharp.

When the zinc can has been exposed (it may be covered with a layer of paper or black pitch), secure the cell in a vise and use a hacksaw to remove the top half-inch of the cell. Remove the carbon rod, the filler compound and the electrolyte-impregnated paper liner from inside the can. The carbon in the compound will stain clothing, so be careful. Watch out for the sharp edges of the zinc can and clean any remaining compound from the can with detergent, water and an old toothbrush.

When the zinc is clean, use a file to remove the sharp edges left by the hacksaw. Then cut the can into strips with shears or a nibbling tool. Remove any corrosion from the strips with sandpaper.

Homemade "Moist" Cells. Figure 4 shows a simple 1.45-volt cell made from a 1/4-inch wide strip of magnesium or zinc wrapped with two layers of paper towel previously dipped in a solution of salt and water. A piece of copper foil (available at

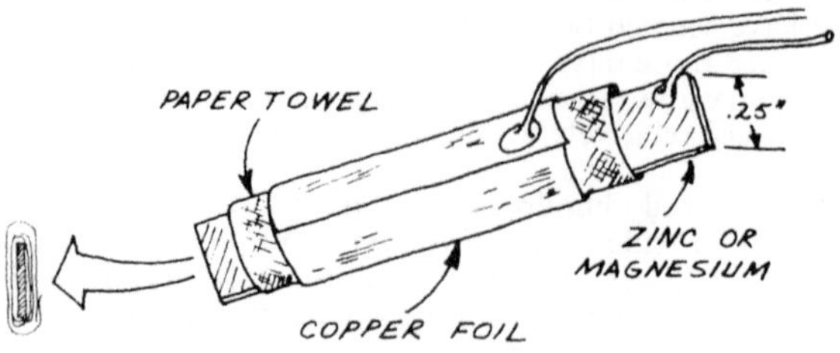

Fig. 4. How to make a simple 1.45-volt "moist" cell.

craft and hobby shops) the size of a postage stamp is wrapped over the paper towel.

For best results, the paper towel should be dried before the cell is assembled. When the cell is to be used, it can be activated by dipping it in water. Alternatively, a few drops of water can be applied to the exposed ends of the paper towel.

The cell shown in Fig. 4 is merely one of many possible configurations. You can make round, square or triangular cells. You can even cut the zinc or magnesium anode into long, narrow strips and make ultra-thin, cylindrical cells. You can increase the current capacity of a cell by increasing the area of its electrodes. Two or more cells can be connected in series to achieve higher voltages.

If a discharged cell is disconnected, in time it will gradually recover. Add moisture, and the cell will again deliver power. After several discharge cycles, you can rejuvenate a cell by unwrapping the copper foil and cleaning both the anode and cathode with steel wool. Reassemble the cell with a fresh, salt-impregnated separator layer.

There are two ways to attach wires to the cell. The simplest is to use miniature clip leads. I prefer to solder short lengths of wrapping wire to the electrodes prior to assembly. Copper foil is easily soldered. Zinc must be sanded for best results. Solder will not adhere to magnesium, so you will have to use a clip lead if you use this anode material.

Homemade Stacked Batteries. You can assemble a miniature version of Volta's stacked battery, which was called a *Voltaic pile*, with the help of a 1/4-inch paper punch. Punch a dozen or so holes in a piece of *thin* cardboard like that used for shoe boxes and soak the cardboard disks in salt water or lemon juice. Then punch an identical number of disks out of sheets of copper foil and magnesium or zinc. Solder a six-inch length of wrapping wire to one copper disk and one zinc disk. If you use magnesium, make an extra copper disk and solder a wire to it.

For best results, assemble the cell inside a hollow plastic tube. I used a small tube which originally contained the point of a drafting pen. Flexible tubing can also be used.

Install the copper disk with an attached wire lead first. Then, install alternating disks of cardboard, zinc (or magnesium), copper, cardboard, etc. The final disk should be zinc with an attached wire. If magnesium is used in place of zinc, top off the stack with the other copper disk to which a lead has been attached.

You may need to press the disks lightly against the end of the tube to achieve maximum output from the battery. Too

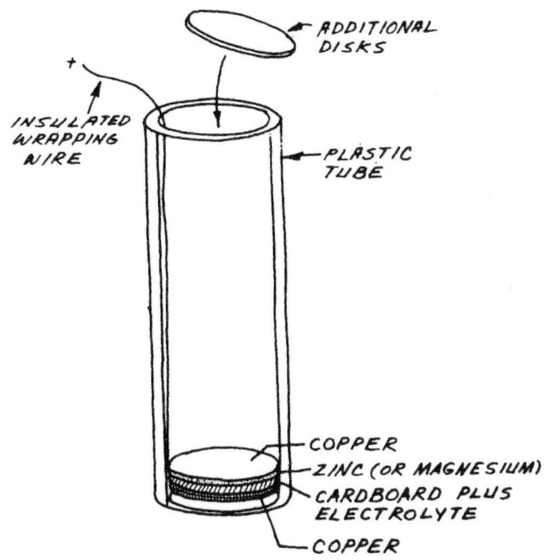

Fig. 5. Construction of a multi-cell stacked battery using zinc or magnesium.

much pressure, however, will squeeze electrolyte from the cardboard disks. Free electrolyte can short adjacent cells and reduce the battery's output voltage.

Incidentally, you will find it very helpful to use pointed tweezers when assembling a battery like this. Also, be sure to blot excess electrolyte from the cardboard disks before installing them in the tube.

Figure 5 shows how a 12-cell stack is assembled. This battery delivers 5.8 volts open circuit and is able to drive a LED with built-in flasher. The load of the LED and flasher circuit drops the voltage from the battery to a few volts, so the LED is not very bright. I used zinc and copper disks and lemon juice electrolyte.

Unfortunately, the battery in Fig. 5 is not suitable for post-assembly water activation. Adding water to the battery would short all the cells together. ■

Six

Digital Circuits

IN LAST MONTH'S column, we discussed the analog comparator and illustrated some applications for it. This month, we're going to look at its digital counterpart, the *magnitude comparator.* Figure 1 is the logic diagram for a simple magnitude comparator. The circuit, which can be made using a single 7400 quad NAND gate, compares two logic signals applied to its inputs. When both input signals are at the same logic level, a logic 0 appears at the output. When the inputs are at different logic levels, the output is a logic 1.

If this behavior sounds familiar, you're probably acquainted

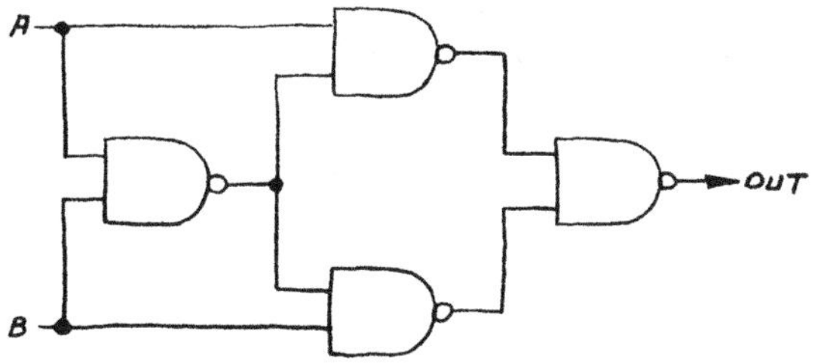

Fig. 1. Logic diagram for an Exclusive-OR gate synthesized from NAND gates.

with the Exclusive-OR gate, one of the most important combinational logic circuits. Figure 1 is, in fact, an Exclusive-OR gate synthesized from NAND gates. The logic diagram and truth table for this important circuit are shown in Fig. 2.

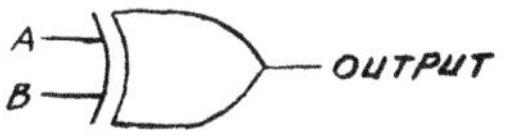

A	B	OUTPUT
0	0	0
0	1	1
1	0	1
1	1	0

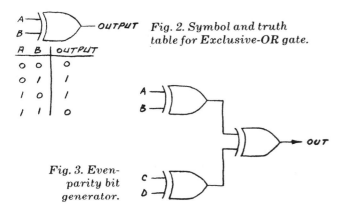

Fig. 2. Symbol and truth table for Exclusive-OR gate.

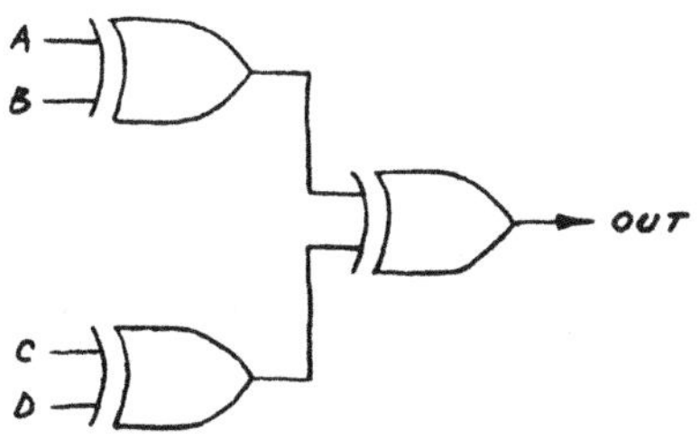

Fig. 3. Even-parity bit generator.

Exclusive-OR Parity Generator. When large quantities of digital information are transmitted from one point to another, it is not uncommon to lose (or "drop") a few bits. Although the level of error might be extremely low, in many applications, it is important to find and correct *any* error that occurs. A single dropped bit, for example, can change a number such as 128 (10000000_2) to 0 (00000000_2). On the other hand, if a leading bit is somehow changed from a 0 to a 1, a transmitted 0_{10} will be received as a 128_{10}. You can imagine the problems that would occur if a few bits are dropped (or a dropped bit or two turn up) in a computerized payroll account!

The exclusive-OR gate gives rise to clever methods of preventing most dropped bits from slipping into a digital system undetected. The even-parity bit generator, shown in Fig. 3, is employed in one common error-detection system. This logic circuit continually monitors each bit in a four-bit nibble. If the nibble has an odd number of 1's, its output is logic 1. If the nibble has an even number of 1's, the circuit generates a logic-0 output. The extra bit is tagged onto the nibble as a fifth bit.

The fifth (parity) bit stays with the nibble while it is transmitted to a data processing circuit anywhere from a few feet to perhaps thousands of miles away. When the nibble is received, it is inspected by a parity detector which, like the parity generator, is made from Exclusive-OR gates. If the parity is correct, the nibble is accepted for processing. If not, an error signal is generated.

This simple parity method is effective, but it is not foolproof. A dropped parity bit, for example, could cause an otherwise perfectly valid nibble to be flagged as erroneous. Nevertheless, the odds for losing a parity bit are much smaller than those for losing a nibble bit since there are four times as many of the latter. In demanding applications, more complicated parity generation and detection methods can be used.

Multiple-Bit Digital Comparators. Many interesting applications are made possible by multiple-bit digital comparators. These comparators use Exclusive-OR circuits to compare respective pairs of bits in each of two words. The outputs of the comparators are applied to a gate that generates a logic 1 at its output when all the bit pairs are equal and a logic 0 when one or more are not.

Additional output bits can indicate which input is larger than the other, should that be the case. Additional inputs can permit two or more multiple-bit comparators to be cascaded so that larger words can be compared.

The 7485 4-Bit Magnitude Comparator. You can use Exclusive-OR gates for comparing one- or two-bit numbers, but the 7485 4-bit magnitude comparator is by far the best solution in more advanced applications. The 7485 provides three fully decoded outputs that indicate which of two input nibbles is larger than the other or if both nibbles are equal. It also includes three cascade inputs that permit two or more 7485's to compare words having eight or more bits.

Figure 4 is the pin diagram of the 7485. Note the reasonably consistent placement of the cascade inputs and comparator outputs, a feature you will find useful when you're working with this chip. Although the 7485 is rarely used in experimenter and hobbyist projects, it is a very versatile chip with some powerful applications. Just compare some of its decision making features with a pocket programmable calculator or a microcomputer and you'll see why. Indeed, you might be able to solve some fairly complex circuit requirements with a 7485 comparator rather than a more complicated approach such as using a microprocessor.

7485 Demonstration Circuit. A good way to learn how the 7485 works is to assemble the test circuit shown in Fig. 5. This circuit allows you to manually apply two 4-bit nibbles to a 7485 and simultaneously monitor three LEDs that indicate the status of the chip's outputs. Because the activated output of the 7485 is high, the LED connected to it is *off* and the other two LEDs are on. This might seem a little confusing at first, but you'll soon become accustomed to it. If you prefer that the LED connected to the activated output glow and the other two LEDs remain off, you can use the 7485 as a current source

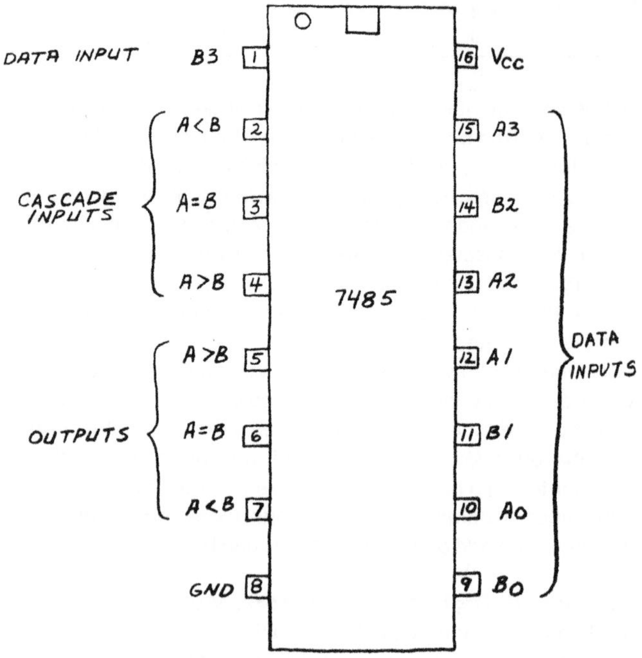

Fig. 4. Pin diagram for 7485 4-bit magnitude comparator.

rather than a current sink. We'll see how this is done later.

Although the circuit shown in Fig. 5 specifies two 4-position DIP switches, you can use a single 8-position switch if you prefer. In operation, a switch that has been closed applies a logic 0 to the comparator input, and an open switch applies a logic 1. For this reason, it's best to install the DIP switches upside down. While experimenting with this circuit, you might find that the DIP switch occasionally fails to register properly. Some DIP switches are less reliable than conventional toggle switches, and you might have to press a switch into position more than once before it makes or breaks contact. You will also, no doubt, find that this circuit poses a test of your ability to manipulate binary numbers. We'll explore this aspect of using the 7485 in more detail later.

Finally, you'll probably begin to develop some applications of your own for the 7485. For example, how about a binary combination lock? One switch holds the combination, which you can easily change in only a few seconds. The second switch becomes the combination dial. Connect a solenoid lock to the A=B output (pin 6) through an SCR or power transistor and your combination lock is complete.

You can make a more sophisticated lock by inserting a normally-open pushbutton switch between the 7485 and the positive power supply and using the two A≠B outputs to activate an alarm bell. An incorrect combination entry followed by a press of the switch to activate the lock will ring the bell and discourage further tampering.

How to Cascade Two 7485's. The circuit in Fig. 5 has only sixteen possible combinations, but you can increase the number to 128 by adding a second 7485 and using an 8-bit combination. Cascaded comparators have other applications as well, particularly in light of the fact that most microprocessor and controller circuits work with words having eight or more bits.

Figure 6 shows how to cascade two 7485's. Although this circuit includes DIP switches to allow you to make manual entries, it can be readily adapted to receive bit patterns from a pair of 8-bit buses. Be sure to compare Fig. 6 with the pinout in Fig. 4 to see how the outputs of one 7485 are connected to the cascade inputs of the second 7485. Also, note how the indicator LEDs are sourced by the 7485 in this circuit. This means the LED connected to the activated output glows and the other two are off.

It's possible to cascade more than two 7485's so that words having more than 8 bits can be compared. The Texas Instruments *TTL Data Book*, for example, shows how to connect six 7485's to compare two 24-bit words (see page 7-64). *Five* 24-bit comparator circuits can be used with a single 7485 to compare two 120-bit words.

A 120-bit comparator made from 7485's would be very fast, but would require 31 chips. If speed is not essential, a much more practical way to compare very long words is to break

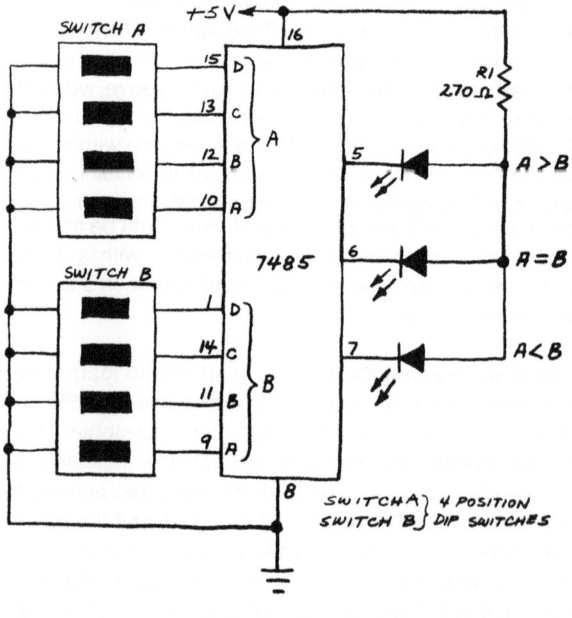

Fig. 5. Demonstration circuit using a 7485.

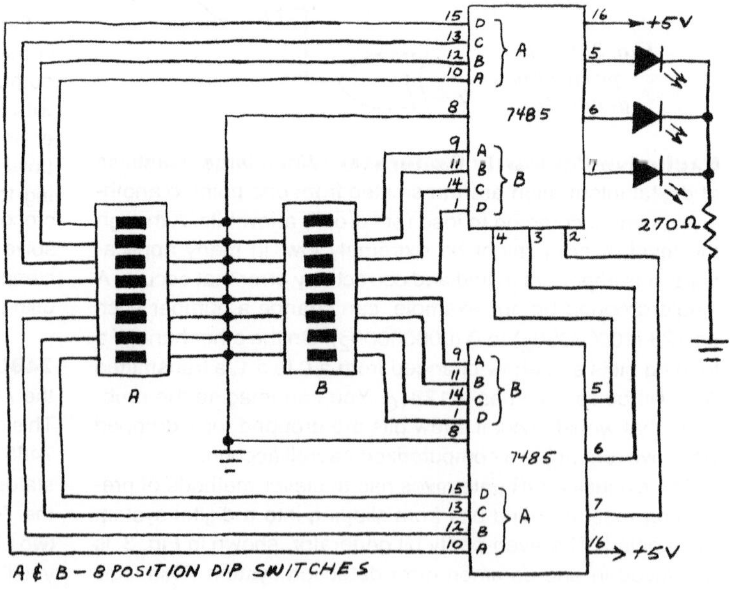

Fig. 6. Two 7485 comparators can be cascaded as shown.

them up into 4-bit nibbles or 8-bit bytes and compare respective nibbles or bytes a pair at a time in sequential fashion. That's a comparatively easy task for a microprocessor.

Using the 7485 in a Programmable Counter. Figure 7 shows how to make a programmable 4-bit counter with the help of a 7485. The desired count, which can be any number between 0001 and 1111, is entered into the DIP switch. The 555 functions as a clock that drives the 74193 through its count sequence. In operation, the 7485 continually interrogates the count and compares it with the number loaded into the DIP switch. When the two are equal, the comparator sends a pulse to the CLEAR input of the 74193 which resets the count to 0000. The cycle then repeats.

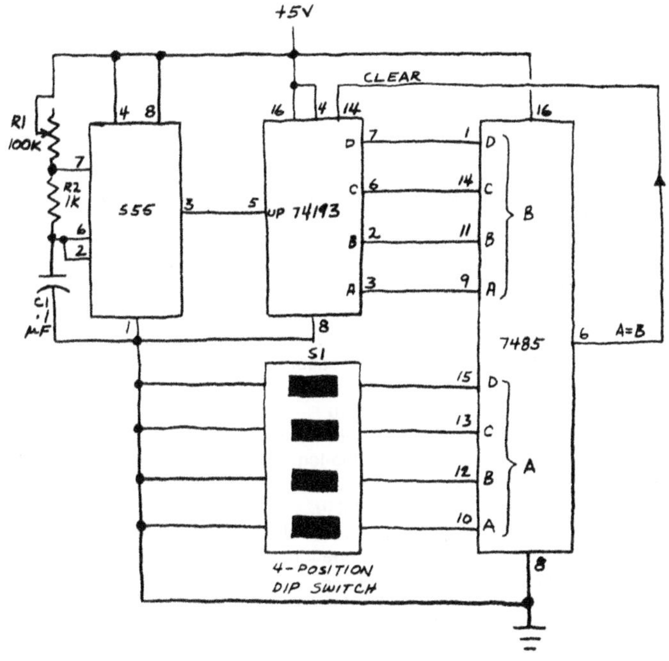

Fig. 7. A programmable 4-bit counter using a 7485 with a 74193 and 555 clock.

This basic circuit is easily modified for more sophisticated applications. Cascading counters and comparators, for example, will permit higher counts. By adjusting *R1* and *C1* to provide one clock pulse at a known interval, the circuit can be used as a programmable digital timer. Increasing *R1* or *C1* or both will slow down the clock rate.

The A≠B outputs of the 7485 provide a number of other application possibilities. For example, by connecting the A<B output to the CLEAR input of the 74193, the counter will be reset upon the next clock pulse *after* the programmed number has been reached. If you build this circuit, be sure to experiment with this and other modes of operation.

BCD Trainer. Once you have assembled any of the 7485 circuits described thus far, you will realize how important the ability to think in binary is. The circuit shown in Fig. 8 will help you quickly learn the first ten binary numbers (0000-1001). These numbers are used in most if not all logic circuits that employ decimal readouts and are collectively known as the *binary-coded decimal* or *BCD* system.

The circuit is operated by pressing pushbutton *S1* for a second or two. This allows a fast stream of clock pulses to enter the 74192 BCD counter. When the switch is released, the last count attained by the 74192 remains in the counter register. This number is for practical purposes random because the clock frequency is far too high to allow anyone to second-guess what the count is when *S1* is released.

The BCD number stored in the 74192 is decoded by the 7447, which lights the appropriate segments of a LED readout to display the decimal equivalent. The way to use this training circuit is to set up on the DIP switch the BCD equivalent of the

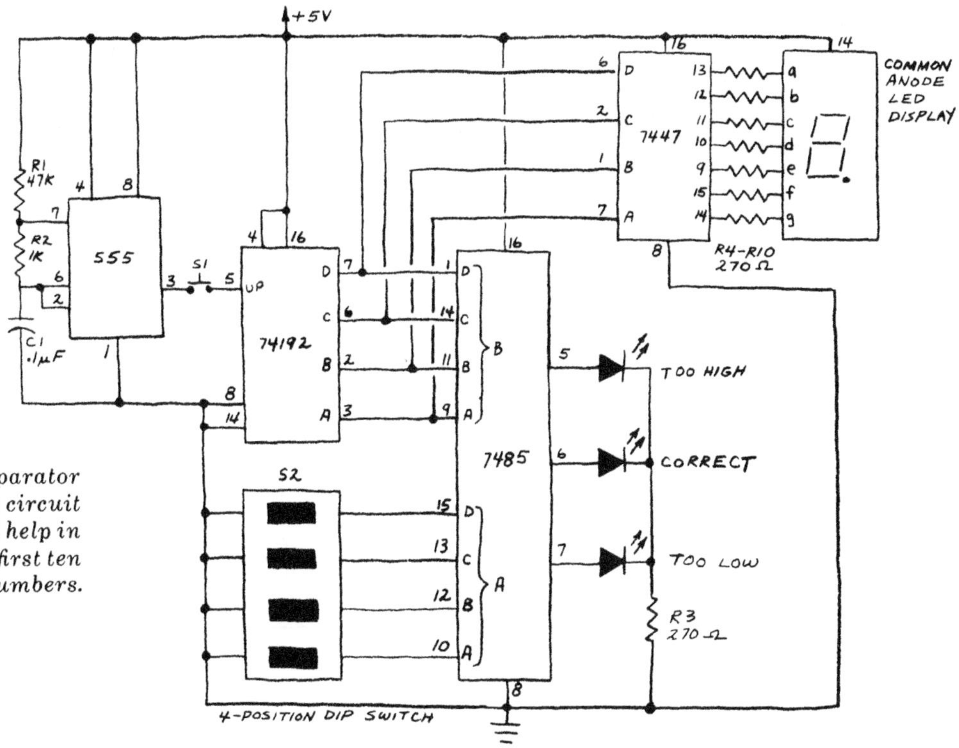

Fig. 8. A 7485 comparator is used to form a circuit that will help in learning the first ten binary numbers.

displayed digit. If the switch entry is correct, the appropriate LED will glow. If not, one of the other two LEDs will indicate that the switch entry is either too high or too low.

As with all the other 7485 circuits we've looked at so far, this one is easily modified. You can use a clock of your own design. You can also substitute a 7448 decoder and common cathode display in place of the readout circuit shown if you remember to connect the display's common cathode pin to ground instead of to +5 volts.

Another possibility is to eliminate all three response LEDs and use the decimal-point LED in the readout in place of the LED that indicates a correct response.

Number Guessing Game. The circuit in Figure 8 can be easily converted into a guessing game that is both educational and entertaining. Simply connect pin 6 of the 7485 to pin 4, the blanking input, of the 7447. The display will then stay extinguished *until* the correct BCD number is entered on the DIP switch. The two A≠B LEDs will indicate if your guesses are too high or too low. ■

2. CMOS Tone Sequencer

TONE sequencers can be used to generate electronic music, as toys, annunciators, or warning alarms, and in remote signalling. The sequencer shown in Fig. 1 requires only three CMOS chips to provide a sequence of up to ten individually programmable tones.

In operation, two of the inverters in a 74C04 are cross-coupled to form an astable multivibrator that generates clock pulses for a 74C192 BCD counter. The rate at which clock pulses are delivered to the counter is determined by the values of *R1* and *C1*.

The tone generator is made from two of the remaining inverters in the 74C04. Its frequency is determined by *C2* and the effective resistance between pins 5 and 9. This resistance is provided by one of the eight resistors (*R2* through *R9*) connected to the outputs of the 4051 analog multiplexer/demultiplexer. At any given instant, only one resistor is selected by the 4051 in response to the address applied to its inputs. This resistor is automatically connected between pins 5 and 9 of the 74C04. The resistors are selected in sequence as the 74C192 counts incoming pulses, and this steps the frequency of the tone generator through the range of tones determined by the individual resistors.

The 74C192 is a BCD counter (0000–1001) with a 4-bit output word. Because the 4051 accepts a 3-bit address, the highest-order bit from the counter is ignored. (This is why pin 7 of the 74C912 is not connected.) The net result is that the sequencer repeats the two lowest order tones (addresses 000 and 001) once each cycle when the *full* counter output is 1000 and 1001. If a full 4-bit counter such as the 74C193 is used, no repetition occurs because the counter will present all eight address combinations to the 4051 twice during each 16-step sequence.

The circuit's tone generator can drive a small, high-impedance earphone directly. For more volume, connect the circuit to a small speaker using a driver transistor or unused inverter as shown in Fig. 2. Alternatively, connect the circuit to a small amplifier through *C3* and *R10* as shown in Fig. 1.

When the value of *C2* is 0.1 µF, the frequency of the tone generator ranges from 34 to 6500 Hz for a resistance range between pins 5 and 9 of the 74C04 of 220,000 to 470 ohms. The output frequencies I measured for a selection of standard resistance values are shown in the table above.

These measurements are plotted in Fig. 3 on a log-log graph to help you select the resistance values for tones not listed in the table.

If you prefer to select the tones experimentally, connect a 100,000-ohm potentiometer between pin 13 of the 4051 and pin 5 of the 74C04. Leave the remaining 4051 outputs unconnected. Slow down the clock rate by adjusting *R1*. When a tone is heard, quickly disable the clock by shorting *C1* with a short jumper wire. Then rotate the shaft of the potentiometer until the desired tone is heard. Remove the pot, measure its resistance and record the value next to *R2* on a notepad

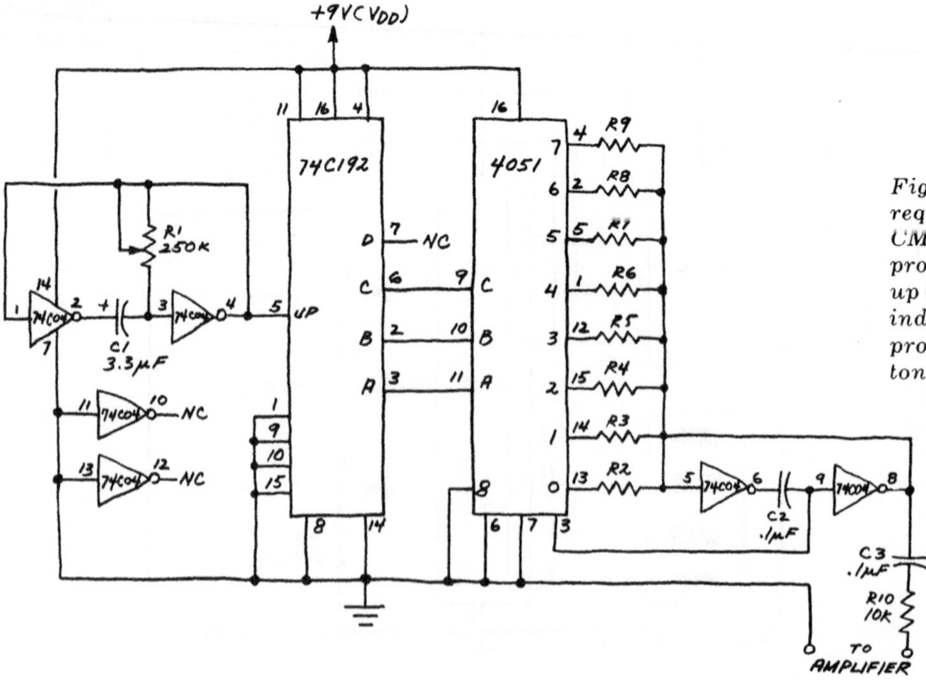

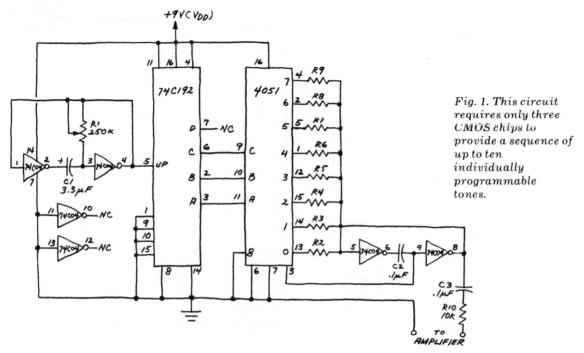

Fig. 1. This circuit requires only three CMOS chips to provide a sequence of up to ten individually programmable tones.

OUTPUT FREQUENCY VS. RESISTANCE

Resistance (ohms)	Tone frequency (Hz)
470	6,481
680	5,201
1,000	3,937
1,500	3,081
2,200	2,320
3,300	1,792
4,700	1,192
6,800	938
10,000	629
15,000	441
22,000	330
33,000	226
47,000	160
68,000	110
100,000	70
150,000	52
220,000	34

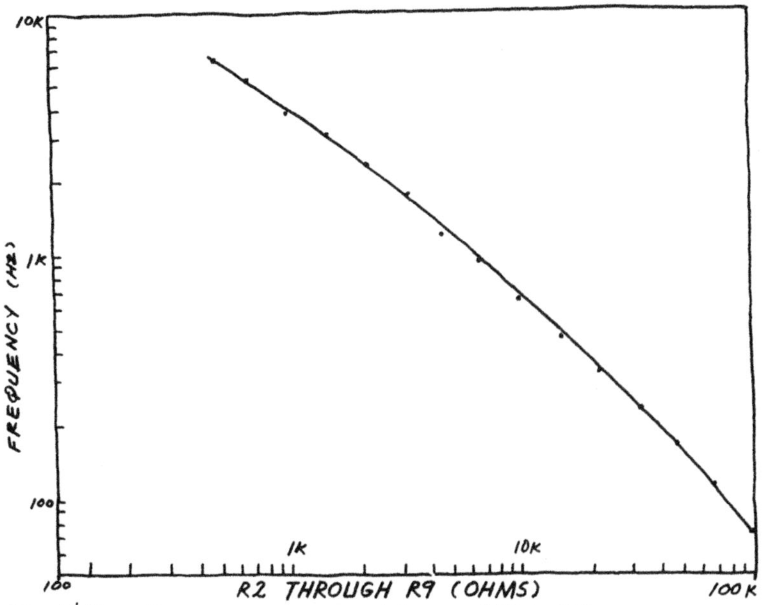

Fig. 3. Tone frequency versus resistance of R2 to R9.

which lists component designations *R2* through *R9*.

Next, connect the pot between pin 14 of the 4051 and pin 5 of the 74C04 to select the resistance of *R3*. Continue this procedure until the values of all eight resistors have been selected. Then install fixed resistors having resistances close to the measured values.

Modifying the Circuit. This project is ideal for experimenters because it is easily modified. The simplest modification is to replace *R2* through *R9* with trimmer resistors to permit the tone frequencies to be easily changed. For higher tone frequencies, reduce *C2* to 0.01 μF.

Adding a Second Clock. Referring to Fig. 1, note that there are two unused inverters in the 74C04. Normally, the inputs of the unused inverters and the unused data inputs of the 74C192 must be tied to ground or V_{DD} since these are CMOS chips.

These two gates can, however, be used to make a second clock circuit which is connected to the DOWN input of the 74C192. Use a 3.3-μF timing capacitor and a 1-megohm potentiometer for the timing resistor. Disconnect pin 4 of the 74C192 from V_{DD} and reconnect it to the output of the second clock. This dual UP-DOWN clock arrangement will liven up the otherwise predictable tone sequence.

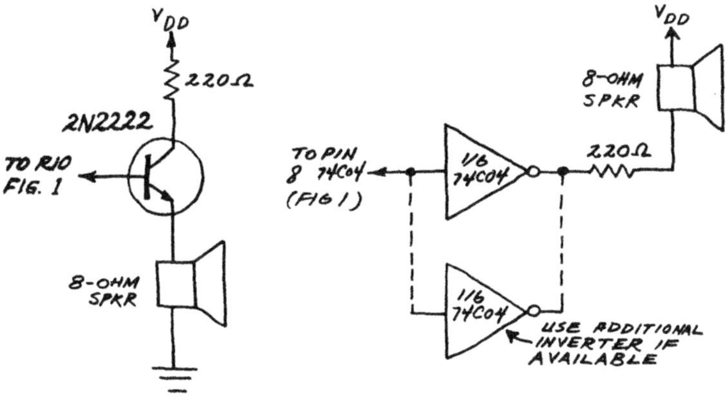

Fig. 2. Two simple ways to add a small speaker to sequencer.

Single Sequence Operation. It's possible to modify the circuit to emit a single sequence of seven tones each time a switch is toggled.

Referring to Fig. 1, here's how the circuit is modified:

1. Connect pins 1, 9, 10 and 15 of the 74C192 to V_{DD} instead of ground.
2. Disconnect pin 11 of the 74C04 from ground and connect it to pin 7 of the 74C192.
3. Disconnect pin 11 of the 74C192 from V_{DD} and connect it to pin 10 of the 74C04.
4. Disconnect pin 14 of the 74C192 from ground and connect it to the pole of a spdt switch. Connect one position of the switch to V_{DD} and the other to ground.
5. Remove *R9* from the circuit (unless you want a continuous output tone while the circuit is in standby condition awaiting the switch to be toggled).

To operate the circuit, throw the switch so that its pole is grounded. Then throw the switch so the center pole is connected to V_{DD}. The speaker will emit a steady tone indicating the 74C192 is cleared to address 0000 and *R2* is connected to the tone generator. When the switch is thrown once more, the complete tone sequence will be generated. After the seventh note, the counter will inhibit itself (can you figure out why?) and the circuit will be ready for another tone cycle to be initiated. ■

NOW THAT "four-banger" electronic calculators are so inexpensive, modifying them for special-purpose applications is an attractive and cost-effective possibility. This month, we'll examine several ways of adding external features to a four-function calculator incorporating an automatic constant. To determine if a calculator you are thinking of modifying has an automatic constant, enter the following keystroke sequence: 1; +; =; =; =. If the display reads 3, the calculator is equipped with an automatic constant feature.

A Calculator Event-Counter. As you discovered when you ran through the simple keystroke sequence given above, a calculator with an automatic constant can count the number of times the "=" key is pressed. To automate this counting ability, it's necessary to connect an external switch across the contacts of the "=" key.

The calculator I modified, a Texas Instruments TI-1200 purchased new for about $6.00, has a 5×4 matrix keyboard. This keyboard is readily accessible by removing the four screws which hold the calculator's front cover in place. It has nine flexible leads, four of which address the vertical columns of keys and five the horizontal rows. The "=" key is accessed by the first and eighth wires from the top left of the keyboard.

Other calculators have different keyboard arrangements, and some recent models do not have a separate keyboard at all. Unless both sides of the circuit board are visible, you'll have to determine which wires access the "=" key by trial and error. Simply enter the sequence: 1; +; =; and, with the help of a jumper, begin shorting pairs of wires or foil conductors leading to the keyboard. Shorting digit keys may overwrite the 1 in the display. If so, reenter the 1; +; = sequence before trying again.

When you find the conductors that lead to the "=" key, carefully solder an insulated wire lead to each of them using a grounded or battery-powered iron. There is room in the TI-1200 and some other calculators for one or more miniature phone jacks. If your calculator has this extra space, drill a mounting hole, install a jack and solder the leads to it. Once the calculator has been reassembled, it can be used for both calculating and event counting.

Many different devices can be used to actuate the "=" function. For manual operation, an ordinary spst pushbutton switch connected to a two-conductor cable and plug is sufficient. For automatic counting, a magnetic reed switch or phototransistor can be used.

Fig. 1. A phototransistor connected to a calculator.

Figure 1 shows how a common npn phototransistor can be connected directly across the "=" key. Flashes of light will then actuate the "=" function. This permits moving objects to be counted without the need for mechanical contact. It also permits such novel applications as counting nearby lightning strokes during a nighttime storm.

A standard npn transistor can also be used as a switch. For example, to determine the maximum count rate for a TI-1200 calculator, I connected a 2N2222 across the "=" key and applied pulses from a variable-rate pulse generator as shown in Fig. 2.

The maximum usable count rate of each of these add-on circuits will be limited by the rate at which the calculator scans its keyboard to detect key closures. The TI-1200 that I modified has a multiplex rate of about 360 Hz, but that doesn't mean the unit will accept 360 closures of the "=" key each second. All twenty key locations are scanned one at a time by the multiplex circuit so, it would at first appear, the maximum number of counts per second is 20. Actually, my unit will accept a maximum of only

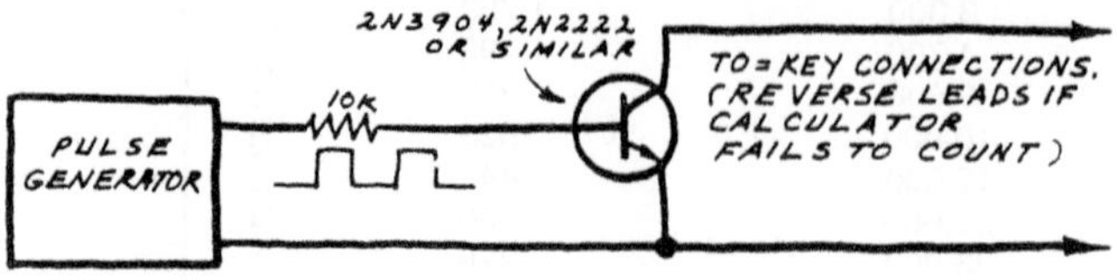

Fig. 2. Using a pulse generator with a calculator.

13.44 counts per second. That's because only those input signals present when the "=" key is in the process of being scanned are accepted. Those which arrive *and* depart between scans are *not* detected.

This can cause problems in applications where the pulse to be counted is very brief. For example, the reason I decided to modify my calculator was to count the number of times the front wheel of my bicycle rotated during specific time intervals (to determine the average speed of the bike) and during various trips (to determine the total distance travelled). A magnetic reed switch secured to the front fork was connected in parallel with the "=" key. A magnet attached to the wheel rim served as its actuator. It didn't take me long to discover that at speeds greater than about 5 mph some wheel rotations were not counted because the switch closed and opened again *between* the time intervals when the calculator was scanning its "=" key. This problem can be remedied by moving the magnet and switch closer to the hub assembly (using care to keep these components and the connecting wires away from the spokes!) or by adding a one-shot between the reed switch and the calculator to stretch out the pulses generated by the switch.

Calculator as a Timer. The addition of a simple timebase permits the TI-1200 or other low-cost calculator to function as a programmable timer. Figure 3, for example, is a simple CMOS timebase that can be assembled on a small circuit board to be tucked either between the display and keyboard or below the battery compartment of a TI-1200.

Two of the gates in a 4011 are connected as an astable multivibrator that delivers a stream of pulses to the LED in an op-

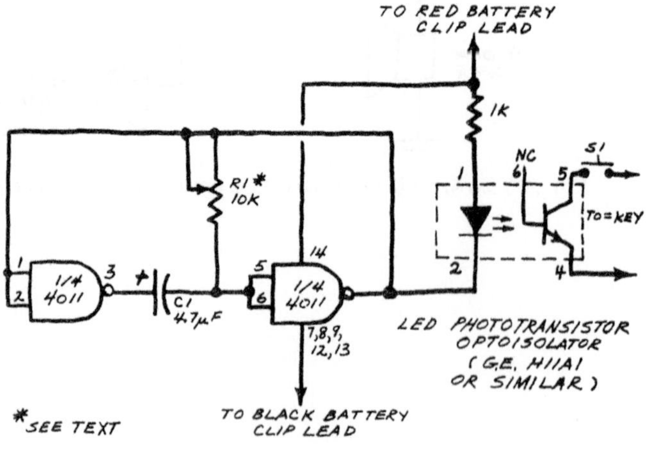

Fig. 3. CMOS time base converts calculator into timer.

toisolator. The collector and emitter of the phototransistor in the optoisolator are connected directly across the "=" key.

For 0.1-second resolution, it's necessary to calibrate the timebase so that it generates pulses at a rate of 10 Hz. This can be done by using a physically small trimmer potentiometer for *R1* and connecting a frequency counter to the output of the timebase. The prototype timebase that I assembled generated a 10-Hz output when *C1* was nominally 4.7 μF and *R1* was adjusted to 2270 ohms.

To operate the calculator as a timer, enter the sequence: .; 1; + and then close *S1* (Fig. 3) to allow the timebase to feed pulses to the "=" key. Release *S1* when the event being timed is over. Read the elapsed time to the nearest one-tenth of a second from the display. You can then use the calculator to convert the time, which is displayed in seconds, into minutes or hours.

Adding an Output Port. Upon the addition of an output port, a low-cost calculator can become a primitive, but useful, digital controller. Microprocessor chips usually have one or more pins designated as ports. The ports permit external devices to influence the microprocessor when they are functioning as input ports or to be controlled by the microprocessor when they are acting as output ports.

There are several ways to add one or more output ports to a calculator. So far, the two simplest methods I've identified are monitoring the minus sign and the decimal point in the display. Let's see how the decimal point can be monitored.

If you enter in the keystroke sequence: 10.0; −; 1.0 on a TI-1200 or similar calculator, the display will be decremented by 1.0 each time the "=" key is pressed. That is, the display will read 10.0; 9.0; 8.0; . . . 2.0; 1.0; 0.; −1.0; etc. Notice that when the count reaches zero the decimal point moves one place to the right. When the count is above or below zero, however, the decimal point stays at least one place to the left of the lowest-order digit in the display.

This makes possible the use of the lowest-order decimal point as an output port. All you have to do is find the contacts on the display that lead to the lowest-order digit and the decimal point. The TI-1200 display has 17 connection tabs. Tab 9 is connected to the common cathode of the lowest-order digit and tab 13 is connected to the decimal point.

Figure 4 shows one way to interface an external circuit to a decimal-point output port. The LED/LASCR optoisolator provides a latching action that keeps a LED or other output device continuously on once it has been triggered. The calculator display will keep a record of the number of trigger events that occur after the LASCR has fired.

A typical application for a calculator modified to include an output port is a programmable timer (such as one for darkroom use) that is capable of controlling an external device. If the timebase is delivering pulses at a rate of 10 Hz, a maximum delay of up to 9,999,999.0 seconds (more than 115 days!) is available, assuming that the calculator is programmed to decrement the total by 0.1 per clock pulse *and* that the power supply does not fail. Using a slower clock rate or reducing the tally in much smaller increments can easily increase the longest possible time delay to *years*!

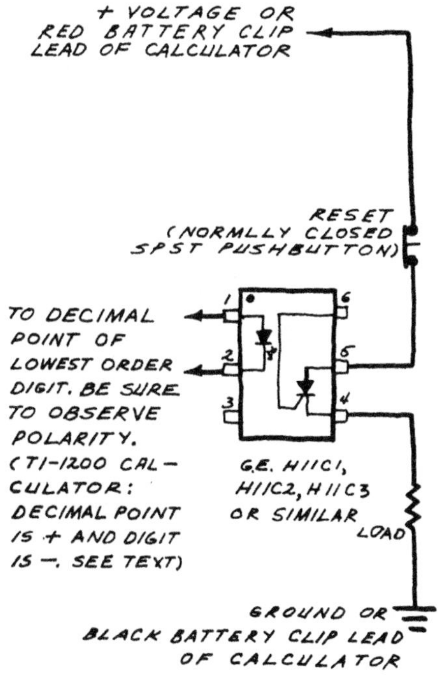

Fig. 4 Adding an output port to a calculator.

With a little care, you should be able to fit both the timebase and output-port circuits into the vacant space inside a TI-1200 or similar calculator. Alternatively, the additional circuits can be installed in a small enclosure and interfaced with the calculator using miniature phone plugs.

The output port has a number of applications other than timing. For example, you can program the calculator to count a given number of events (revolutions of a wheel, openings of a door, passing of cars, etc.) and then generate an output signal when the desired number have taken place. Unless you make special modifications which affect the use of the keyboard, the calculator can be used for its normal purpose when it's not being used for special applications.

Going Further. With a little experimentation, you will be able to come up with some clever applications of your own for modified calculators. For starters, you can remotely actuate any key on a calculator keyboard using the same techniques we've described in this column to actuate the "=" key. Keep in mind that the automatic constant feature of the TI-1200 and many other calculators works for all four primary arithmetic functions.

For advanced applications, consider modifying more powerful calculators. Some programmable calculators are now available for under $50. If you're not concerned about voiding the warranty of a programmable (or if it has expired), you might consider adding external circuits employing some of the methods described in this column. One possibility is a beeper that's automatically actuated when a long program is completed. Automatic data entry at a specified point in a program is another. ■

4. Digital Stopwatch

HERE'S a two-digit counter with a crystal-controlled timebase that you can either use as a stopwatch or modify for other timing applications. The timebase (see Fig. 1) uses an MM5369 crystal-controlled oscillator/divider. This chip generates a stable 60-Hz reference when connected to a standard 3.58-MHz color TV oscillator crystal. A miniature trimmer capacitor (*C2*, 1.5 to 20 pF or similar) permits you to tune the oscillator to precisely the right frequency. For best results, connect an accurate frequency counter to pin 7 and adjust *C2* until the counter indicates a frequency of 3.579545 MHz.

If you don't have a miniature trimmer capacitor, substitute a 5-pf capacitor for *C2*. Then twist two lengths (approximately 2" or 5.1 cm long) of wrapping wire together and solder one end of each wire to each lead of *C2*, leaving the other end of each

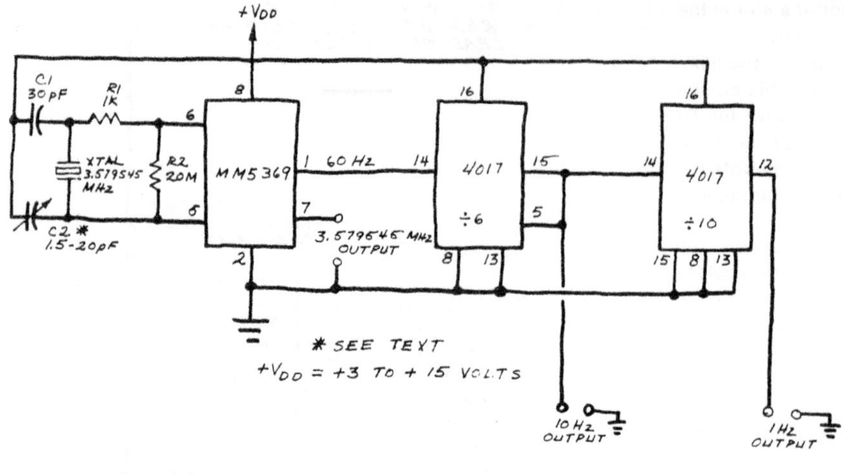

Fig. 1. Schematic of a precision crystal-controlled timebase.

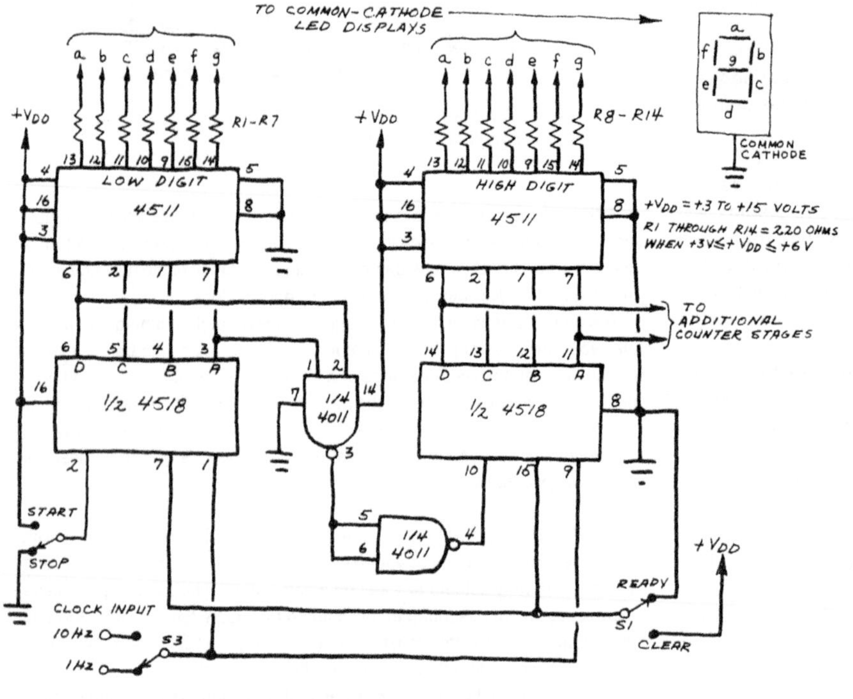

Fig. 2. Circuit for a two-digit digital counter/stopwatch.

wire unconnected. The two leads form a "gimmick" capacitor in parallel with *C2*. Now you can trim the oscillator by snipping off short sections at the free ends of the twisted wires until the desired frequency is obtained.

The 60-Hz output of the MM5369 is divided down to 10 Hz by a 4017 CMOS counter and is further divided to 1 Hz by a second 4017. These two output signals are the project's timebase.

Figure 2 is the schematic of the counter portion of the Digital Stopwatch. Depending on the position of *S3*, either 1-Hz or 10-Hz square waves are applied to the input of the 4518 CMOS dual BCD up-counter. To prepare the project for the timing of an event, place *S2* in its STOP position and toggle *S1* from CLEAR to READY. Then place *S2* in its START position at the beginning of the interval to be timed. When the event is over, place *S2* in its STOP position. The elapsed time will be frozen and displayed on the LED readout.

This circuit can be modified to take advantage of the latch built into the 4511 CMOS decorder/driver ICs. When a logic one is applied to pin 5 of these chips, the data present at their BCD inputs are stored. This permits the continuing display of the result of one timing sequence while the dual counter is timing a later event.

The basic two-digit counter shown in Fig. 2 will increment from 0.0 to 9.9 seconds in 0.1-second steps when the 10-Hz timebase is selected, and from 0 to 99 seconds in 1-second steps when the 1-Hz pulse train is routed to the counter. To increase the maximum timing intervals, you can add more counter/decoder/display stages. For more resolution (e.g., time measurements to the nearest one hundredth of a second), design a circuit that will multiply the frequency of the 10-Hz pulse train by a factor of ten. Then use the 100-Hz output as the timebase. ■

5. The Digital Phase-Locked Loop, Part 1

RECENTLY, I was talking to the parts buyer for an electronics supplier about sales volumes of various integrated circuits. The most surprising thing I learned was that sales of the 4046 digital CMOS phase-locked loop (PLL) are only a trickle compared to those of other ICs.

This is puzzling, because the 4046 is is one of the most versatile CMOS chips. It is also unfortunate—the 4046 is very handy if you know how to use it. Among the many applications of the 4046 are those in frequency modulation and demodulation, voltage-to-frequency conversion, frequency synthesis, tone decoding, FSK demodulation, and frequency multiplication.

One possible reason for the low sales volume of the 4046 is that little descriptive or applications information about this chip has appeared in electronics magazines and books. To rectify this situation, we will unravel some of the mysteries surrounding the digital PLL and present some basic circuits. By the time you finish experimenting with some of the more advanced application circuits, you'll be well acquainted with the operating principles of the digital PLL, an exceptionally versatile CMOS chip.

Phase-Locked Loop Basics. The simplest PLL consists of a phase comparator, a voltage-controlled oscillator (vco),

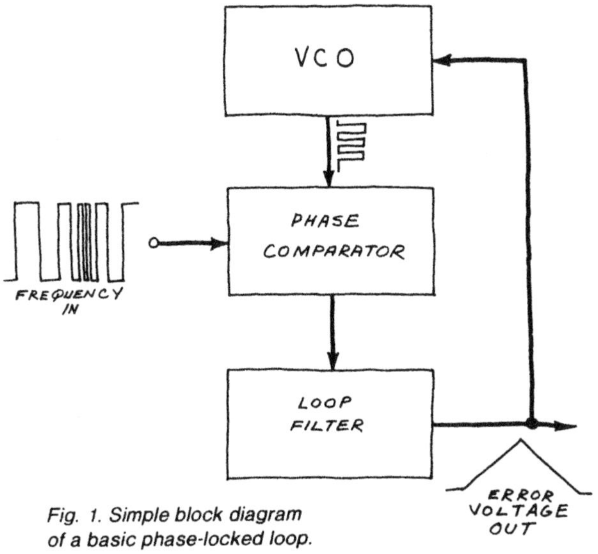

Fig. 1. Simple block diagram of a basic phase-locked loop.

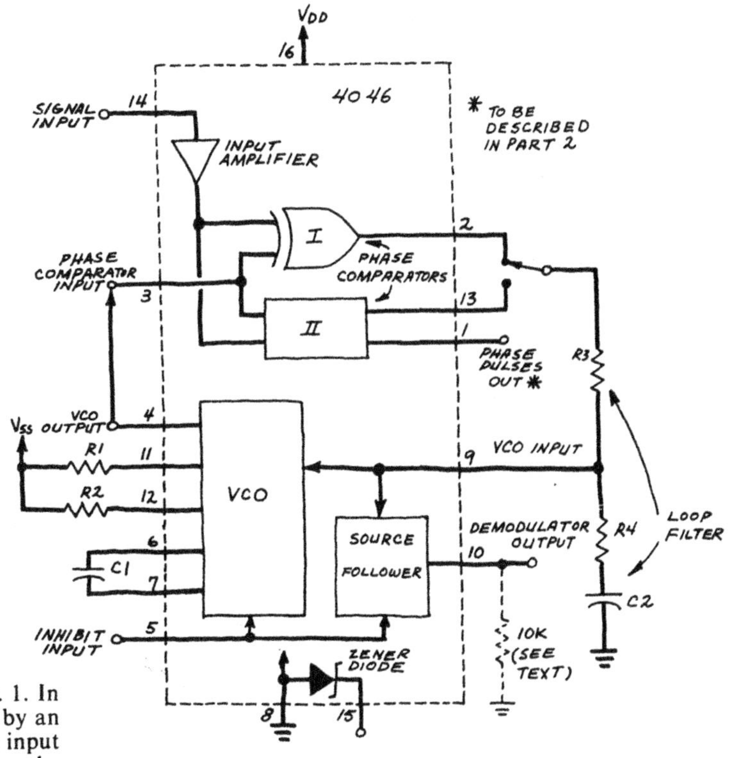

Fig. 2. Block diagram of the 4046 CMOS micropower phase-locked loop.

and a low-pass loop filter, all arranged as shown in Fig. 1. In operation, the vco oscillates at a frequency determined by an external *RC* network. This frequency is applied to one input of the phase comparator. An external signal applied to the second input of the phase comparator causes it to generate an *error voltage* whose magnitude is proportional to the difference between the external source and vco frequencies.

The low-pass loop filter smooths the pulsating error voltage into a dc level which is applied to the control input of the vco. The vco responds to the error voltage by moving its frequency of oscillation toward that of the input signal. This *capture* process continues until the vco frequency equals the input frequency. When this occurs, the PLL is said to be *locked* or *phase-locked* to the input signal.

When the PLL is locked to the input frequency, the vco automatically tracks any changes in the input frequency that fall within a window called the *lock range*. The lock range is always greater than the *capture range*, the band of frequencies over which the PLL can hunt for and "capture" an incoming signal.

It is important to understand that, although the loop filter is essential for proper operation of the PLL, its time constant limits the speed with which the system can track changes in the input frequency. It also limits the capture range. On the other hand, the loop filter helps prevent noise voltages from adversely affecting loop operation. The charge stored in the loop filter's capacitor helps the quick recapture of a signal temporarily lost because of a noise spike or other transient.

In short, the loop filter is a necessary part of the PLL, but it imposes certain operating restraints and tradeoffs. Be sure to keep this in mind when you experiment with PLL circuits, because optimizing PLL performance often requires experimentation with loop-filter component values.

Inside the 4046 PLL. Figure 2 is a block diagram of the 4046 CMOS micropower PLL. One of the most obvious features of this chip is that it includes *two* phase comparators. Phase Comparator I is an exclusive-OR gate that provides a high degree of noise immunity. Unfortunately, this comparator has a tendency to lock onto input signals having frequencies close to harmonics of the vco frequency. Also, it requires a square-wave input with a 50% duty cycle.

Phase Comparator II is a relatively complex network of four edge-triggered flip-flops with control gates and a 3-state output stage. While this detector is less susceptible to the harmonic problem that plagues Phase Comparator I, it is much more sensitive to noise.

Both phase comparators are simultaneously driven by an input amplifier which will be described later. Their outputs, however, are brought out to separate pins (2 and 13). This means that the user can select either comparator for a specific application by simply connecting its output pin to the vco through the loop filter.

Because the flip-flop comparator has a frequency-tracking range of more than 1000:1, it is often a better choice than the exclusive-OR comparator which tracks over a range of only ±30 percent. Another advantage of the flip-flop comparator is that it can accept input pulses of any duty cycle (for example, very narrow pulses).

The vco incorporates an NMOS input stage that provides an input impedance of 10^{12} ohms. Its linearity ranges from 0.1 percent (V_{DD} = +5 V) to 0.8 percent (V_{DD} = +15 V). The oscillator's maximum operating frequency typically ranges from 0.7 MHz (V_{DD} = +5 V) to 1.9 MHz (V_{DD} = +15 V).

Figure 2 shows a source follower connected to the vco input. This buffer stage is intended specifically for frequency-demodulation applications. It allows an external amplifier or other circuit to be driven by the output signal from the loop filter (the filtered error voltage) without loading down the filter. When the DEMODULATOR output (pin 10) of the source follower is used, a load resistor of at least 10,000 ohms must be connected between pin 10 and ground (V_{SS}). Otherwise pin 10 should be left floating.

Both the vco and source follower are provided with a common INHIBIT terminal (pin 5) to reduce standby power consumption. A logic 0 (V_{SS}) at pin 5 enables the vco and follower, and a logic 1 (V_{DD}) inhibits them.

The final component in the 4046 is a 5.2-volt zener diode. This zener is intended for voltage-regulation applications, and its use is optional.

Using the 4046. The 4046 requires a power supply that can furnish from 3 to 18 volts at modest current levels. Power consumption depends upon both the vco frequency and what percentage of time the vco is enabled. For example, at a frequency of 10 kHz, the 4046 consumes only 600 microwatts—about 1/160th the power required by a typical analog bipolar PLL such as the 565. Suffice it to say that the 4046 is ideally suited for battery-powered operation!

A minimum number of external components is required to use the 4046. The center frequency of the vco is determined by one capacitor (*C1*) and one or two resistors (*R1* and *R2*) as shown in Fig. 2. When only *R1* is used, the vco frequency can be varied from 0 Hz when the control voltage at pin 9 is V_{SS} to a maximum frequency given by the equation: $f_{max} = 1/R1\,(C1 + 32\text{ pF})$ when the control voltage is V_{DD}. For proper operation, the resistance of *R1* should be between 10,000 ohms and 10 megohms.

Resistor *R2* is included when it is desirable to move the minimum vco frequency to some point above 0 Hz. For this reason, it is called the *offset resistor*. The minimum frequency resulting from the inclusion of *R2* is determined by solving the equation: $f_{min} = 1/\,R2\,(C1 + 32\text{ pF})$ when the control voltage at pin 9 is V_{SS}. When *R2* is used, the maximum vco frequency when the control voltage is V_{CC} is found by adding f_{min} to the f_{max} obtained from the previous equation.

These vco design equations are extracted from Motorola's MC14046B specifications sheet. They apply only when the values of *R1* and *R2* are between 10,000 ohms and one megohm and when that of *C1* is between 100 pF and 0.01 μF. Nevertheless, the manufacturer's specifications sheet observes that experimentation is in order to determine the exact component values required for a particular application because, ". . . calculated component values may be in error by as much as a factor of 4." This poses no problem because it's a simple matter to use trimmer potentiometers for *R1* and *R2* and to adjust them to get the desired frequency range.

The loop filter, like the vco, also requires a capacitor (*C2*) and one or two resistors (*R3* and optional *R4*). The best explanation of this rather touchy circuit that I have found is in Don Lancaster's *CMOS Cookbook* (Howard W. Sams, 1977, pp. 363-364).

Earlier, we briefly covered some of the loop-filter design tradeoffs. Don, who seems to know more about the real-world idiosyncrasies of the 4046 than the data-sheet authors, says that both *R3* and *R4* are necessary to avoid driving the loop into near-oscillation. He reports that best operation is obtained when the resistance of *R4* is from 10 to 30 percent of that of *R3*. This provides enough damping to eliminate loop overshoot and oscillation, but still ensures a reasonably quick response to changes in the input frequency.

Don recommends nominal values of 470,000 ohms for *R3*, 47,000 ohms for *R4*, and 0.1 μF for C2. A longer *RC* time constant means excessive delay when the loop is tracking quickly changing input voltages. A smaller *RC* product can cause erratic changes in the vco frequency as the loop tracks a rapidly changing signal.

I said that we would have more to say about the 4046 input amplifier later. Don comes directly to the point on this subject, so let's hear from him again. "*The linear amplifier operation of pin 14 is an unmitigated disaster when the wideband phase detector is being driven. Don't use it this way!* Linear operation causes extra amplitude-variation sensitivity, jitter, tearing and generally poor noise immunity" (*CMOS Cookbook*, p. 363).

One solution to this problem is to apply only full logic levels to the input. If this isn't possible or practical, pin 14 should be pulled up with a 10,000-ohm resistor to V_{DD}. The input signal can then be coupled into pin 14 by means of a 0.1-μF capacitor. In any event, if the input is a low-frequency train of slowly rising and falling pulses, the pulses must be conditioned with an appropriate pulse-shaping circuit.

VCO Application Circuits. An important feature of the 4046 is that the vco section can be used on its own for many practical applications, several of which will now be described. Experimenting with them will provide important experience for working with the chip as a complete PLL.

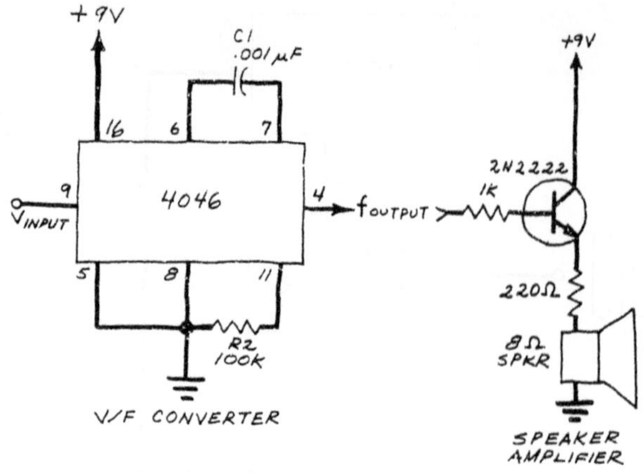

Fig. 3. A basic 4046 vco circuit used as a V/F converter with a speaker amplifier.

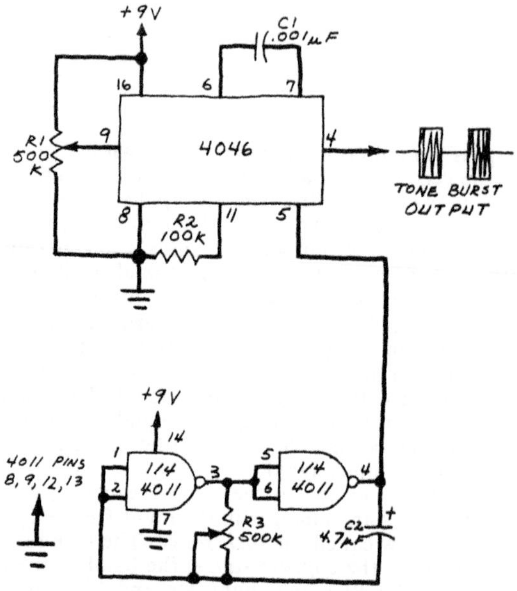

Fig. 4. A tone-burst generator in which R1 controls frequency and R3 burst rate.

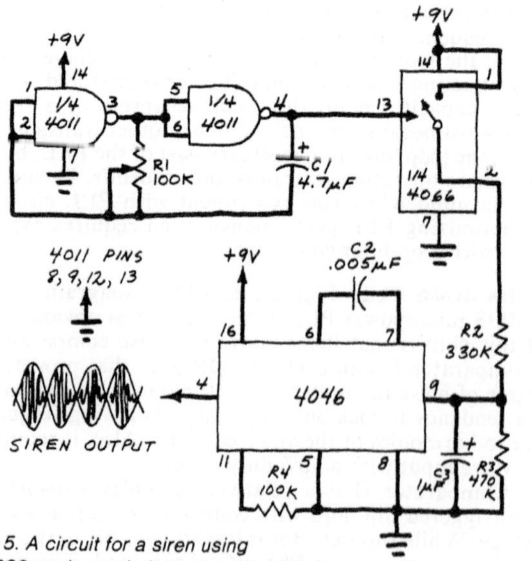

Fig. 5. A circuit for a siren using a 4066 analog switch to vary the sound.

Voltage-to-Frequency Converter. Figure 3 shows the most basic 4046 vco circuit possible, a simple V/F converter. Varying the input voltage from V_{SS} (ground) to V_{DD} will shift the output frequency over a range of 0 Hz to 18.5 kHz. You can use this circuit as a tunable oscillator by connecting the opposite ends of a 500,000-ohm potentiometer to V_{DD} and ground and by connecting the rotor to pin 9.

Figure 3 also includes a basic speaker amplifier that can be used with this and other 4046 circuits.

Tone-Burst Generator. Figure 4 is a simple tone-burst generator. Potentiometer *R1* controls the tone frequency and *R3* controls the burst rate.

Siren. The operation of the siren shown schematically in Fig. 5 is controlled by a 4066 analog switch. When the super-low-frequency NAND gate oscillator closes the switch, capacitor *C3* charges to V_{DD} through *R2*. When the analog switch is opened, *C3* discharges through *R3*. Because the voltage across *C3* controls the vco frequency, the result is an up-down siren effect.

Experiment with the various *RC* time constants to alter the sound of the siren. Components *R1* and *C1* control the cycle time, *R4* and *C2* control the frequency, and *R3* and *C3* control the wail. ■

The Digital Phase-Locked Loop, Part 2

PLL Lock Indicator. It's often difficult to determine whether or not a PLL is out of lock, particularly if a scope is not available. RCA application note ICAN-6101 recommends a simple NOR gate lock-detection circuit, a slightly modified version of which is shown in Fig. 1.

The lock indicator monitors the *phase pulses* output (pin 1) of phase comparator I and the output (pin 2) of phase comparator II. The output of the second NOR gate goes high and extinguishes the red LED when the loop is locked. When the loop is out of lock, the red LED flashes or appears to glow continuously. The optional NOR gate causes the green LED to glow when the loop is locked and go dark when the loop is out of lock.

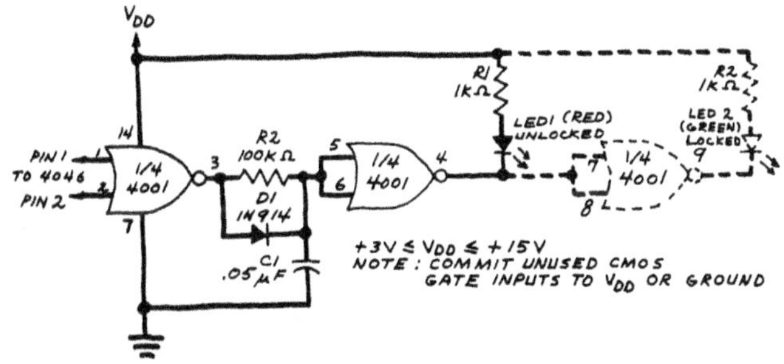

Fig. 1. A simple NOR-gate lock-detection circuit using a 4046 micropower PLL.

This simple circuit is a handy addition to any PLL since it provides an instant indication of a possible malfunction. It can also be used as an active part of a frequency detector or FSK demodulator. In the latter application, a binary signal is converted into a dual-frequency audio tone for remote transmission or storage on magnetic tape. A familiar example of FSK among computer hobbyists, the Kansas City Cassette Tape Standard, assigns a frequency of 1200 Hz to logic 0 and 2400 Hz to logic 1.

The PLL lock indicator can detect the presence of a 0 or 1 if the vco is adjusted so its minimum and maximum frequencies (see Part 1) encompass one of the two frequencies. The lock indicator will then go high for one tone and low for the second tone.

FSK Detector. A 4046 circuit designed specifically for Kansas City FSK detection is shown in Fig. 2. The vco is tuned by selecting *R1* and *R2* to give a capture bandpass from 2100 to 2700 Hz with a peak response of 2400 Hz. Frequencies outside the capture window are not detected; hence the lock detector goes high for a 2400-Hz input signal and low for a 1200-Hz input signal. These logic states can be reversed by adding a third gate in the 4001 to the output of the detector.

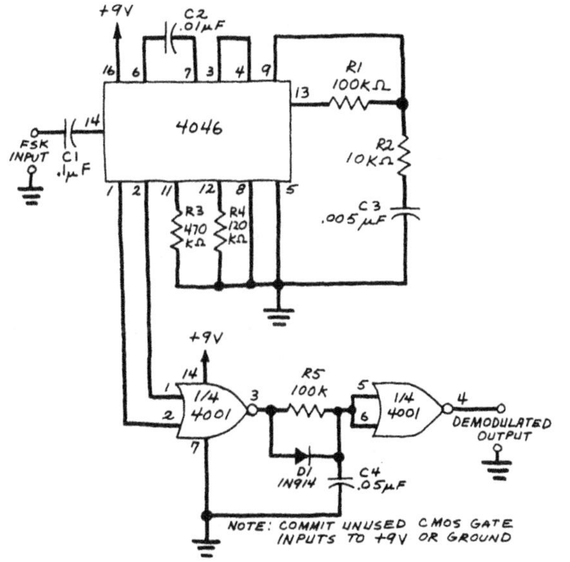

Fig. 2. A frequency-shift-keying (FSK) detector circuit.

Tone Detector. The circuit in Fig. 2 can be used to detect a wide range of input frequencies. For example, when *R4* is 1 kΩ and *R3* is 33 kΩ, the circuit responds to an incoming frequency of 48.775 kHz. The capture window with these values is *very* narrow (48.76-48.80 kHz). This demonstrates the possibility of using the 4046 as a precision tone detector. This application is normally reserved for the 567, a bipolar chip that uses considerably more power than the 4046.

It's easy to alter the response of the loop by substituting potentiometers for *R1* and *R2*. Or you can calculate the resistances required to define a specified frequency window (or loop capture range) by using the equations given in Part 1 or in the 4046 data sheet.

A commercial function generator or a simple dual-gate astable can be used to provide a variable-frequency input signal. A digital frequency meter to measure the peak detection frequency and the capture range is very helpful, but you can design a working circuit without one.

Analog Frequency Meter. In Part 1 we experimented with the 4046 vco as a voltage-to-frequency (V/F) converter. Thanks to the filtered error voltage available from the source follower (pin 10), the 4046 can also be used as a micropower frequency-to-voltage (F/V) converter. Figure 3 shows one of many possible 4046 F/V application circuits: an analog frequency meter. The input frequency is read out on a 0-1-mA

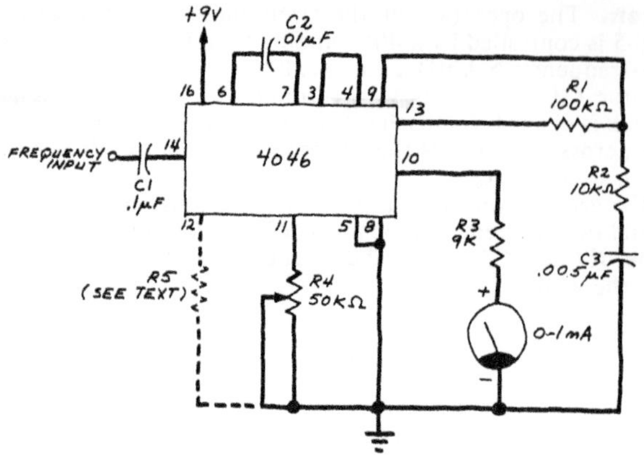

Fig. 3. Frequency-to-voltage converter as an analog meter.

meter connected in series with pin 10 and a 9 kΩ load resistor (which doubles as a current limiter).

With the values shown, the frequency meter has a full-scale response of 100 to 8000 Hz. Below 100 Hz, the meter's needle will oscillate or indicate an erroneous reading.

The circuit is calibrated by applying a 5-kHz input signal and adjusting *R4* to produce a meter indication of 0.5 mA. Since the circuit does not have a perfectly linear response, you will need to make a new meter scale or conversion table if you want to use it as a practical frequency counter.

Resistor *R5* is used only to add an offset to the lower end of the frequency measurement scale. For example, when *R5* is 100 kΩ and R4 is adjusted to give an output of 0.5 mA at an input frequency of 5 kHz, the frequency measurement range is 2.3 to 7.0 kHz.

The lock indicator shown in Fig. 1 is a particularly handy addition to this circuit since it provides immediate warning when the circuit is out of range. This prevents erroneous frequency measurements.

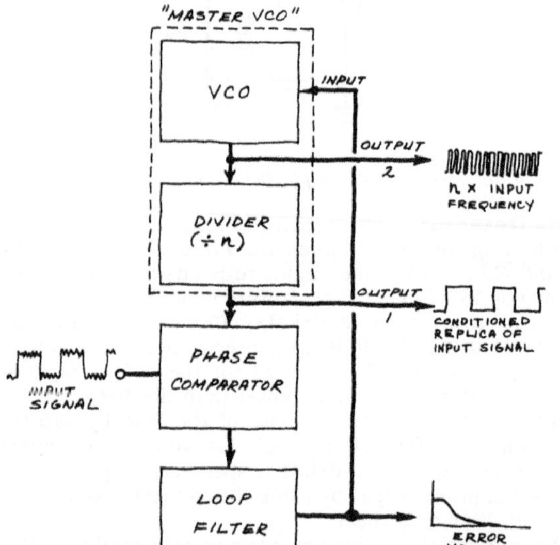

Fig. 4. A phase-locked-loop circuit with a divider.

Frequency Synthesis. Figure 4 shows how to synthesize exact multiples of a specified input frequency by inserting a divide-by-*n* counter between the vco and phase comparator of a PLL. You can understand how this arrangement works by thinking of the vco and divider as a single functional block or *master vco* instead of two separate circuits. The input (error voltage) and output (conditioned replica of the input signal) of this master vco are indistinguishable from those of a loop without the divider. The only difference is a second output connected directly to the internal vco having a frequency of *n* times the input frequency.

PLLs with dividers are used in many kinds of frequency synthesizers and function generators. They are particularly important in CB radios and other multiple-channel telecommunications equipment since they provide a wide range of precise output frequencies from a single crystal-controlled reference oscillator.

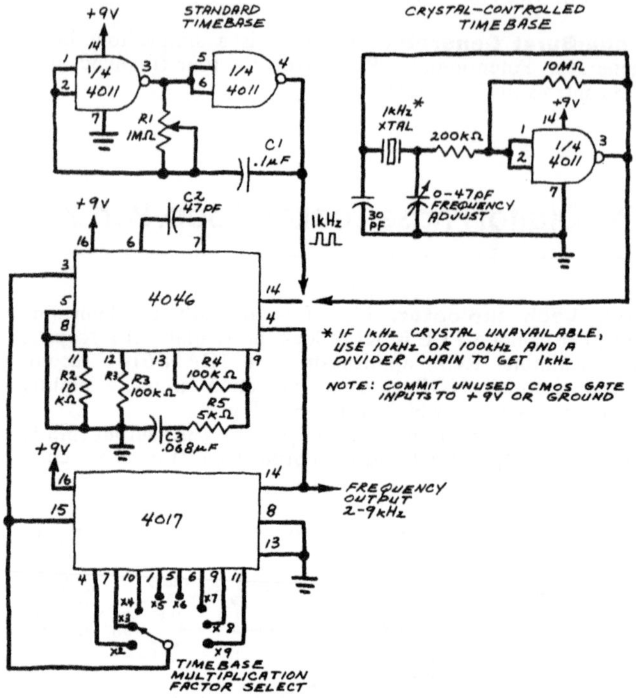

Fig. 5. Basic 2-to-9-kHz frequency synthesizer circuit.

Basic Frequency Synthesizer. One way to make practical use of a PLL with a divider inserted in the feedback loop is shown in Fig. 5. In this circuit, a 4017 counter is connected as a programmable divide-by-*n* counter where *n* is 2 to 9.

In operation, the NAND gate oscillator serves as a time-base which supplies a reference frequency of 1 kHz to the 4046 input. An 8-position selector switch connects the 4017 reset input to one of the eight count outputs. When the selected count is reached, the 4017 is reset and a new count cycle begins. This provides eight frequency steps ranging from two to nine times the time-base frequency. Each is a precise multiple of the time-base frequency.

For best results, especially in precision applications, use the crystal-controlled time base (also shown in Fig. 5). For non-precision applications or preliminary tests while you are awaiting arrival of the crystal, use the version without crystal control.

By using a string of divide-by-ten counters in place of the 4017 you can assemble a wide-range 10-Hz to 1-MHz synthesizer. You can achieve the same result by using programmable counters (*e.g.* 4522, 4018, etc.).

Pulse Frequency Modulator. The vco section of the 4046 can be used to make a simple pulse-frequency modulator which can be adjusted to provide a carrier frequency of 1 MHz or more. A simple pulse-frequency-modulated (pfm) lightwave voice transmitter, complete with a microphone preamp designed around a 3130 BiMOS op amp, is illustrated in Fig. 6. This circuit will also work with a 741 or other standard op amp. (Omit *C1* if you substitute op amp for the 3130 that has an internal compensation capacitor.)

With the values shown for *C3* and *R5*, the vco oscillates at a carrier frequency of about 100 kHz. This, and the circuit's modulation bandwidth, ensures reasonably good transmission of audio-frequency signals. The frequency-modulated signal drives an LED through *Q1* with *R6* limiting LED current.

The easiest way to test this circuit in conjunction with the receiver described next is to disconnect the microphone from *R1* and connect the output of a transistor radio to *R1* through a 0.1-μF capacitor.

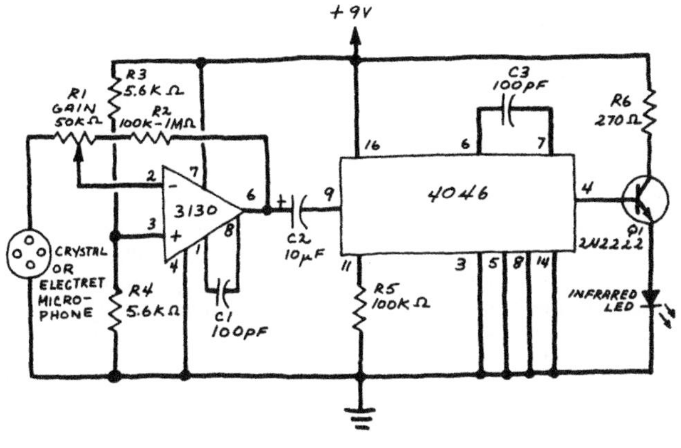

Fig. 6. Pulse-frequency-modulated lightwave voice transmitter.

Pulse Frequency Demodulator. Figure 7 shows a receiver system suitable for detecting and demodulating the pfm signal from the transmitter of Fig. 6. In operation, the infrared signal from the transmitter LED is detected by a phototransistor and coupled into a 3130 BiMOS op amp. This is the same op amp used in the transmitter and it, too, may be replaced with a 741 or other standard op amp. (Don't forget to omit *C2* if you use a 741.)

The amplified signal from the 3130 is ac-coupled via *C3* to the phase comparator input (pin 14) of a 4046. The vco is adjusted by *R5* and *C4* to create a center frequency identical to that of the transmitter (about 100 kHz). Components *C5* and *R7* form the loop filter that determines capture range. In this case, the resistor normally placed in series with the loop filter capacitor (see Part 1) has been omitted. This greatly simplifies the formula for determining the loop capture range:

$$f_c = \pm(1/2\pi)\sqrt{2\pi f_L / R7\ C5}$$

where f_c is the capture range and f_L is half the frequency lock range or, in this case, the vco center frequency. Substituting the values of *R7* and *C5* shown in Fig. 7 gives a capture range or bandwidth of ± 12.6 kHz.

For best results you should connect a lock detector like the one shown in Fig. 1 to the receiver's demodulator. You can use a single red LED to indicate loss of the signal or a single green LED to indicate acquisition of the signal. I prefer to use both the red and green LEDs for a clear go/no-go signal.

For preliminary tests, disconnect the transmitter's microphone from *R1* and connect a radio to *R1* as previously described. When the transmitter's LED is pointed at the receiver's phototransistor, the 4046 in the receiver should quickly lock onto the signal. You should then be able to hear the demodulated signal by means of an audio amplifier connected to the receiver's output.

If the receiver fails to capture the signal, tune its vco by adjusting *R5* until lock is established. If this fails, check the wiring of both the transmitter and receiver. If you've made no wiring errors, experiment with the radio's volume setting until lock is established.

When the receiver's demodulator has captured the input signal, block the beam and note that the signal is *sharply* cut off. This full-on/full-off reception is characteristic of FM transmission systems. It means that the signal from the receiver's demodulator has constant amplitude as long as the received signal has enough amplitude to be captured by the phase-locked loop.

You may notice that the receiver sometimes faithfully reproduces sound sent by the transmitter even when the receiver's 4046 is out of lock. This usually occurs when the signal level is weak. In such cases, the PLL is so close to establishing lock that the sound quality is unaffected.

Though the transmission range of these circuits is only several inches, external lenses or an optical fiber can substantially improve the range. For best results with free-space links, use GaAs:Si LEDs emitting at 940 nm. Suitable LEDs include the TIL-32 (Texas Instruments), OP-195 (Optron), 1N6266 (General Electric), etc. ■

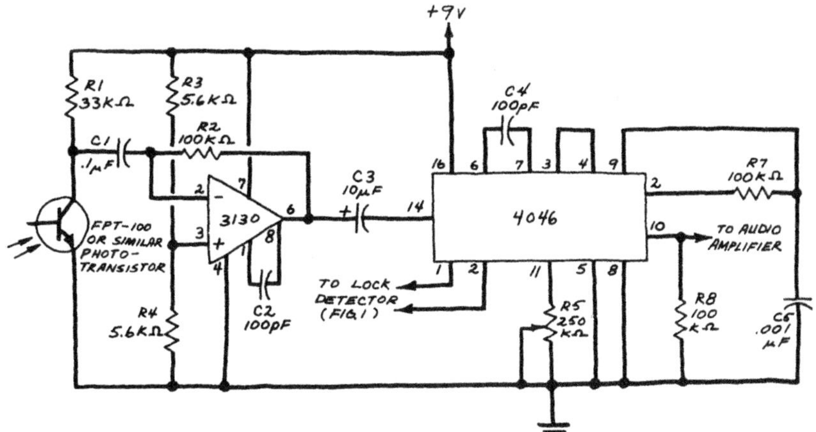

Fig. 7. A receiver system to be used for detecting and demodulating the pulse-frequency-modulated signal from Fig. 6.

6. Digital Color Organ

The schematic diagram here is for one of several digital color organs with which I've been experimenting for several years. In operation, signals generated by a transistor radio are coupled via its phone jack directly to the input of an LM3915 LED dot/bar display driver.

The sensitivity of the LM3915 is adjusted by means of potentiometer *R2* so that radio signals having maximum amplitude activate the green LEDs while low and intermediate levels activate the red and yellow LEDs, respectively. Potentiometer *R2* might have to be readjusted if the radio's volume is changed.

The total display consists of ten tricolor clusters, each of which is scanned by a 4017 counter/decoder at a rate determined by a clock made from two of the gates in a 4011. In accord with good design practices, the inputs of the remaining two 4011 gates are grounded. When the radio is silent, the display is blank. Sound from the radio causes a flurry of scintillating activity, as lights appear to bounce up and down while racing across the display in bursts and filaments. Slow, smooth music with restricted dynamic range does not cause a dramatic display, but a strong beat gives a very flashy show.

Before making a permanent version of the circuit, build a test circuit on a pair of solderless breadboards (one for the circuit and one for the display). This will allow you to evaluate both the circuit's operation and the relative brightness of the LEDs. For best results, operate the circuit in a darkened room and select the brightest LEDs you can find. Once the circuit is operating, experiment with the values of *R4* and *R5* to balance the brightness of the LEDs. You can initially use 1000-ohm potentiometers for *R4* and *R5*, and then substitute fixed resistors having the appropriate values after achieving a good brightness match. Also, be sure to experiment with the clock frequency by adjusting *R6* to find the most interesting level of display activity.

If you build a permanent version of the circuit, don't be afraid to experiment! Arranging the LEDs in concentric circles (with red LEDs the innermost) will produce starburst patterns. Finally, you can expand the display by adding additional 4017s and LEDs. ■

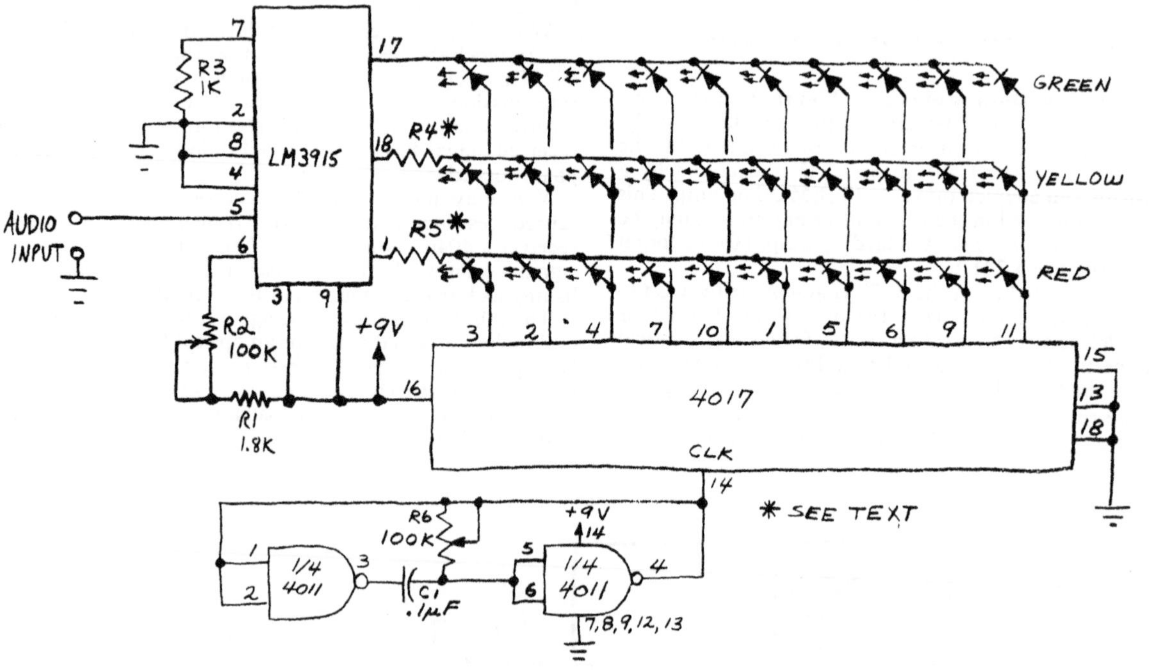

Schematic of a color organ having a display of ten tricolor clusters.

7. Experimenting with Shift Registers

The Basic Shift Register. Figure 1 is a block diagram of a very simple 4-bit shift register made from four D flip-flops connected in series. To understand the operation of this circuit, assume that the Q output of each flip-flop is at logic 0. When a clock pulse is applied to the SHIFT line, the logic level at the D input of each flip-flop is loaded into the corresponding flip-flop. Thus, if all of the Q outputs are initially at logic 0, the status of the four outputs (0000) will not be changed after the arrival of the clock pulse.

If a logic 1 is applied to the SERIAL INPUT, a logic 1 will be loaded into the first flip-flop when the next clock pulse arrives. The four-bit output nibble appearing at the PARALLEL OUTPUTS will then be 1000. If the logic level applied to the SERIAL INPUT is then changed to logic 0, the logic 1 will move one position to the right when the next clock pulse arrives. The four-bit nibble stored in the register will then be 0100. The rightward movement of the logic 1 will continue as additional clock pulses are received. The nibble changes to 0010 and then to 0001. Upon receipt of the fifth clock pulse, the logic 1 is pushed entirely out of the register and replaced by a logic 0. The register will then again contain the nibble 0000.

Several significant things have occurred during the course

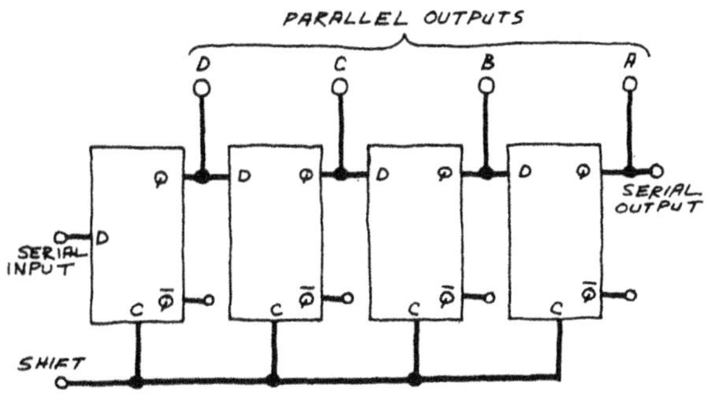

Fig. 1. Basic block diagram of a D flip-flop shift register.

of applying five clock pulses to the basic shift register. First, the logic 1 applied to the SERIAL INPUT appeared at the SERIAL OUTPUT only after the arrival of four clock pulses. Therefore, the shift register has functioned as a *digital delay line*. Secondly, the logic 1 migrated through the register, appearing at one of the four Q outputs at any given time in a sequence controlled by the clock rate. Taken together, the PARALLEL OUTPUTS can be used to actuate sequentially or *strobe* a series of external circuits in accordance with any pattern of bits presented to the SERIAL INPUT. In general, the ENABLE inputs of many logic ICs are active when a logic 0 is applied to them, so a logic 0 would usually be used as an activating strobe bit.

Thirdly, the bit pattern appearing at the four PARALLEL OUTPUTS can be considered a binary word. As the logic 1 moved from left to right, the magnitude of the word was halved at each clock pulse (1000 = 8; 0100 = 4; 0010 = 2 and 0001 = 1). Thus, the shift register performed a numerical divide-by-two operation. Finally, between clock pulses, the shift register has acted as a conventional data storage register. The register stored data without changing or modifying them, and data were always available at the PARALLEL OUTPUTS

Experimental Shift Register. Many different kinds of integrated shift registers are available, and we'll examine several of them next month. However, if you would like to build and experiment with your own flip-flop shift register, you can try the circuit shown in Fig.2. It is made from a pair of CMOS dual D flip-flops, and does everything the basic register of Fig. 1 does.

Two NAND gates (*IC3A* and *IC3B*) connected as a bistable latch provide a bounce-free pulse to the clock inputs of each flip-flop when *S2* is placed in its LOGIC 1 position. This switch and the INPUT DATA SELECT switch allow you to cycle the shift register and change the input data in any fashion you choose. The logic level of each Q output is indicated by a LED.

If you prefer, you can use flip-flops other than those contained in the 4013 to make a shift register. For example, the 7474 is a TTL dual D flip-flop. The 74175 contains four D flip-flops in a single DIP and, as you can see in Fig. 3, readily lends itself to use as 4-bit TTL shift register.

Incidentally, if you don't have any D flip-flops on hand, but do have some JK flip-flops (such as the 4027, 7473, 7476, etc.), you can convert the JK units into D flip-flops. Simply connect the input and output of an inverter to the J and K inputs, respectively. The node comprising the flip-flop's J input and the inverter input behaves as a data (D) input.

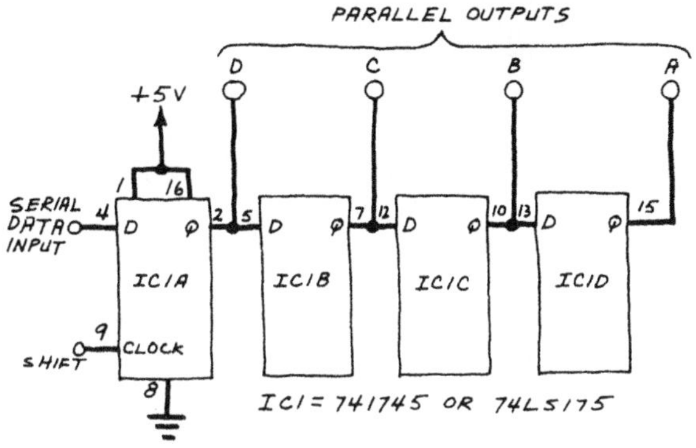

Fig. 3. A four-bit shift register made from a single quad D flip-flop.

Shift Register Types. Now that we've seen what a basic shift register can do and how it does it, let's examine some of the technical jargon used to characterize various types of shift registers. First, that shown in Fig. 1 is called a *serial-in/parallel-out and serial-out* shift register. It is a *serial-in* register because data can be entered bit by bit (serially) into the input of only the first flip-flop. It is a *parallel-out* register since all four outputs are simultaneously available. Because the final output is always available, the circuit also provides a *serial-out* capability. A *parallel-in* capability is not available with the

Fig. 2. Schematic diagram for an experimental CMOS shift register made from D flip-flops.

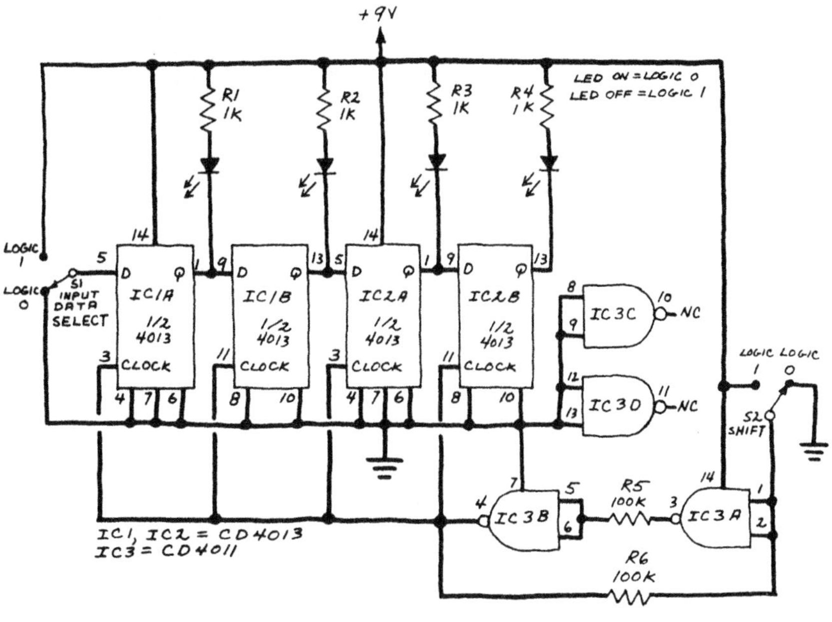

circuit in Fig. 1, but can be added with the help of a suitable logic network.

These descriptive terms allow us to specify the most important kinds of shift registers:

Serial-In/Serial-Out. This is the basic shift register. It can be as simple as a 2-bit register or as complex as a million-bit bubble memory.

Serial-In/Parallel-Out. This register is more flexible than a simple *serial-out* register because all of the contents of the register are always available.

Parallel-In and Serial-In/Serial-Out. Such a register allows all of the bits in a complete digital word to be loaded simultaneously and then clocked out one bit at a time.

Parallel-In and Serial-In/Parallel-Out and Serial-Out. This is the "complete" shift register. It can be used as a conventional data register or as a universal shift register.

Although the basic register shifts bits only to the right, some registers can shift bits in *both* directions. These are the most versatile of all shift registers. Figure 4 shows all the input, output and control lines of a 4-bit universal shift register.

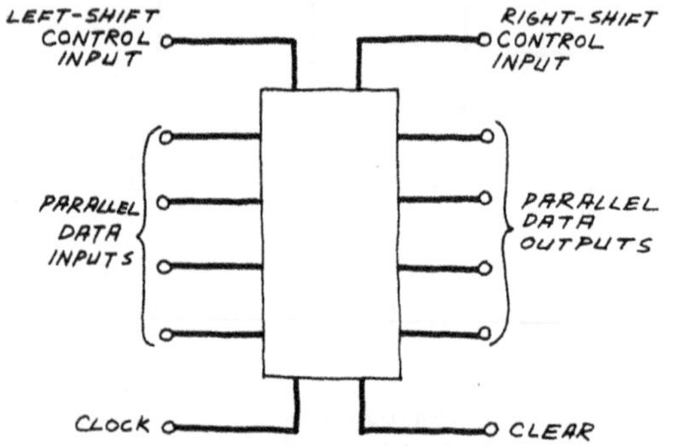

Fig. 4. Inputs and outputs for a hypothtical universal four-bit shift register.

Shift Register Applications. Shift registers have literally dozens of applications. In the remainder of this column we'll examine several important applications conceptually. We'll experiment with some specific circuits next month.

Multiplication. Shift registers are vital components in many digital computing circuits. Consider, for example, this problem in binary multiplication: Multiply 110_2 by 101_2.

```
  110 multiplicand
  101 multiplier
  110
 000     partial products
110
11110      final product.
```

The rules for binary multiplication are: (0) (0) = 0; (0) (1) = 0; (1) (0) = 0; and (1) (1) = 1. The rules for binary addition are: 0 + 0 = 0; 0 + 1 = 1; 1 + 0 = 1; and 1 + 1 = 0, carry 1, or 10.

Refer again to the multiplication problem above and you'll discover a binary-multiplication shortcut: When one bit in the multiplier is 0, its partial product is 000; when the bit is 1 the partial product equals the multiplicand. Therefore, to multiply two binary numbers, inspect the least significant bit in the multiplier. If it is 0, write down a string of 0s equal in length to the number of bits in the multiplicand. If it is 1, write down the multiplicand. This entry becomes the first partial product.

Next, move to the second-most significant bit in the multiplier. Repeat the foregoing procedure to arrive at the second partial product. Then shift the result one bit position to the left and add the two partial products.

Continue inspecting, shifting and adding until all the bits in the multiplier have been accounted for. The sum of the last two partial products becomes the final product.

This exercise illustrates a very important characteristic of digital arithmetic—binary multiplication can be accomplished by shifting left and adding. The arithmetic-logic unit (ALU) in virtually every microprocessor includes a logical comparator, an adder and a shift register. Multiplication can be performed by a relatively straightforward program that makes alternate comparisons, shifts and additions. If you would like to know more, Lou Frenzel has written a very clear explanation of this procedure in an excellent book, *Getting Acquainted with Microcomputers* (Howard W. Sams & Co., 1978, pp. 197-203).

Multiplication and Division by Two. Another neat binary-arithmetic trick that shift registers can perform is multiplication or division by a factor of two. As we have already observed, shifting any binary word one bit to the right divides the word integrally by two. For example, 1110 (14) shifted right one bit is 0111 (7). Similarly, shifting any binary word one bit position to the left multiplies the word by two. For example, 1001 (9) shifted left one bit is 10010 (18).

Serial Addition. A binary *full adder* is a straightforward combinational circuit made from two exclusive -OR gates and several additional gates connected as shown in Fig. 5. The circuit is called a full adder since it can both accept and generate carry bits.

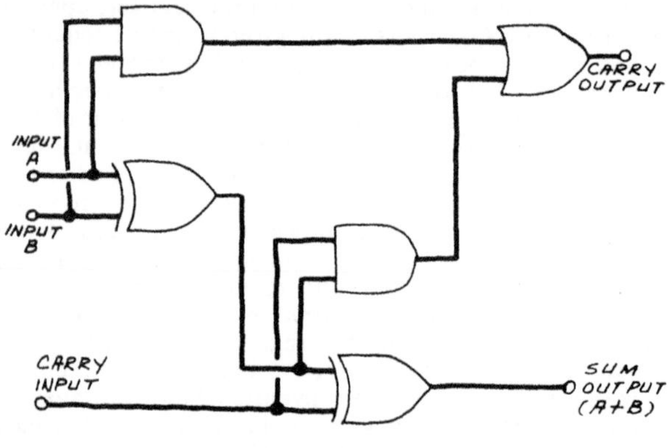

Fig. 5. Logic diagram of a binary full adder.

A single full adder can add only two data bits plus one carry bit. Therefore a number of adders arranged in parallel are required to simultaneously add all of the bits in two data words. For example, the simultaneous addition of all of the bits in two bytes requires a parallel array of eight full adders.

It's possible to add two data words using just one adder if the addition is performed one bit position at a time. Two shift registers are required to store the words being added, and a third is required to store the sum. A single D flip-flop is needed to store the carry bit which will result when the two bits to be added are both 1 or if the sum of the two bits and carry bit which might be present is 10 or 11. Figure 6 is a block diagram of a serial-shift-register adder.

The operation of a shift-register serial adder is a very good example of a sequential logic circuit. Referring to Fig. 6, the two words to be added are loaded into shift registers A and B. They are then clocked through the adder a pair of bits at a time and the resulting partial sums are loaded into shift register C. The complete addition requires only four clock cycles.

Can you think of a way to simplify the serial adder in Fig. 6? Shift register C can be eliminated entirely by feeding the output of the adder back to the input of Shift Register A, which then becomes an *accumulator*.

Although the operation of the serial adder seems simple enough, a control circuit is required to prevent the application of any more clock pulses once the addition has been completed. Otherwise, any new data that happens to be at the inputs of Shift Registers A and B will be cycled through the adder, and the sum stored in Register C will be pushed out and lost.

You can learn about an important aspect of the operation of the control section of a microprocessor or digital computer by designing a simple circuit. The circuit should monitor the operation of a serial adder and save the final sum by either dis-

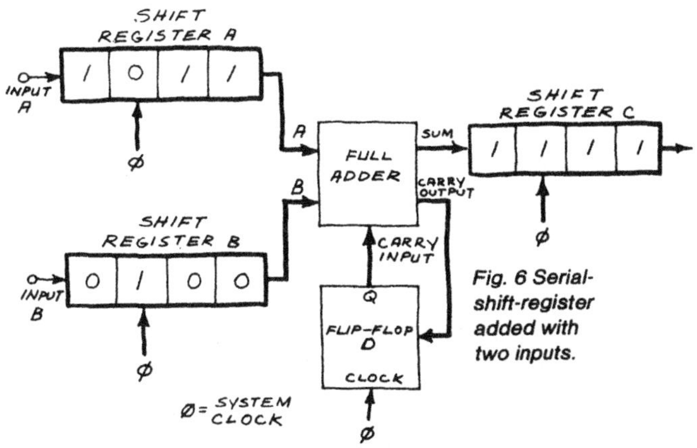

Fig. 6 Serial-shift-register added with two inputs.

abling the clock pulses or moving the sum into still another register. Hint—the use of a 2-bit counter offers one solution.

Data Transmission. Computer data is usually transmitted in serial fashion one bit at a time. A shift register at the transmitting end reduces each word to be transmitted into its component bits, and one bit is transmitted each time a clock pulse arrives. A second shift register at the receiving end reconstructs the transmitted words bit by bit. It passes them to a storage register each time a complete word has been received and reconstructed. Figure 7 summarizes a shift-register data-transmission system.

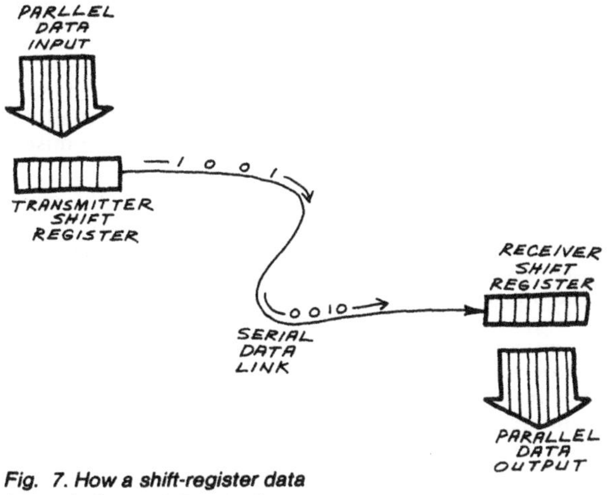

Fig. 7. How a shift-register data transmission system operates.

A shift register that transmits a word one bit at a time is called a *parallel-to-serial converter.* A shift register that assembles data words from a stream of incoming bits is called a *serial-to-parallel converter.* Both applications find use in many operations other than data transmission. Closely related to data transmission are applications in which a shift register acts as a temporary storage register or delays the arrival of a data word by a preselected number of clock pulses.

Memory Stack. A *memory stack* consists of two or more data registers used to hold temporary data. In a microcomputer, a stack is implemented within the main memory (RAM) or by a special set of data registers. A *pointer register* keeps track of where data is stored in the stack.

Shift registers can be used to make a memory stack. In the version shown in Fig. 8, four shift registers are arranged in

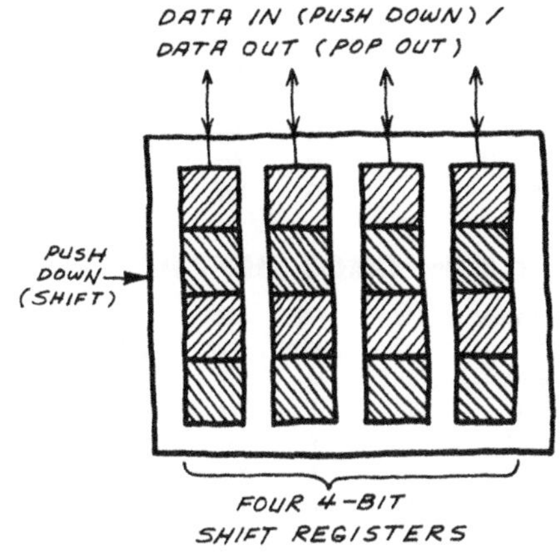

Fig. 8. Four 4-bit words can be stored in this memory stack.

parallel so that up to four 4-bit data words or nibbles (half of an 8-bit byte) can be stored. The clock (shift) lines of all four registers are tied together so that a 4-bit nibble can be loaded into the stack in one operation. The nibble can then be pushed down into the stack as more are loaded.

If the nibble moves in only one direction through the stack, the first nibble to enter is the first to exit. This is a FIFO (first in/first out) stack. Several variations are possible. For example, the capability of shifting in both directions means that a nibble can be pushed into and popped out of the stack. In a LIFO (last in/first out) stack, the last nibble pushed into the stack is the first to be popped out of the stack. ■

8. More on Shift Registers

Tables I and II list most of the TTL and CMOS shift registers available to experimenters. These tables, which were compiled from Motorola, National, RCA and Texas Instruments data books, list only some of the more important shift register parameters. Also, some specifications (such as maximum clock frequency) can vary somewhat within the same chip type if different manufacturers are involved. Therefore, you should check the specifications provided by the manufacturer of the chip you are thinking of using in a project for more specific information.

Here is a brief explanation of the table headings:

Function. The function of each shift register is identified by a four- or five-letter code. S means serial, P parallel, I in, O out, and B bidirectional. Therefore, a shift register listed as B/PIPO can provide bidirectional, parallel-in/parallel-out operation. Incidentally, many of the shift registers listed in the tables will provide more than one operating function. Most PIPO registers, for instance, also provide SISO operation.

Bits. This parameter specifies the number of register elements in the listed device.

Frequency. The maximum shift (clock) frequency is given. Frequency specifications for CMOS shift registers vary with both the V_{DD} supply voltage and the manufacturer. Consult the manufacturer's literature for the specific operating frequencies for a given chip.

Modes. Four operating modes are given. Those whose functions are obvious are *shift left* and *shift right. Load* is normally found on parallel-input shift registers. Depending upon the status of the load input, upon receipt of the next clock pulse, the register will either ignore or accept the data

TABLE I—TTL SHIFT REGISTERS

Type	Function	Bits	Freq. (MHz)	Shift Right	Shift Left	Load	Hold
7491	SISO	8	10	Yes	No	No	No
7494	SISO	4	10	Yes	No	Yes	No
7495	PIPO	4	36	Yes	No	Yes	No
7496	PIPO	5	10	Yes	No	Yes	No
74L99	PIPO	4	3	Yes	No	Yes	No
74164	SIPO	8	25	Yes	No	No	No
74165	PISO	8	25	Yes	No	Yes	Yes
74166	PISO	8	20	Yes	No	Yes	Yes
74178	PIPO	4	25	Yes	No	Yes	Yes
74179	PIPO	4	25	Yes	No	Yes	Yes
74194	B/PIPO	4	25	Yes	Yes	Yes	Yes
74195	PIPO	4	30	Yes	No	Yes	No
74198	B/PIPO	8	25	Yes	Yes	Yes	Yes
74199	PIPO	8	25	Yes	No	Yes	Yes
74LS295	PIPO	4	25	Yes	No	Yes	No
74LS299	B/PIPO	8	35	Yes	Yes	Yes	Yes
74LS323	B/PIPO	8	35	Yes	Yes	Yes	Yes
74LS395	PIPO	4	25	Yes	No	Yes	No

TABLE II—CMOS SHIFT REGISTERS

Type	Function	Bits	Freq. (MHz)	Shift Right	Shift Left	Load	Hold
4006	SISO	18	10	Yes	No	No	No
4014	PISO	8	5	Yes	No	Yes	No
4015	SIPO	8	9	Yes	No	No	No
4021	PISO	8	5	Yes	No	Yes	No
4031	SISO	64	8	Yes	No	No	No
4034	B/PIPO	8	10	Yes	Yes	Yes	Yes
4035	PIPO	4	5	Yes	No	Yes	No
4094	SIPO	8	5	Yes	No	No	Yes
40100	B/SISO	32	3	Yes	Yes	No	Yes
40104	B/PIPO	4	9	Yes	Yes	Yes	No
40194	B/PIPO	4	9	Yes	Yes	Yes	Yes

present at its input(s). *Hold* means that the clock input can be inhibited or disabled so that the shift register will store its contents like a memory register.

Several other operating modes not listed in the tables may also be available. *Preset* or *clear* enables all the register elements to be cleared to logic 0. *Recirculate* causes the data at the output of a shift register to be cycled back to the input.

Exotic Shift Registers. Several kinds of elaborate integrated shift registers are available for such serial-memory applications as refreshing cathode-ray tube traces and storing characters to be printed by high-speed printers. A typical example is Synertek's SY1404A 1024-bit MOS dynamic shift register. This family also includes a dual 512-bit register (SY1403A) and quad 256-bit register (SY1402A). The maximum clock rate for these chips is a relatively slow 2.5 MHz, but by means of a multiplexing technique, data can be accepted at a 5-MHz rate. Even more capability is provided by 2048-bit shift registers such as the SY2401 and SY2827.

Bubble memories and charge-coupled devices (CCDs) are among the most exotic shift registers available. Bubble-memory capacity can exceed a *million bits per chip*, making this exotic register a strong contender in the search for a solid-state replacement for the disk memory.

Application Circuits. Now let's try experimenting with some readily available shift registers to see how they work and what they can do. The circuits that follow use both CMOS and TTL shift registers. Once you see how easy it is to use these various chips, you'll want to consider experimenting with other shift registers as well.

Pseudorandom Sequencer. Figure 1 is the schematic of a pseudorandom generator whose operation is patterned after the S2688/MM5837 shift-register noise generator that was described in the March 1980 installment of this column. The circuit consists of two 8-bit 4021 CMOS shift registers (*IC2* and *IC4*) cascaded to form a single 16-bit register.

The 4021 is a PISO register with the bonus feature that the outputs of the 6th, 7th and 8th stages are available. This means it can be used as a 6-, 7- or 8-bit shift register and makes possible the pseudorandom sequencer circuit. Such a circuit requires that the outputs of two stages in a shift register be coupled back to the input via an exclusive-OR gate.

By connecting the exclusive-OR gate's inputs to the final two outputs of the shift register, the sequencer will generate a pseudorandom sequence that recycles after 255 clock pulses. The bit pattern within a single 255-bit cycle is essentially random, but the periodic recycling of the pattern compromises its randomness over the long term.

Two of the gates in *IC1* form a simple clock for this circuit. This is a CMOS chip, so be sure to connect the unused inputs (pins 8, 9, 12 and 13) to V_{DD} or V_{SS} to keep the chip from drawing excess current and overheating. You should also connect the unused inputs of *IC3* (pins 5, 6, 8, 9, 12 and 13) to either V_{DD} or V_{SS}.

This circuit has several useful applications. Connect its output to an audio amplifier through the low-pass filter composed of *R2* and *C2* to convert the pseudorandom bit pattern into audible noise. For best results, speed up the clock by reducing the value of *C1* to 0.01 μF or so. At very fast clock speeds, the output will sound like a pure tone. At slower frequencies, the repetitious pattern present in the output signal can be heard.

An interesting LED flasher can be made by connecting the output of the circuit to the cathode of a LED whose anode is

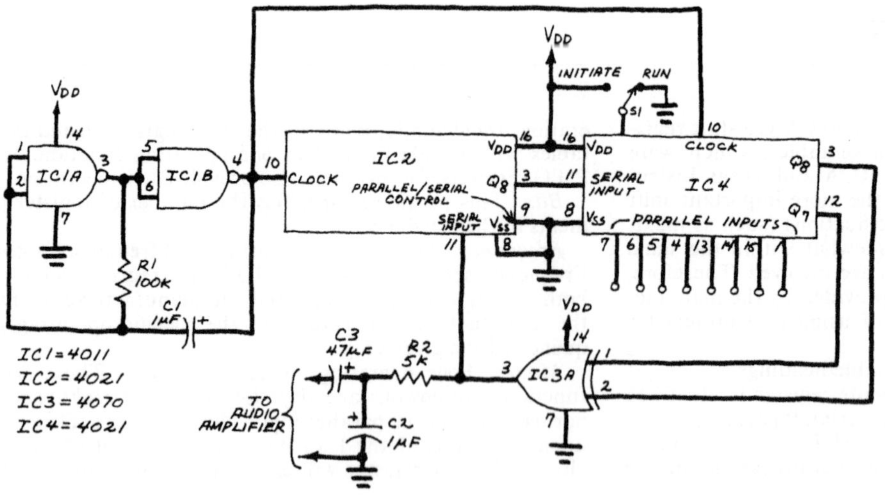

Fig. 1. Schematic diagram of a pseudo-random sequencer made from CMOS shift registers.

connected to V_{DD} through a 1000-ohm series resistor. Slow the clock rate to a few hertz by increasing the value of *C1* to 10 μF or so. The LED will then flash on and off in an irregular, seemingly random pattern. This LED-flasher application illustrates how the circuit can be used to strobe (actuate) another circuit at a pseudorandom rate.

As you alter the clock rate or change the connections to the shift registers, the pseudorandom sequence generator might shut down. Make sure the clock is working by monitoring its output with an audio amplifier and loudspeaker (a tone should be heard) or with an oscilloscope (a square wave should be displayed). If the clock is running, try resetting each shift register by switching its pin 9 (the PARALLEL/SERIAL CONTROL input) from V_{SS} to V_{DD} and then back to V_{SS}. The circuit should then resume normal operation.

You can control the pattern of bits moving through the shift registers. Connect the PARALLEL DATA inputs of *IC4* to V_{DD} or V_{SS} to set up any desired sequence of logic 0's and 1's. Then toggle *S1* from RUN to INITIATE and then back to RUN to load the PARALLEL DATA inputs. Repeat this procedure as desired to create many different bit sequences.

Incidentally, if you intend to use this circuit for commercial purposes, you should first write the U.S. Patent and Trademark Office (2021 Jefferson Davis Highway, Arlington, VA 22202) and request a copy of U.S. Patent 4,191.175. The fee for a single copy of a patent is 50¢. I've not yet seen this patent myself. However, after the March 1980 "Experimenter's Corner" appeared, William L. Nagle, president of Paratronic Systems, Inc. (Honeybrook, PA 91344) wrote to this magazine that, ". . . your readers should be cautioned that use of this for any other than private purposes would be an infringement of the patent our company holds for such devices and their applications." The complete letter was reproduced in the *Letters* column of the July 1980 issue.

I am interested in seeing this patent because its number indicates an issuance date late in 1979. Publications describing pseudorandom sequence generators date back to at least 1973 when Fairchild Semiconductor's *The TTL Applications Handbook* (a truly outstanding book) described two such sequencers on pages 8 through 21.

Incidentally, one of the Fairchild circuits employed one 4-bit and seven 8-bit shift registers to generate a truly *long*, nonrepetitive output bit sequence. According to the descriptive text, ". . . even at a frequency of 20-million states per second the counter would not repeat until more than 18 centuries had elapsed." (!) For more information about pseudorandom sequencers, see Don Lancaster's indispensable *CMOS Cookbook* (Howard Sams & Co., 1977). Don discusses the topic and presents two circuits on pages 318 through 323.

Pseudorandom Voltage Generator. Connect a suitable resistive ladder network to the outputs of a SIPO or PIPO shift register set up as a pseudorandom sequencer and you get a pseudorandom voltage generator. The resistor network serves as a digital-to-analog converter.

Figure 2 shows such a circuit designed around a 74164 or 74LS164 SIPO 8-bit shift register. The four NAND gates (*IC2A* through *IC2D*) form an exclusive-OR gate. You can substitute one-fourth of a 7486 quad exclusive-OR gate in the circuit if you prefer.

Use a 555 timer IC connected as an astable oscillator like the one shown in Fig. 3 to provide clock pulses for the circuit. The amplitude of the pseudorandom, stepped output voltage can be changed by connecting pin 2 of *IC2A* to any of the six other output pins of *IC1*.

An interesting application for this circuit is a pseudorandom tone sequencer that can be made by connecting its output to the control input of a voltage-controlled oscillator or voltage-to-frequency converter. One suitable vco is the 4046, a chip that was highlighted in "Experimenter's Corner" for July and August 1980. Suitable V/F converters include the

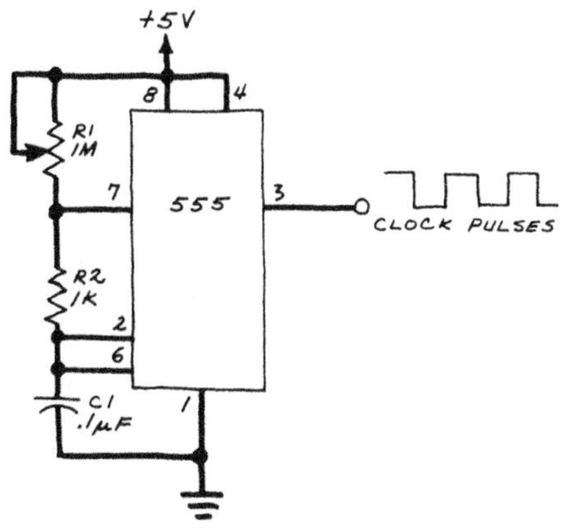

Fig. 3. A 555 clock pulse generator.

9400 and LM331—see "Experimenter's Corner," for October 1979 for details.

8-Bit Serial-to-Parallel Data Converter. Here's a circuit that you can use to load serial data onto an 8-bit data bus. Referring to Fig. 4, clock pulses are applied simultaneously to counter *IC1* and shift register *IC3*. The data to be loaded on the bus should be applied at the clock rate to the serial input of the shift register.

The 7490 (*IC1*) is configured as a divide-by-eight counter. It generates a load pulse that causes *IC2*, a 74LS374 octal D flip-flop, to accept the data appearing at the parallel outputs of shift register *IC3*.

Summarizing, after 8 bits have been loaded into the shift register, a pulse from the counter causes the data to be transferred to the octal flip-flop. The flip-flop is updated with new data from the register after eight additional clock pulses.

You can use the 555 oscillator shown schematically in Fig. 3 as a clock circuit for the data converter. If you want to experiment with the circuit as I did when working with the prototype, connect the cathode of a LED to each output pin of *IC2*. Connect the anodes of the LEDs to 1000-ohm resistors which are in turn connected to +5 volts. You can use smaller resistance values (as small as 270 ohms) for the current-limiting resistors if you want the LEDs to glow brighter. I prefer 1000 ohms to keep current consumption down to 2 or 3 milliamperes per LED.

An additional LED and series resistor connected between

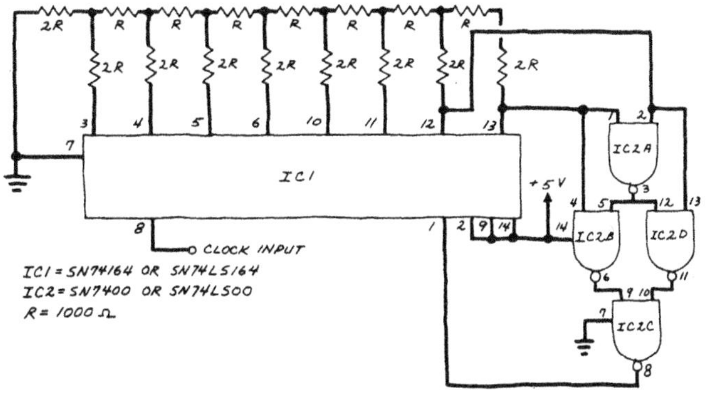

Fig. 2. Schematic of a pseudo-random voltage generator.

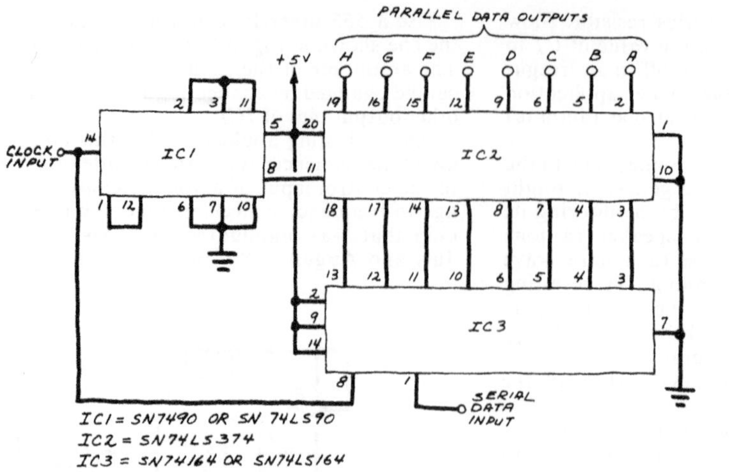

Fig. 4. Schematic of an 8-bit serial-to-parallel data converter.

pin 8 of *IC1* and ground will allow you to monitor the status of the shift register. You can then manually test the circuit by slowing down the clock to about one pulse per second and applying input data bits to pin 1 of *IC3*.

This circuit can be used as the receiving portion of a digital data-transmission system. A good design exercise would be to devise a suitable transmitter for converting 8-bit bytes into a serial bit stream. Hint—use a PISO shift register and a divide-by-8 counter.

Programmable Sequence Generator. Figure 5 schematically shows an 8-bit programmable sequence generator made from two series-connected 74194 shift registers (*IC1* and *IC2*). In operation, any desired bit pattern is first selected by means of switches *S1* through *S8*. A normally closed pushbutton switch, *S9*, is momentarily opened to load the selected bit pattern into the two shift registers. When clock signals are applied simultaneously to pin 11 of both shift registers, the bit pattern will be shifted one position for each clock pulse. Because the serial output of the second shift register (pin 12 of *IC2*) is connected to the serial input of the first shift register (pin 2 of *IC1*), so long as the clock pulses are received the bit pattern will recirculate through the registers. It will remain unchanged until it is modified by means of the data-select and load switches.

Use the oscillator shown in Fig. 3 to generate clock pulses for this circuit. The output LEDs shown in Fig. 5 are optional, but including them allows the circuit to double as an attention-getting programmable light flasher. The LEDs can be arranged in various patterns to enhance the effect.

A more practical application can be accomplished by connecting a resistor ladder network (see Fig. 2) to the outputs of the shift registers. The circuit then will function as a programmable waveform generator. Without the ladder network, the circuit can be used to strobe various circuits in any programmed sequence. More shift registers can be added for even longer sequences.

If you build this circuit, be sure to include the 0.1-μF power-supply decoupling capacitors. Without these capacitors, the shift registers will be affected by power-supply transients that can arise during the switching sequence. A typical effect of such a transient is an unwanted change in the sequence.

Bargraph Generator. Figure 6 is the schematic diagram of an unusual bargraph generator made from a pair of 74194 shift registers. In operation, the circuit's ENABLE INPUT is brought momentarily to logic 0 to start the circuit. This causes seven of the eight bit positions to be loaded with logic 1's, because their inputs are left unconnected and therefore assume a high state. The output of the first stage of *IC1* goes to logic 0 because its input (pin 3) is grounded.

The bits in the first and last stages, logic 0 and logic 1, respectively, are combined in AND fashion by *IC3A* and *IC3B*. The result is presented to the serial input of the first shift register. On the first clock pulse, therefore, the outputs of the first two positions go to logic 0 while all of the other outputs stay at logic 1. This pattern continues as subsequent clock pulses are received until each of the shift-register outputs switches in turn from logic 1 to logic 0. When the final stage goes low, the bit resulting from the gating of the first

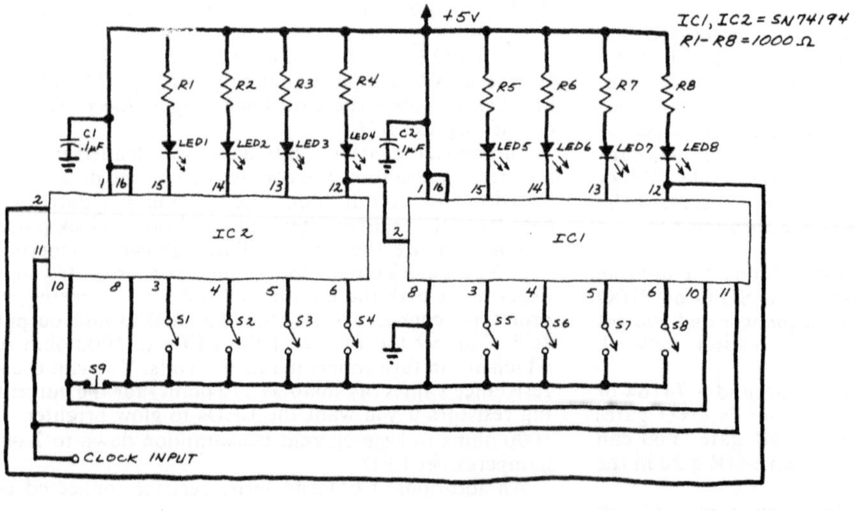

Fig. 5. A shift-register 8-bit programmable sequence generator.

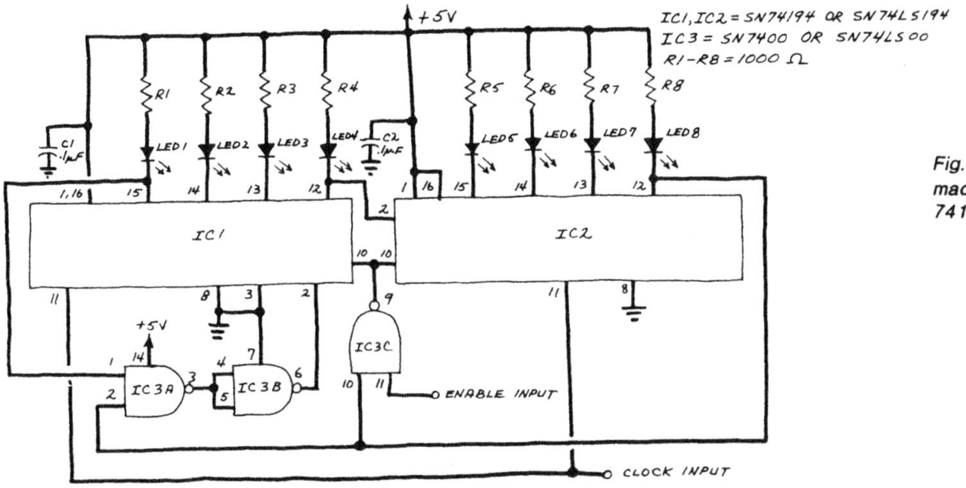

Fig. 6. A bargraph generator made from a pair of 74194 shift registers

and last bits is a logic 0.

Because this bit is fed back to the input of the first register (pin 2 of *IC1*), it might at first glance appear that all of the register outputs would remain at logic 0 after the first cycle of clock pulses. Note, however, that the output of gate *IC3C* is connected to pin 10 of both registers. When the output of this gate is logic 0, the shift registers ignore the data presented to their parallel inputs. When its output switches to logic 1, the shift registers load the data present at their inputs. This, of course, is what occurs when the ENABLE INPUT is brought to logic 0 to start the circuit.

This circuit has such practical applications as strobing various external circuits in a sequential fashion. It also makes a very interesting visual display.

Going Further. Shift registers are ideally suited for experimentation. The bargraph generator shown in Fig. 6 is a good example of this. I began with the remnants of the circuit in Fig. 5 (the input switches were removed) and tinkered with the basic circuit by adding a single 7400 quad NAND gate IC. Within a few minutes, the bargraph effect was achieved. You can do the same kind of experimentation on your own by selecting different register outputs to be connected to NAND gate *IC3A* in Fig. 6. You can also try adding additional stages for more complex effects. The cost of integrated shift registers is very reasonable, and you'll learn a good deal about these very versatile logic circuits by experimenting with them. ■

9. CMOS Basics: The 4011 Quad NAND Gate

IT'S common knowledge that CMOS integrated circuits lacking internal zener protection are vulnerable to electrostatic damage. That's why many experimenters are reluctant to use them. This month we're going to try to convince those of you who shy away from CMOS to get your feet wet by experimenting with one of the most basic CMOS chips, the 4011 quad 2-input NAND gate. In doing this, we hope to make clear the advantages of CMOS to those of you who still use TTL exclusively in your projects.

As you'll see by building the circuits that follow, CMOS is in many respects much more versatile and easier to use than TTL. Its principal advantages are its wide operating voltage range, low current consumption, and high input impedance. CMOS chips also provide very good noise immunity and large fan-out (that is, the number of gate inputs that can be driven by a single output). The only major advantage TTL has over CMOS is considerably greater switching speed. Typical maximum speeds for CMOS logic range from 1 to 5 MHz.

Using CMOS Chips. The circuits described here will enable you to learn firsthand about the operating characteristics and requirements which distinguish CMOS from TTL. Probably the most important requirement is that *every* input pin of a CMOS chip *must* go somewhere. If the pin is unused, it must be connected to V_{DD} (the positive supply voltage) or V_{SS} (the negative supply voltage, which is usually ground). Otherwise, stray signals can enter the device via the unused pin, turn on a gate, and cause a very large current to flow. This can cause the chip to overheat or exhibit erratic, unpredictable operation.

Another important difference is power-supply voltage. TTL must be operated within half a volt of +5 volts. Most CMOS chips can be operated with a potential difference between the V_{DD} and V_{SS} terminals of from +3 to +15 volts. Although the wide operating voltage range of CMOS is very desirable, its ultra-low current consumption is even more important. At a switching speed of 1 MHz, for example, a typical CMOS gate draws only 0.1 milliampere when the supply voltage is +5 volts and only 0.2 milliampere when it is +10 volts.

Today's CMOS chips are less vulnerable to electrostatic-discharge damage than those of older vintage, thanks to protective zener diodes diffused across their inputs. These diodes shunt high voltages (generated as static electricity by simple physical handling) away from the delicate gate structures of CMOS devices. Nevertheless, always play it safe and follow standard CMOS handling precautions.

- Never store CMOS chips in *nonconductive* plastic trays, bags or foam.
- Always place CMOS chips *pins down* on a conductive aluminum foil sheet or tray when the chips are not being stored in conductive foam.
- Avoid touching the pins of CMOS chips.
- Use a grounded or battery-powered soldering iron to solder the pins of CMOS chips. Better yet, employ IC sockets or Molex Soldercons.

Recently, there have been reports that some so-called conductive plastic bags and foams intended for the storage of CMOS chips are not nearly as effective as they are supposed to be at protecting the chips from static discharge damage.

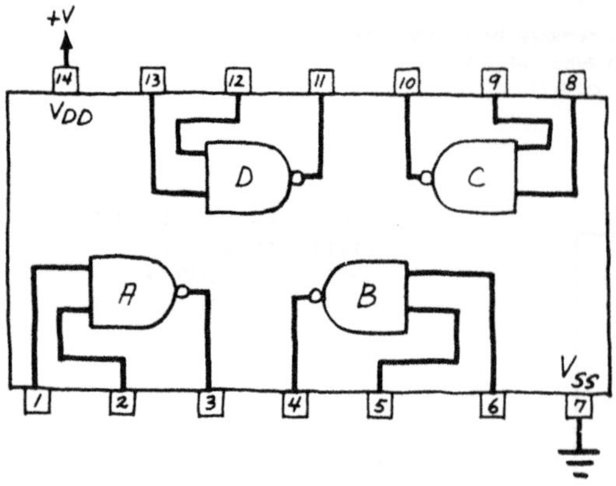

Fig. 1. Pin outline for a 4011 quad 2-input NAND gate.

You can avoid this potential problem by plugging CMOS chips you want to store into a flat piece of foam plastic wrapped with aluminum foil.

Figure 1 is the pinout of one of the simplest CMOS gate packages, the 4011 quad 2-input NAND gate. If you're a veteran TTL user, you will immediately notice that the gates connected to pins 4, 5 and 6 and pins 8, 9 and 10 are oriented in the opposite direction as compared to the corresponding gates in the 7400, the TTL counterpart to the 4011. Be sure to keep that in mind when you use the 4011 in the circuits here.

Audio Oscillator. Figure 2 shows a simple tone generator

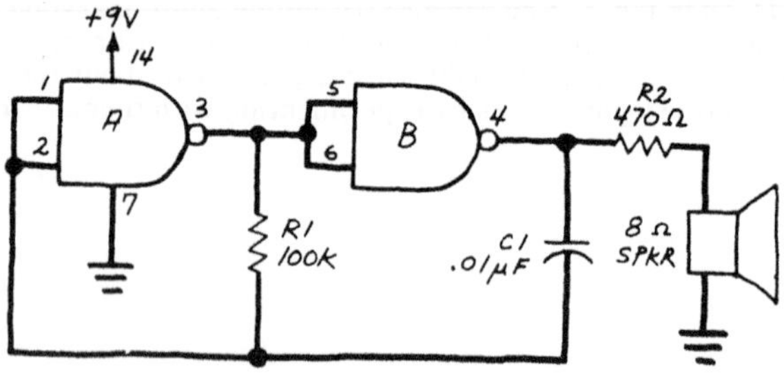

Fig. 2. Circuit for an ultra-simple CMOS tone generator.

made from half of a 4011. The circuit delivers a 1-kHz square wave to a miniature 8-ohm speaker. The frequency of the output signal can be increased by decreasing the value of *C1*.

The square-wave output can be made more symmetrical by inserting a 1-megohm resistor between pins 1 and 2 of the 4011 and the common connection of *R1* and *C1*. For increased drive capability, connect together all four inputs and both outputs of the two unused gates (*C* and *D*) to form a buffer which is then inserted between pin 4 and the speaker.

You can turn the tone generator on and off with an external logic signal by disconnecting one of the inputs of the first gate and using it as an enable input. The circuit will oscillate when the enable input is at logic 1.

LED Flasher. Figure 3 schematically shows an LED flasher patterned after the basic oscillator circuit of Fig. 2. The LED will flash once or twice each second. The flash rate can be reduced by increasing the value of *C1*. To use the circuit as a 1-kHz LED tone transmitter, use 0.01-μF for *C1*.

Simple Touch Switch. A single 4011 provides the nuclei oı up to four momentary touch switches. The switch shown in Fig. 4 includes an ENABLE input. Touch switches make ideal replacements for pushbutton switches in many circuits. The touch wires can be any pair of closely spaced contacts, terminals or exposed wires.

One-Shot Touch Switch. Figure 5 shows a one-shot touch switch which provides a one-second output pulse when actuated. The circuit consists of two cross-coupled gates which, together with *C1* and resistors *R1*, *R2* and *R3*, form a mono-

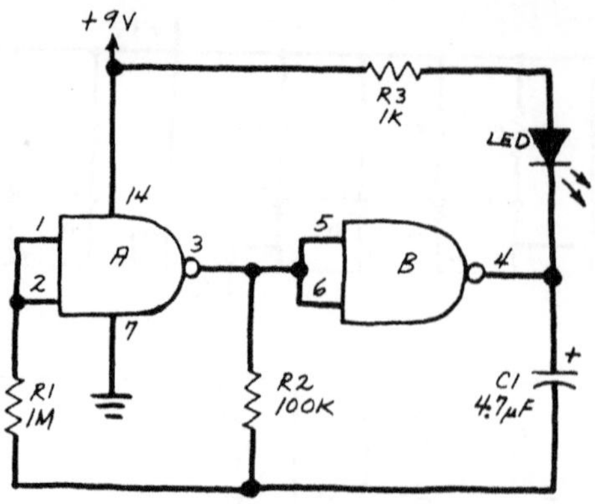

Fig. 3. Schematic of a CMOS LED flasher.

stable multivibrator. For an output pulse of greater duration, increase the value of *C1*.

A one-shot touch switch has many applications. One possibility is to connect its output to the ENABLE input of a CMOS tone generator such as the one shown in Fig. 2. The tone generator will then issue a one-second burst of sound when the TOUCH terminals are bridged by a finger.

Bounceless Switch. The bounceless switch is essential to digital experimentation. A single 4011, two spdt switches and four 100-kilohm resistors can form *two* independent bounceless-switch circuits.

Figure 6 is the circuit for one bounceless switch. Resistors *R1* and *R2* reduce the current spikes that occur when the circuit changes states. Actuating the switch forces the latch formed by the two gates to assume the appropriate logic state irrespective of any contact bounces which the switch produces. Without the latch, the bounces of the switch mechanism would be interpreted as individual input pulses by the logic circuits that the switch is intended to control.

X10 Linear Amplifier. A unique application for CMOS gates which has no TTL counterpart is *linear amplification.* Figure 7, for example, shows an ultra-simple ×10 voltage amplifier made from a single gate in a 4011. The voltage gain of the circuit is determined by the ratio of the value of *R2* to

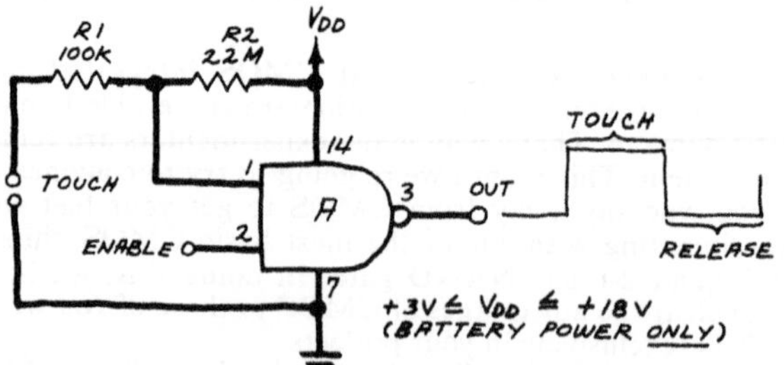

Fig. 4. Simple touch switch includes ENABLE input.

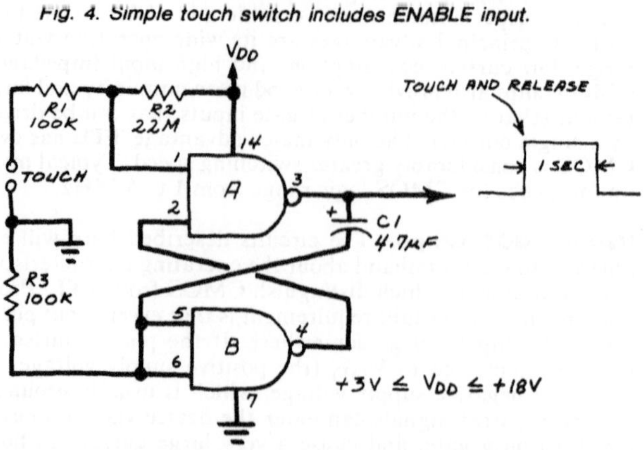

Fig. 5. One-shot touch switch provides 1-second output.

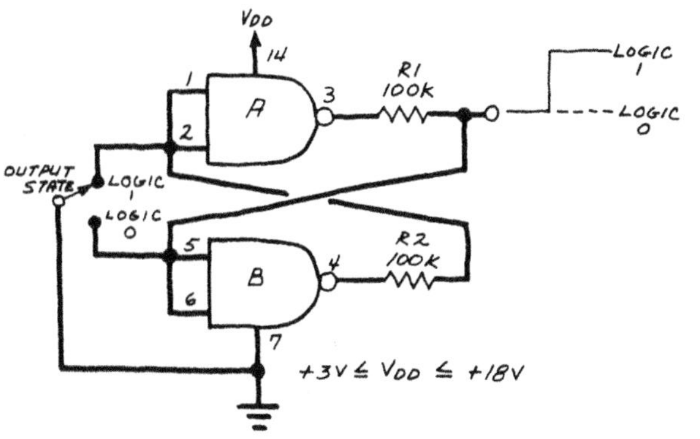

Fig. 6. Bounceless switch necessary for digital experiments.

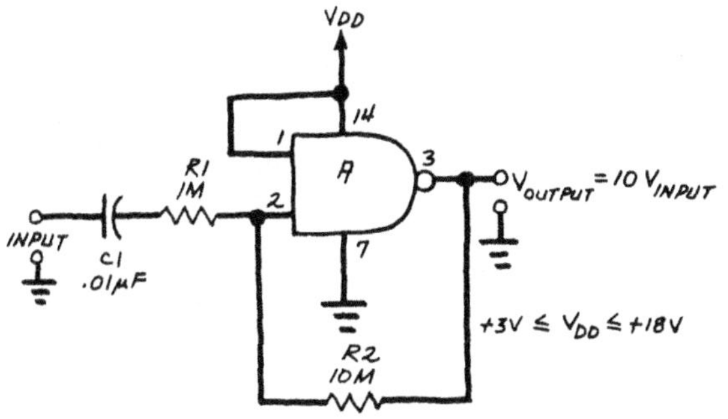

Fig. 7. A X10 linear amplifier using one gate in a 4011.

that of $R1$. In this circuit, therefore, the gain is 10.

The chief advantages of CMOS gate amplifiers are convenience, simplicity, and high input impedance. They can easily replace op amps in such applications as gain blocks for frequency counters and other circuits that require input buffering and amplification. High-frequency response is about 1 MHz. The voltage gain can be set as high as 50 by increasing the value of the feedback resistor ($R2$ in Fig. 7).

Dual-LED Flasher. The dual-LED flasher in Figure 8 illustrates several important CMOS applications. The first is the use of cross-coupled gates to form an astable multivibrator. This configuration is very similar to the latch shown in Fig. 5. The capacitors have been added to provide astable operation. The second application is the use of very large capacitances to give a time constant or flash rate of about 1 Hz. The rate can be slowed even more by further increasing the values of $C1$ and $C2$.

The third application that is illustrated by this circuit is the use of output buffers to interface LEDs to the circuit. Without the buffers, the LEDs might disrupt the operation of the oscillator. The buffers, of course, are formed from the two unused gates in the 4011.

Miscellaneous Logic Functions. Figure 9 sums up several of the fundamental logic functions which can be achieved by interconnecting the gates in a single 4011. Perhaps these functions can save a chip or two in a CMOS design you have in mind if you have some unused 4011 gates on your breadboard. Or, they might come in handy when you need a specific logic function but don't have the chip that incorporates the gate to perform it. Indeed, with the exception of three-state logic and transmission gates, one or more 4011's can be used to form virtually *any* conceivable logic function, including that of a flip-flop, a latch, a counter, a decoder or even a memory circuit. ■

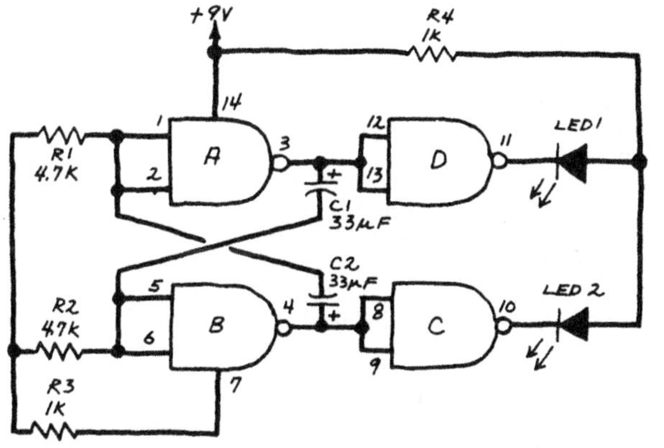

Fig. 8. Dual-LED flasher uses cross-coupled gates.

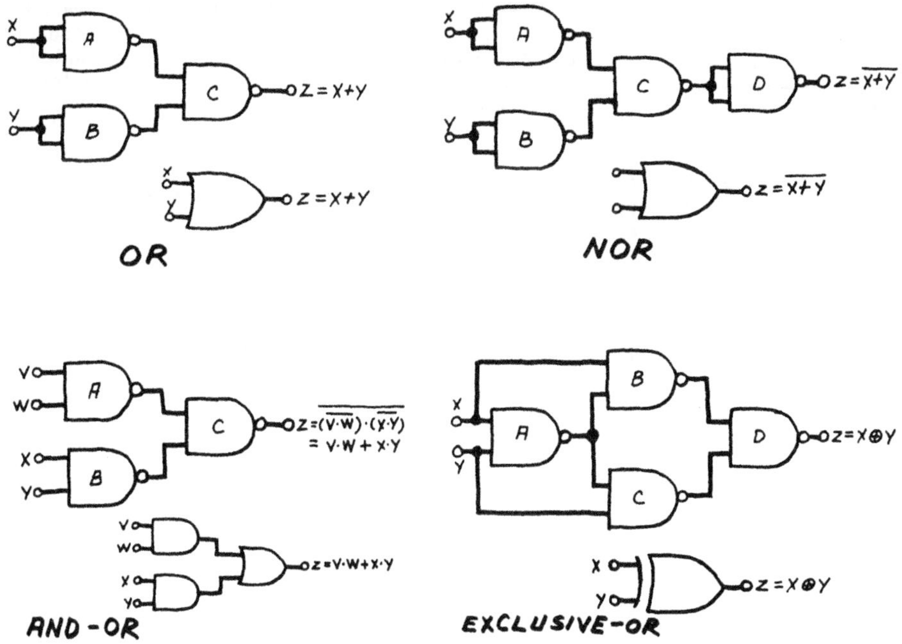

Fig. 9 Logic diagrams of some of the fundamental functions that can be achieved by interconnecting gates of a single 4011.

MODEL railroad enthusiasts can add a touch of realism to their layout with this flashing signal light. It includes two red LEDs which begin flashing alternately when a train approaches and continue flashing until the train has passed. The signal does so no matter which direction the train is moving.

Circuit Description. Figure 1 is the schematic diagram of the crossing light. Phototransistors *Q1* and *Q2* are installed on either side of the protected crossing as shown in Fig. 2. Both transistors are illuminated by a dc-powered light source. A single incandescent lamp can illuminate both *Q1* and *Q2*, or individual lamps or infrared LEDs can be used. Fig. 3 shows how to connect a pair of infrared emitters so that a forward current of 30 mA will flow.

When both *Q1* and *Q2* are illuminated, the inverting inputs of both op amps are slightly above ground. The outputs of *IC1* and *IC2* are at +9 volts. These outputs are applied to NAND *IC3A* whose output is at logic 0 under these conditions. This disables the dual LED flasher comprising gates *IC3B* and *IC3C*. Gate *IC3D* turns off *LED2* when the flasher is disabled. If *LED2* were connected directly to the output of *IC3C*, it would glow continuously when the flasher was disabled.

Should the light path between the light source(s) and either *Q1* or *Q2* be blocked by an arriving train, the output of gate *IC3A* goes high. This allows the dual flasher to oscillate, and LEDs 1 and 2 begin flashing alternately. The flasher continues to oscillate as long as either *Q1* and *Q2* or both are blocked. When the train has passed and light again strikes both *Q1* and *Q2*, the flasher circuit is again disabled.

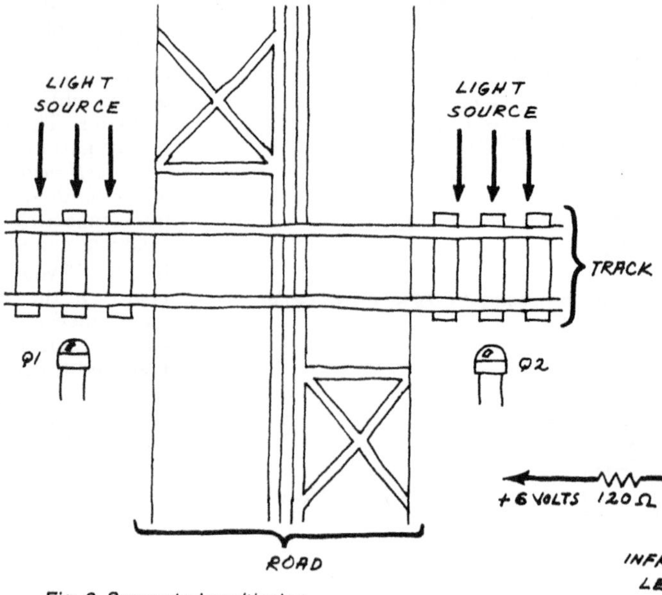

Fig. 2. Suggested positioning for phototransistors for railroad crossing lights.

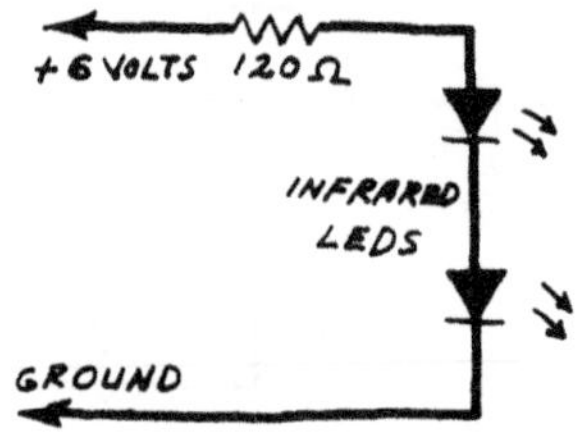

Fig. 3. A possible infrared illuminator circuit.

Using the Circuit. Operation of the circuit is fairly straightforward. If the flasher fails to operate properly, substitute a 0.1-μF disc capacitor for *C1*. If the LEDs then flicker rapidly when the flasher is enabled, the electrolytic capacitor you previously used as *C1* is the culprit. Reconnect it to the circuit, making sure to observe its polarity. If the LEDs still fail to flash, use another electrolytic capacitor.

If your train layout is brightly lighted, you may need to place light shields or infrared bandpass filters over the apertures of the phototransistors. One-inch lengths of quarter-inch black heat-shrinkable tubing work well. Unexposed, processed pieces of Kodachrome or Agfachrome color slide film make excellent, inexpensive infrared bandbass optical filters. They can be secured with glue.

Be sure the distance between the two phototransistors is either greater or less than the length of one of your pieces of rolling stock. Otherwise, the warning signal might be deactivated if both phototransistors are simultaneously illuminated by light passing between adjacent cars. Also, install the phototransistors at heights low enough so that the light paths will be blocked by low-lying rolling stock such as flatcars.

Going Further. For best results, you should install the two LEDs in a miniature replica of a signal-crossing light. Avoid the temptation of increasing the brightness of the LEDs by reducing resistance of *R5*. If you want the LEDs to flash more brightly, parallel two of the gates in a second 4011 to drive each LED. You can

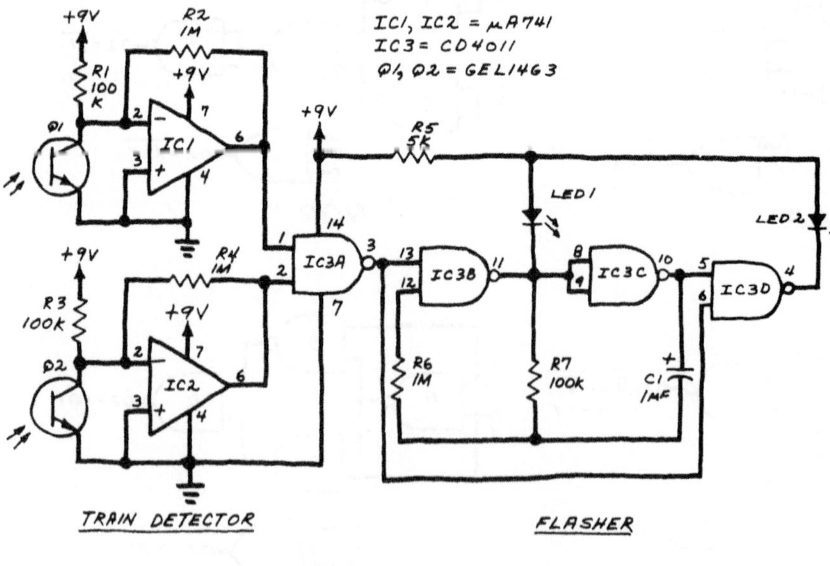

Fig. 1. Schematic diagram of a circuit for a model railroad crossing signal light.

then use series resistors of, say, 270 ohms to limit the forward current through each LED to approximately 25 mA. This will result in plenty of brightness.

Referring again to Fig. 1, note that the outputs of *Q1* and *Q2* are compared and then the result put through a NAND gate. Separate op amps are used to reduce the chance of stray noise pick-up. You can combine the two op amps into a single chip to reduce the parts count by using a dual op amp. A more efficient modification, however, is to eliminate one op amp entirely by wire-ANDing the outputs of the phototransistors.

Fig. 4 shows how this is done. Phototransistors *Q1* and *Q2* are connected in series. NAND gate *IC3A* of Fig. 1 is converted to an inverter to achieve the NAND operation required for proper control of the flasher circuit.

Flip-flops and timer chips can also be used to make railroad crossing lights. Perhaps you can use them to design custom circuits on your own. ■

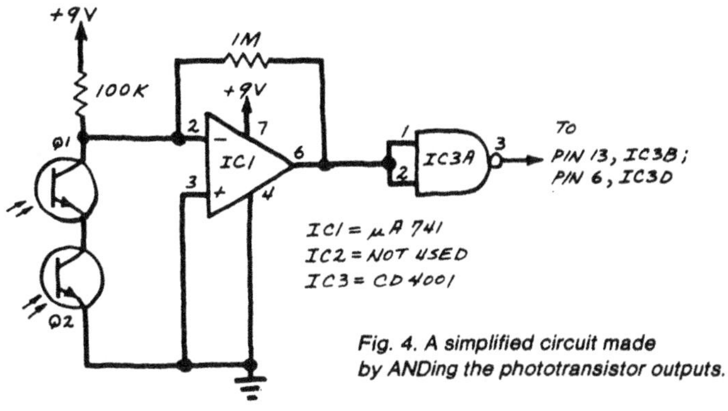

Fig. 4. A simplified circuit made by ANDing the phototransistor outputs.

11. Simple BCD Keyboard Encoder

THE November 1978 "Project of the Month" was a hexadecimal keyboard encoder assembled from four TTL chips (a 7400, a 74173 and a 74193). Many circuits require only one decade of decimal entry (0 through 9). The circuit in Fig. 1 implements this function with only three readily available CMOS chips. Because this new circuit is a CMOS design, its power consumption is considerably less than the TTL circuit presented earlier.

In operation, an astable oscillator made from two cross-coupled inverters (*IC1A* and *IC1B*) supplies clock pulses to *IC2*, a 4017 decade counter/decoder, and to *IC3*, a 4518 dual BCD counter. Initially, both counters are disabled by the application of appropriate logic levels to their respective enable inputs (a logic 1 at pin 13 of *IC2* and a low at pin 10 of *IC3*). The LED readout, therefore, displays the status of the outputs of *IC3* immediately after power is applied.

The keyboard is activated by closing any of the ten input switches *S0* through *S9* and then toggling RESET switch *S10* from ground to $+V_{DD}$ and back to ground. If desired, the BCD output can be cleared to 0000 (all LEDs glowing) by toggling RESET switch *S10* prior to selecting a data input switch.

Assume *S3* is closed. All inactive outputs of the 4017 are *low*, so the keyboard (*S0* through *S9*) bus goes low and enables both *IC2* and, via *IC1C*, *IC3*. Both counters then begin a synchronized count of the pulses, applied to their CLOCK inputs.

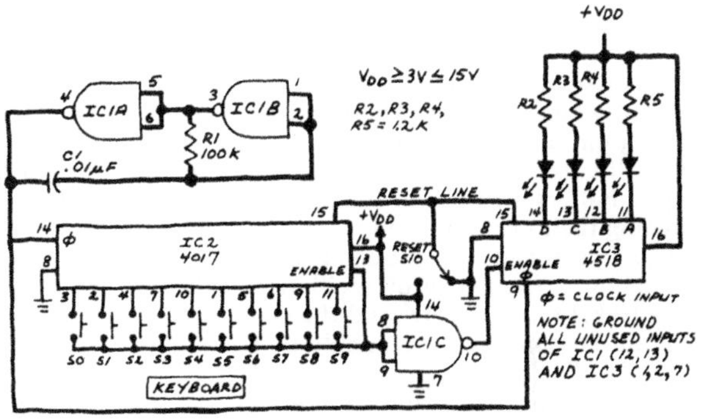

Fig. 1. A BCD keyboard encoder circuit using CMOS chips for low power consumption.

When the fourth clock pulse has been counted, pin 7 of *IC2* goes to logic 1 and, via closed switch *S3*, disables *both* counters. The LEDs then display the BCD equivalent of the selected switch: 0100. Counter *IC3* stores and presents at its outputs the BCD equivalent of the selected switch, even if the selected switch is opened and another is closed. Only after *S10* has been momentarily toggled will a new switch closure be detected and indicated by the output LEDs.

What happens if two or more input switches are closed when *S10* is toggled? The first closed to be scanned by the 4017 is selected. This is a form of *priority encoding*. The October 1978 "Experimenter's Corner" described the operation and use of the 74147 priority encoder. This TTL chip accepts up to ten inputs and presents at its outputs the BCD equivalent of the highest priority or most significant input while ignoring all others.

Going Further. The output LEDs shown in Fig. 1 are optional. They permit the operation of the circuit to be verified but are unnecessary in many practical applications. Of course, they can be retained. Alternatively, the outputs can be decoded by a BCD-to-seven-segment decoder/driver such as the 4511 or 4543 for display on a LED or liquid-crystal readout.

The basic circuit shown in Fig. 1 can also be modified for different applications. For example, recall that the 4518 contains *two* BCD counters, only one of which is used. The second counter can be clocked in parallel with the first (and the 4017) to provide a storage register which can remember a previous keystroke.

Other modifications may require the addition of one or more chips. For example, a 4066 quad bilateral switch can be connected to the outputs of the second counter to provide a 3-state output. ■

12. Precision CMOS Clock Generator

SEQUENTIAL digital logic circuits require one or more clock pulse generators. Microprocessors often include built-in clock generators. Other sequential circuits may use clocks made from 555 timers, a pair of cross-coupled inverters, or a trio of inverters connected as a ring oscillator.

Intersil makes a general-purpose timer chip which, for a CMOS device, has extraordinary specifications. The chip is the ICM7209, available from some electronics mail order suppliers for about $4.00.

The ICM7209 is guaranteed to oscillate at frequencies up to 10 MHz, and it can directly drive as many as five TTL gates. With a 5-volt power supply, the chip typically consumes 11 milliamperes and will operate with a minimum of three external components—two capacitors and a quartz crystal (Fig. 1).

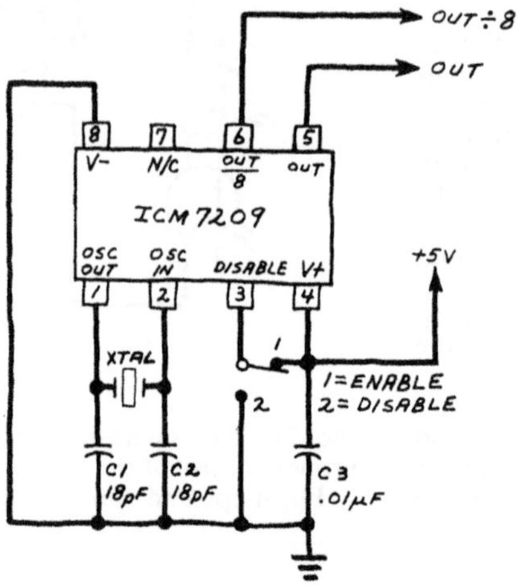

Fig. 1. A clock pulse generator using the ICM7209 chip with just three external components.

The power dissipation of the ICM7209 is directly related to its oscillation frequency. Since the oscillator portion of the chip consumes much less power than its output buffers, power dissipation can be dramatically reduced when the chip is disabled by making pin 3 low. The oscillator portion will continue to operate, but the output buffers will be disabled, thus reducing their current drain.

The crystal can be any quartz crystal having a frequency of oscillation from 10 kHz to 10 MHz, and the circuit can be powered by a supply of 3 to 6 volts. For best results, the crystal should have a load capacitance of 10 pF rather than the usual 30 pF. When *C1* and *C2* are 18 pF, this provides a typical frequency stability of one part per million (ppm) per one volt change in supply voltage.

Note that the ICM7209 includes two output pins. The divide-by-eight output (pin 6) can be used to obtain many combinations.

While Fig. 1 shows the disable input (pin 3) connected to a switch, disable/enable controls can be easily provided by external logic. Pin 3 can also be connected to either the oscillator IN or OUT pins for some interesting results. For example, when pin 3 is connected to pin 2, each of the divide-by-eight pulses appearing at pin 6 are further divided into four separate pulses. This provides a burst output mode not mentioned in the ICM 7209's data sheet.

Figure 2 illustrates this chip's operation at its maximum guaranteed frequency of 10 MHz. Intersil claims typical rise and fall times of 10 nanoseconds (25 nanoseconds maximum) as measured from the 0.5-to-2.4-volt output points. These represent TTL logic levels.

As you see in Fig. 2, the circuit in Fig. 1 has a risetime better than 8 nanoseconds and a falltime faster than 7.5 nanoseconds. The pulse width is 50 nanoseconds FWHM (full width, half maximum). Figure 2 was taken directly from the screen of a 100-MHz oscilloscope. I assembled the oscillator on a standard plastic, solderless breadboard with short, point-to-point connection wires.

The ICM7209 provides an excellent solution to the need for a precision clock generator. Though the circuit shown in Fig. 1 isn't tunable, the oscillation frequency can be quickly altered by changing the quartz crystal used. ■

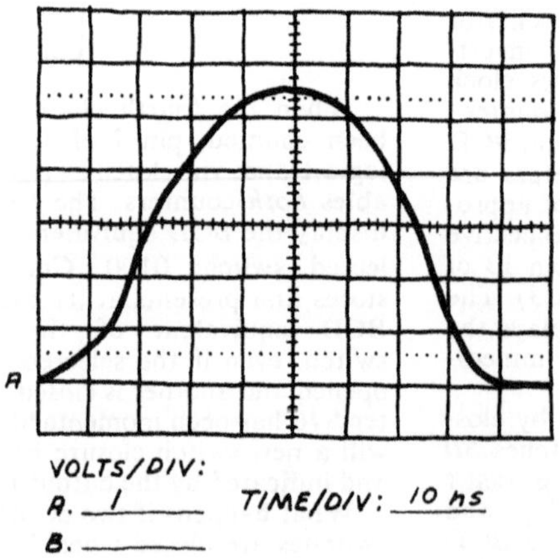

Fig. 2. Output waveform from an ICM7209 with a 10-MHz crystal.

13. Experimenting with High-Speed Logic

HOW WOULD you like a flip-flop that can switch states 500-*million* times in a single second? Flip-flops this fast actually exist and are used in ultrafast computers, communication interfaces for computers, high-speed phase-locked loops, and high-performance controllers.

Ultrafast flip-flops are representative of a family of logic circuits characterized by nanosecond switching speeds. The family is called *emitter-coupled logic* or simply ECL.

I first became interested in ECL while pondering the possibility of measuring the time light takes to travel from a miniaturized laser transmitter to a nearby reflective surface and back. Dividing the elapsed time in half and multiplying the quotient by the speed of light gives the distance from the laser to the surface.

In one second, light travels 299,800,000 meters, or 984,000,000 feet, or 186,280 miles. Put another way, light travels about one foot in one nanosecond (0.000000001 second). Since I wished to measure the distance to objects a few feet, or few tens of feet, distant, nanosecond resolution would be required for successful use of the time-of-flight method.

In a typical time-of-flight optical radar, the transmitter emits a fast-rising, very short light pulse while simultaneously enabling a high-speed counter. Reflected light from the target illuminated by the transmitted pulse is returned to a photodetector, then shaped and amplified. The resultant signal stops the counter. Half the elapsed time stored in the counter provides the time-of-flight from transmitter to target.

The fastest ECL gates change states in a nanosecond; thus ECL is suitable for making the high-speed gate and counter of a time-of-flight optical radar. Though I have not yet designed a practical short-range time-of-flight system, I have experimented with a number of ECL circuits designed around a quad NOR gate. Before having a look at how they work, let's find out more about ECL.

A Typical ECL Gate. The circuit and logic symbol of a typical three-input ECL OR/NOR gate is shown in Fig. 1. Depending upon your point of view, you can think of the circuit as an OR gate with a complementary (NOR) output or a NOR gate with a complementary (OR) output.

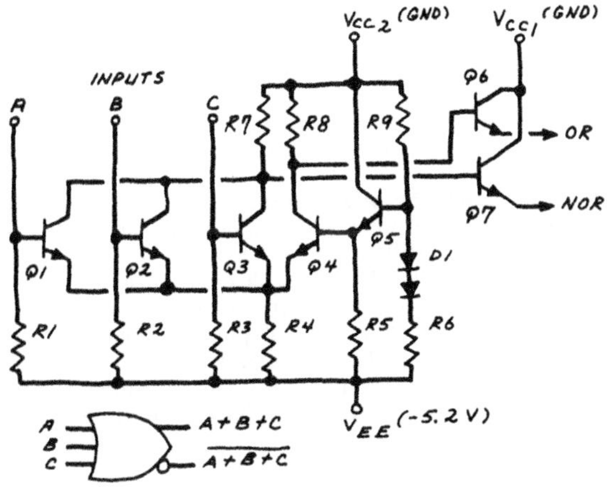

Fig. 1. An emitter-coupled logic (ECL) 3-input OR-NOR gate.

In the instance of the OR gate, the complementary NOR output eliminates the necessity for an external inverter and avoids propagation delays that such an external inverter would add. In either case, the complementary outputs make possible a number of interesting design shortcuts which can reduce circuit complexity and gate count.

In operation, input transistors *Q1-Q3*, together with *Q4*, form a differential amplifier. The bias network composed of *Q5, R5, R6, R9, D1*, and *D2* sets the switching threshold for the differential input amplifier.

If the base voltages at *Q1, Q2* and *Q3* coincide with the voltage at the base of *Q4*, then the current flow between V_{CC} and V_{EE} will divide between the transistors. If, however, the voltage at input A (*Q1*) is increased about half a volt above the reference voltage at the base of *Q4*, then *Q3* will turn on and the current flow will be diverted away from *Q4* and flow through *Q3*. The same applies to inputs B (*Q2*) and C (*Q3*).

Output transistors *Q6* and *Q7* form a complementary pair that monitors each half of the differential amplifier. Should *Q1, Q2* or *Q3* receive an input signal of sufficient amplitude, *Q7* will be turned on. Otherwise, *Q6* is turned on. Since only one side of the differential amplifier can be on at any time, when *Q6* is on, *Q7* is off, and vice versa.

The *transfer curves* for a typical ECL gate are given in Fig. 2. These curves show both the switching thresholds and the high and low logic levels. Note that the difference between an EDL low (−1.75 volts) and high (−0.9 volt) is only 0.85 volt. This means a conventional ECL gate cannot be interfaced directly with TTL logic (where a low is less than 0.8 volt and a high is more than 2 volts). Instead, special ECL circuits called *TTL translators* must be used to interface ECL with TTL.

Note that the ECL logic levels in Fig. 2 are *negative* voltages. This is in accordance with the ECL convention in which

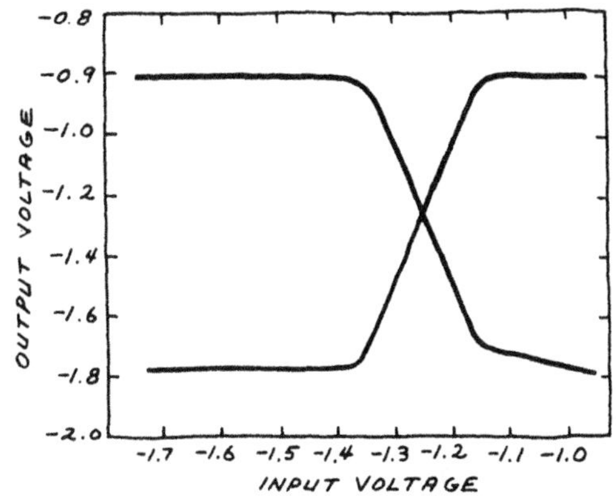

Fig. 2. Transfer curves of a typical ECL gate. The difference between a high and a low is only about 0.85 volt.

V_{CC} is at ground potential and V_{EE} is −5.2 volts. This convention can be reversed so that V_{EE} is at ground potential and V_{CC} is +5.2 volts. However, maintaining V_{CC} at ground potential provides much better noise immunity since any V_{EE} power supply noise becomes a common-mode signal that is cancelled by the differential input amplifier.

ECL Advantages. The principle advantage of ECL is its speed, but it offers other benefits also. One is the very desirable combination of high input impedance and low output impedance. This means a single ECL gate output can drive many ECL inputs. In other words, ECL has a large *fanout* capability.

Another important advantage of ECL is its ability to drive transmission lines and twisted pairs *directly*. This is a result of the open emitter output at an ECL gate (see Fig. 1).

Still another ECL advantage is that unused inputs need not be connected to V_{CC} or V_{EE}. This is because each input is connected internally to V_{EE} via a 50,000-ohm resistor (*R1-R3* in Fig. 1).

Finally, ECL chips have a nearly constant power-supply drain. This greatly simplifies power-supply design and reduces the possibility of noise transients on the supply lines during switching transitions.

Advantages and Drawbacks. ECL circuits have the potential of providing one-nonosecond switching times and propagations delays. Motorola, for example, makes a family of ECL chips called MECL III, having ultrafast operating speeds.

These ultrafast ECL chips require very careful design techniques to avoid uncontrolled oscillation, excessive ringing, and other problems associated with very fast pulses. Wrapped wire interconnections are *not* recommended, and the maximum length of an interconnection should be under one inch.

The 10,000-series ECL made by Fairchild, Motorola, and other companies avoids some of the problems associated with ultrafast ECL by purposely slowing switching times to several nanoseconds and stretching propagation delays to about two nanoseconds. These modifications allow 10,000-series ECL to far exceed the speed of any other logic family while relaxing interconnection requirements. For example, wrapping wire can be used to interconnect 10,00n-series ECL chips so long as connections are less than eight inches in length.

Though 10,000-series ECL is much easier to use than ultra-fast MECL III, attention must still be given to interconnections. Each foot of interconnection inserts a delay of about two nanoseconds. This is approximately equivalent to the propagation delay of an ECL gate.

Transmission lines such as coaxial cables and twisted pairs are ideal for interconnecting 10,000-series ECL over distances of up to 1,000 feet. But if the line is not properly terminated, transmitted pulses will be distorted by considerable leading and trailing edge ringing. Since an ECL output is an uncommitted open emitter, an external resistor to V_{EE} must be added. In a properly terminated transmission line, this resistor is inserted at the *receiving* end rather than the transmitting end. Figure 3 shows the effects on a transmitted pulse under both configurations.

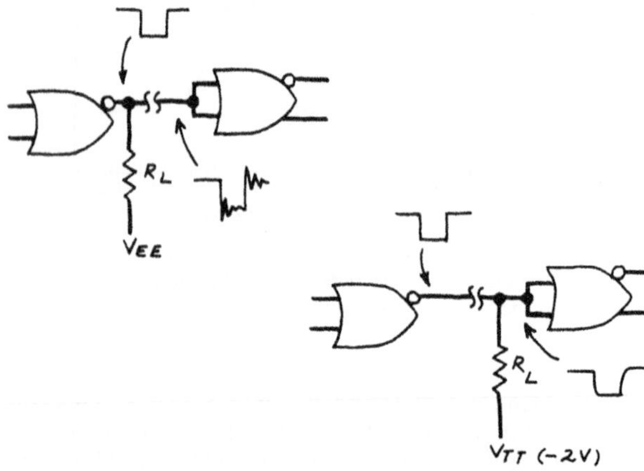

Fig. 3. The effects of an improper (left) and proper termination on a transmission line are evident in the noise on the output signal.

Experimenting with an ECL Quad NOR Gate. A good way to learn about ECL firsthand is to experiment with the 10102 quad 2-input NOR gate. The pin outline for the DIP version of this gate is shown in Fig. 4. As in TTL gate packages, pins 8 and 16 are reserved as power-supply terminals. Pin 1 is also used as a power-supply terminal.

The pin connections to the individual gates are unlike those of any comparable CMOS or TTL gate package. Note in par-

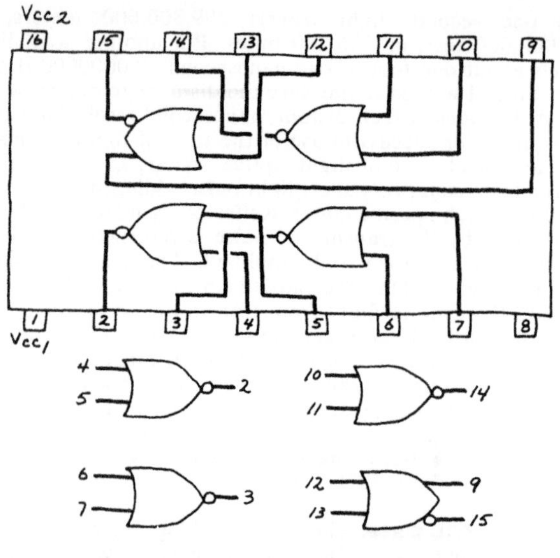

Fig. 4. Pin layout and internal schematic diagrams of the 10102 ECL quad NOR gate.

ticular how the outputs from two gates cross over the inputs of the two adjacent gates.

Finally, note that one of the 10102 gates has complementary outputs. This will give you an opportunity to experiment with this unique feature of ECL gates should you wish to go beyond the simple circuits that follow.

A 78-MHz Oscillator. A straight-forward ECL ring oscillator patterned after similar TTL versions is shown in Fig. 5. The only significant difference is the addition of the required pull-down resistors *(R1-R3)* at each ECL output.

I assembled this simple circuit on a standard solderless breadboard using short lengths of point-to-point connection wire. Power was supplied by a standard TTL power supply.

The output from this oscillator is a 1.6-volt sine wave riding on a 2.6-volt dc level. This means that, while the circuit will easily drive an LED, compensation for the dc level must be provided or the LED will be saturated.

An Ultrafast Schmitt Trigger. The Schmitt trigger is a bistable (two-state) logic circuit with a host of useful applications. Typical uses include threshold detection, signal condi-

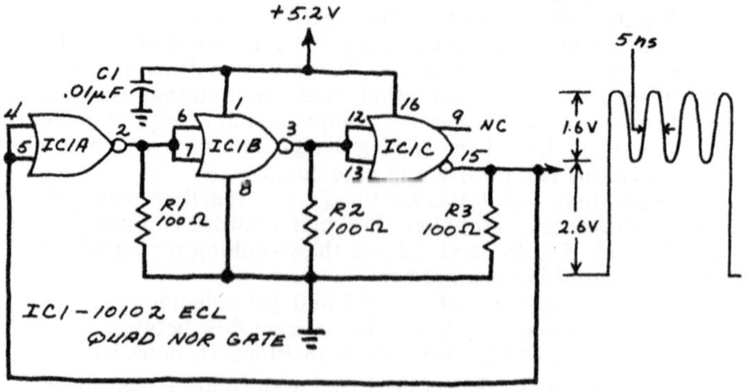

Fig. 5. Schematic of a 78-MHz ring oscillator using ECL. A pull-down resistor is required at each ECL output.

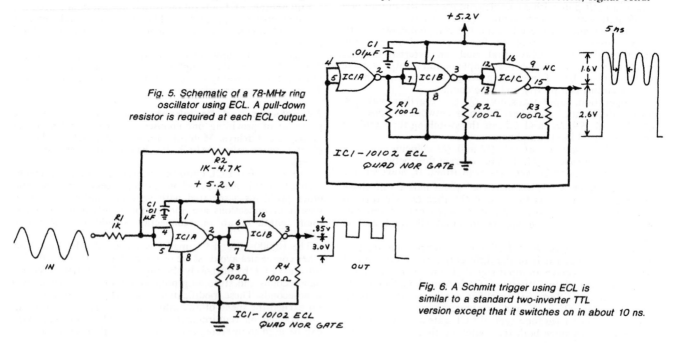

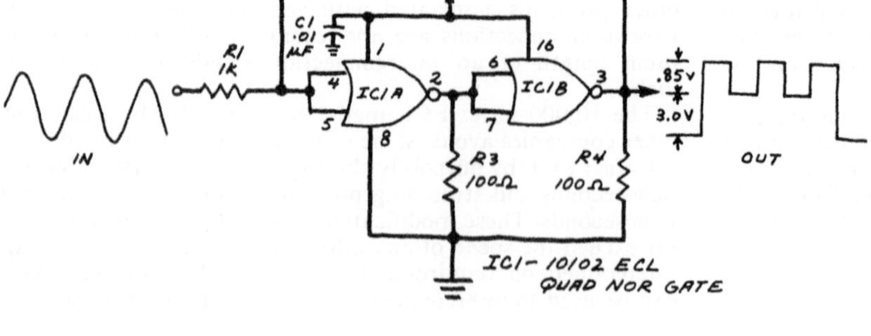

Fig. 6. A Schmitt trigger using ECL is similar to a standard two-inverter TTL version except that it switches on in about 10 ns.

tioning, and sine-to-square-wave conversion. Figure 6 shows a Schmitt trigger designed after a standard two-inverter TTL version. The chief difference is that the ECL version in Fig. 6 switches on in about 10 nanoseconds.

When the signal at the input of the Schmitt trigger is below the circuit's switching threshold, the output is a dc level of 3.0 volts. When the input signal exceeds the circuit's switching threshold of about 3.6 volts, a very fast rising pulse with an amplitude of 0.85 volt is superimposed over the dc output.

Like the oscillator in Fig. 5, the Schmitt trigger was assembled on a standard solderless breadboard using short point-to-point connections. Figure 7 shows the response of the Schmitt trigger to a triangular waveform while Fig. 8 is an expanded view of the Schmitt trigger's output showing a rise and fall time of about 10 nanoseconds at the 10%-90% points.

Other ECL Chips. If you would like to try some more sophisticated ECL circuit designs, a wide variety of standard ECL chips is available. The 10,000 series, for example, includes many different gate packages, flip-flops, decoders, encoders, memories, and other functions.

In the past, some of the parts suppliers who advertise in this magazine have carried some ECL chips. Recently, however, I haven't noticed any ECL chips in their ads. If you have trouble locating a supplier for ECL chips, try manufacturer's representatives. Most big cities have a number of such representatives who can order chips for you. They may even be in stock. Signetics, Motorola, Fairchild, and other companies make ECL chips.

Summing Up ECL. This column provides only a very elementary introduction to ECL. For more information, visit any technical library and review books on digital logic which cover ECL. Even better, get a copy of Fairchild's *The ECL Handbook*. Another excellent manufacturer's handbook is Motorola's *MECL High-Speed Integrated Circuits*. A wide range of ECL application notes is also available from the various ECL manufacturers. ■

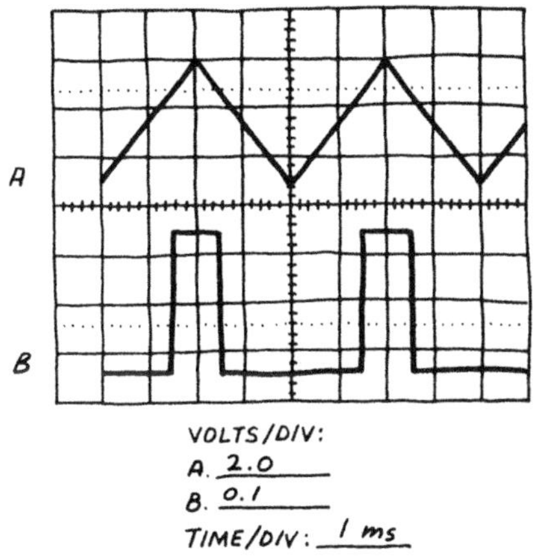

Fig. 7. Response of circuit in Fig. 6 shows fast rise and fall times.

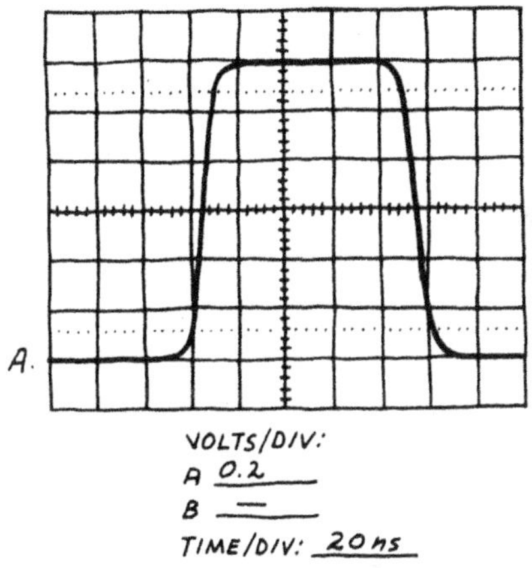

Fig. 8. Expanded view of the output of Fig. 6 with 10-ns rise and fall times.

Seven

Experimenter, Hobby, and Game Circuits

1. Binary Hi-Lo Game

IT'S EASY to learn all sixteen 4-bit binary numbers (0000–1111), but can you *think* in binary? The growing popularity and importance of programmable logic makes the ability to "think binary" very helpful to those who wish to exploit as fully as possible the various tricks and shortcuts made possible by binary number manipulations.

The circuit described here provides a painless way to help learn to "think binary." It's a HI-LO game that uses binary instead of decimal numbers. For consistently good scores, you must know and be able to manipulate binary nibbles.

In operation, the 555 generates a rapid stream of clock pulses at a rate determined by the values of *R1* and *C1*. When normally open pushbutton *S1* is pressed for a few seconds, the 74193 4-bit counter cycles through its count sequence hundreds of times. This means an essentially random number will be stored in the counter when *S1* is released.

After *S1* is released, one of the output LEDs will glow to indicate whether the 4-bit nibble entered on the DIP switch is the same as the nibble in the 74193 or, more likely, if it is too high or too low. If the entry is incorrect, and the odds are 16 to 1 *against* the first entry being correct, a second guess is entered into the DIP switch. This process is continued until the CORRECT LED glows.

The best way to find the correct number with the fewest possible guesses is to make your first entry equal to approximately half the highest possible number. Since there are sixteen possible numbers, your first guess would then be 0111 (seven) or 1000 (eight).

If this guess is too low, then your second guess should be halfway between your first try and 1111. Similarly, if your first guess is too high, your second try should be halfway between your first guess and 0000.

This process is continued until you arrive at the correct number. As you can readily see, the game certainly encourages you to "think binary."

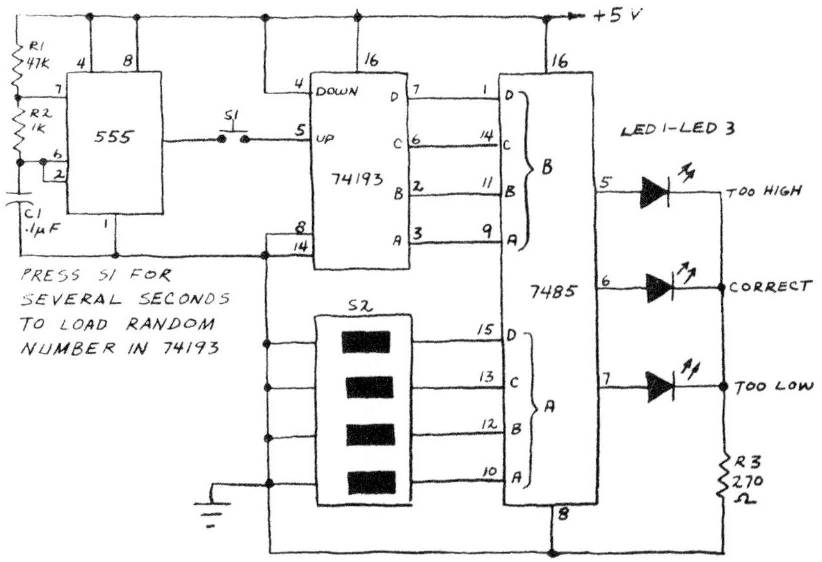

It's easy to make a permanent version of this game on a small circuit board. Use a perforated board with copper solder pads at each hole and connect the components with wrapping wire for fast assembly. Apply wrapping wire directly to the IC and DIP switch pins and use a low-wattage soldering iron and small-diameter solder to secure the wires in place. Power the circuit with four AA cells in a plastic holder. A 1N914 silicon diode connected between the positive battery terminal and the rest of the circuit will drop the voltage down to about 5½ V, the level required by TTL. ■

2. IC Interval Timers

AN INTERVAL timer is a circuit that provides an output pulse of predetermined width at periodic intervals. This can be readily accomplished using any one of several timer ICs available to today's electronic experimenter. Many IC timers, such as the well-known 555, are not only capable of such astable operation but can also function as monostable multivibrators or "one-shots."

Figure 1 is a timing diagram comparing the operation of a monostable to that of an interval timer. Note that a one-shot timer is designed to activate an external device or circuit for or after a fixed period. An interval timer, on the other hand, provides uniform output pulses at an adjustable interval.

You are probably already familiar with numerous applications for conventional one-shot timers. Common examples include automatic switches that extinguish the headlights of a car a minute or so after the ignition is turned off, delayed-action intrusion alarms, switch debouncers, kitchen and darkroom timers, etc.

Although the applications for interval timers are not as numerous, they include two that are particularly interesting: time-lapse photography and time-lapse sound recordings.

You have probably seen many examples of time-lapse photography—the opening of a flower, formation of clouds, construction of a building, etc. Time-lapse sound recordings can store periodic samples of data encoded as an audio tone as well as simply capture ambient sounds. In the latter category, an interesting possibility is to compress a 24-hour history of the sounds at a busy street corner into a one-minute recording. Another is entertaining your family or friends at a party by sampling brief segments of a record, radio program or conversation and playing back the string of sound "snapshots."

Of course, time-lapse photography and sound recordings are not the only applications for interval timers. Before you've finished reading this column, you will probably have thought of several more.

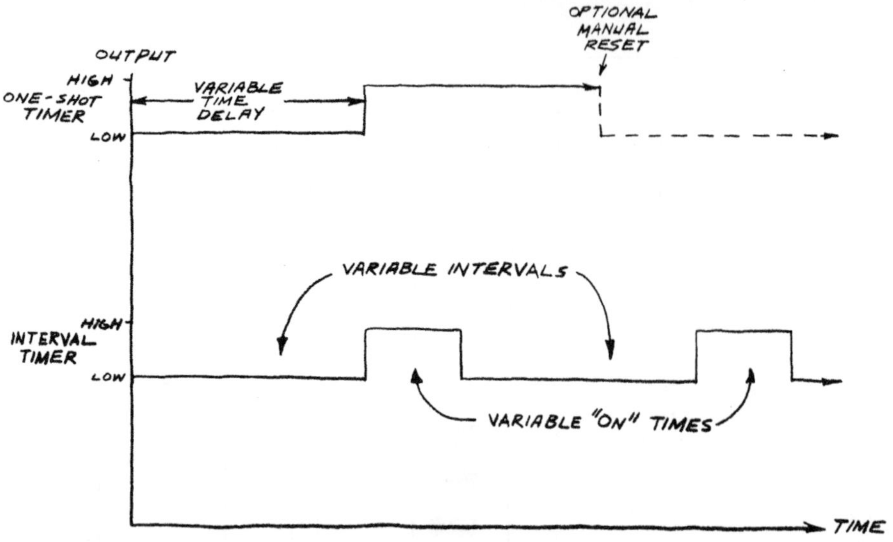

Fig. 1. Timing diagrams of the output waveforms generated by monostable multivibrator or one-shot (top) and interval timer (bottom).

Basic 555 One-Shot. Although most experimenters have assembled either breadboard or permanent circuits that use a 555 timer, many do not fully understand how this IC works. For those of you in this category, the following paragraphs will provide a quick overview of the monostable operation of the 555. If you're already familiar with 555 basics, you can skip ahead to the next section.

Figure 2 is a simplified block diagram of a 555 connected as a monostable or one-shot timer. The key sections of the 555 are the two comparators, *VC1* and *VC2*. They sense when the timing capacitor (*C1*) has charged or discharged to a predetermined level.

To understand how the 555 works, assume the circuit in Figure 2 is "off." This means the control flip-flop is reset and *Q1* is on. Capacitor *C1* is therefore short circuited by *Q1* and cannot charge. The output of the circuit (pin 3) is low. A negative pulse applied to the TRIGGER input (pin 2) momentarily causes the output of comparator *VC2* to go high, setting the control flip-flop. This cuts off *Q1*, which allows *C1* to charge exponentially at a rate determined by the values of *C1* and *R1*. During this period, the output at pin 3 is high.

Notice the three series-connected 5000-ohm resistors in the 555. These resistors form a voltage divider that biases the noninverting input of comparator *VC2* at one-third of the supply voltage and that of comparator *VC1* at two-thirds of the supply voltage. When the voltage across *C1* reaches two-thirds of the supply voltage, the output of comparator *VC1* goes high and resets the control flip-flop. This turns *Q1* on and shorts out *C1*. The output at pin 3 returns to ground and remains there until the entire timing cycle is repeated. This is accomplished by applying a new trigger pulse at pin 2.

This explanation should give you some insight into the operation of the 555 in its monostable mode. It should now be obvious that you can easily select the time delay by the proper choice of components for *R1* and *C1*. If long delays (more than several minutes) are to be obtained, it's important to use

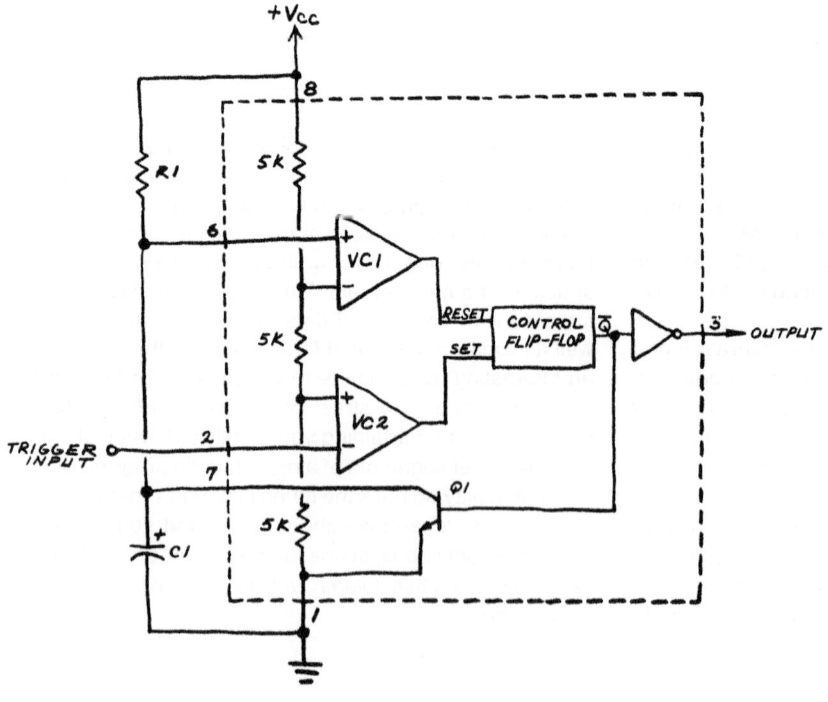

Fig. 2. Simplified functional diagram displays inner workings of a 555 timer IC. External components R1 and C1 control timer's period.

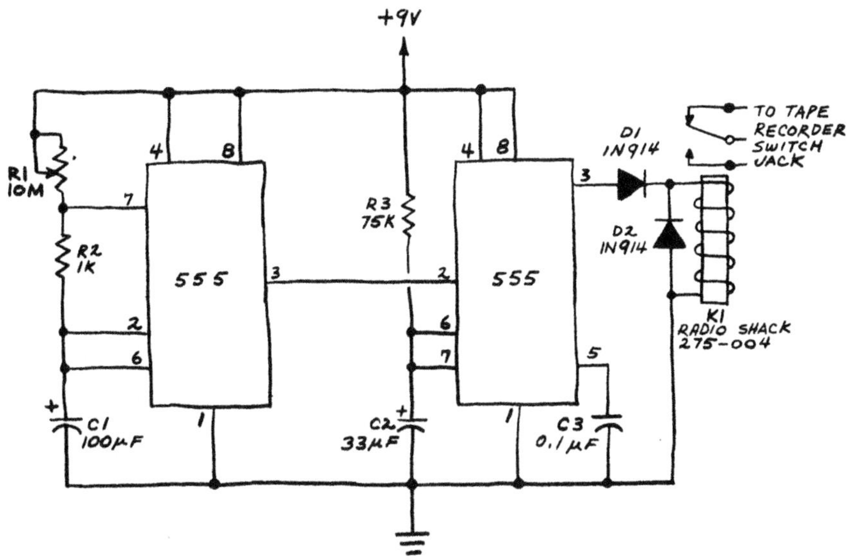

Fig. 3. Interval timer employs one 555 as an astable multivibrator to trigger a second IC operating as a monostable. Relay K1 keys external circuit.

a component with extremely low leakage for *C1*. Otherwise, the capacitor will never be able to charge as it should and the circuit will not function properly.

555 Interval Timer. A 555 monostable can function only as a single-delay timer. A reset pulse is required to initiate a new delay period. An interval timer, however, can be made by connecting the output of a 555 operated as a free-running (astable) oscillator to the TRIGGER input of a 555 monostable. The period of oscillation of the astable will determine the interval time. The *RC* time constant of the monostable will determine the duration of the output pulse that follows each timing interval.

Figure 3 shows the schematic of a working dual-555 interval timer. Interval times (determined by the values of *R1* and *C1*) of up to several minutes are achievable with the values shown. Note that the output pulse from the first 555 is directly coupled into the input of the 555 monostable. The output of the monostable is connected to a low-voltage relay coil through *D1*. Diode *D2* shorts out the powerful inductive kick produced across the relay coil when current to it is interrupted, thereby protecting the 555's output stage from damage.

The values of *R3* and *C2* determine how long the relay is energized after each timing interval. Those specified keep the relay energized for almost exactly 5 seconds (4.98 seconds for the breadboard circuit I built). Change the value of *R3* or *C2* or both to obtain different times.

The relay contacts can be used to switch many different circuits or devices on *or* off. Figure 3, for example, shows the normally-off contacts connected to the switch jack of a tape recorder. This jack is commonly found adjacent to the microphone jack on many cassette recorders. It allows the recorder to be turned on and *off* remotely by means of a small switch such as one mounted on the case of the microphone.

If you want to connect the relay to a tape recorder, use an appropriate plug. You'll have to improvise when connecting the relay contacts to other equipment or circuits. (A few words of caution—*never* connect the relay to a circuit that exceeds the maximum ratings for the relay's contacts. Also, *never* switch ac line power with an unenclosed relay. Personally, I prefer to play it safe with low-voltage applications only.)

556 Interval Timer. The 556 is a pair of 555 timers on a single silicon chip. The circuit in Figure 3, as you might suspect, can be readily assembled with a single 556 dual timer rather than separate 555's. Figure 4 shows the functionally identical circuit.

XR-2240/555 Long-Duration Timer. Because of leakage in the timing capacitor, the maximum period of a 555 operated as an astable oscillator is usually limited to several minutes. The XR-2240 (or XR-2340) is a specialized IC timer that incorporates a self-contained flip-flop divider chain to increase the length of the fundamental time delay by a factor of up to 255. Because the output of each flip-flop in the chain is directly accessible, many different time intervals can be selected without having to alter the values of the circuit's timing capacitor and resistor.

Figure 5 is the schematic of a long-duration, programmable interval timer made from an XR-2240 connected as an astable oscillator and a 555 operated as a monostable. Timing components *R1* and *C1* control the oscillation rate of the XR-2240. The values shown give an adjustable interval *T* of up to about

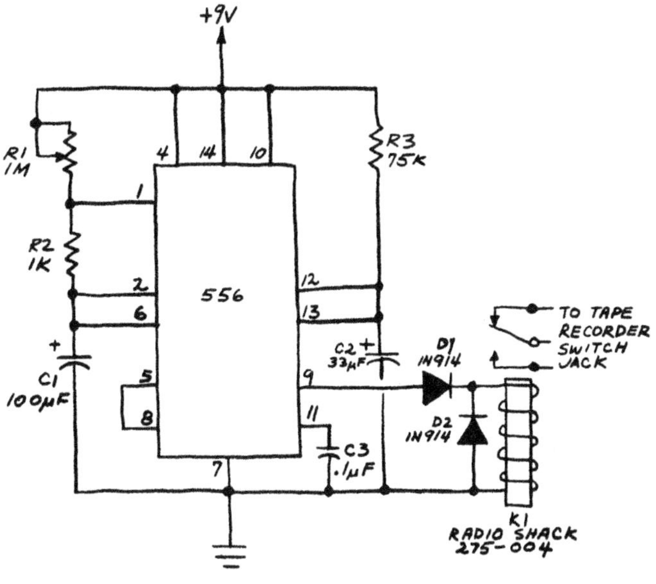

Fig. 4. This circuit, which employs a 556 dual timer, is functionally identical to the one shown in ***Fig. 3.***

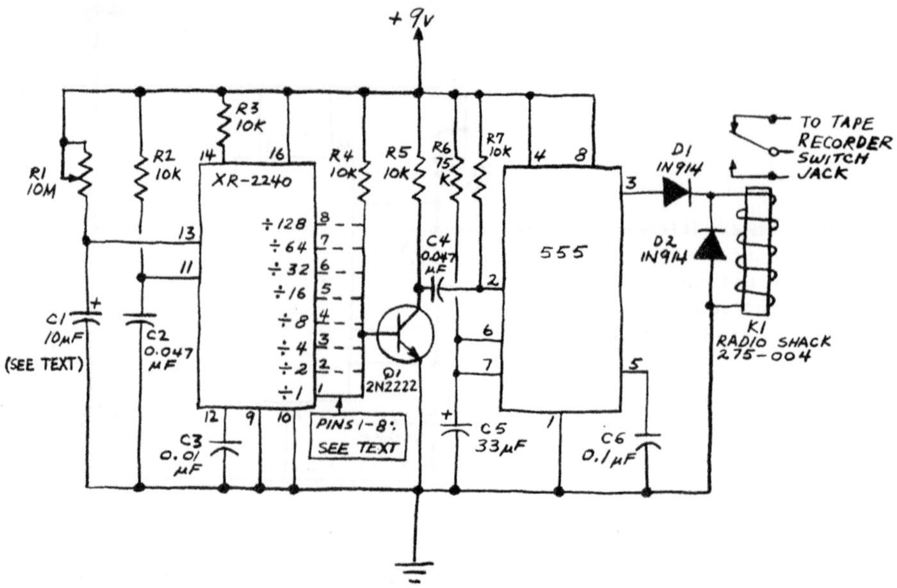

Fig. 5. Long-duration, programmable interval timer employs XR-2240 as an astable and 555 as a monostable. Relay K1 keys external circuit.

100 seconds. The outputs at pins 1 through 8 allow you to select multiples of *T* ranging from 1 to 128. Therefore, selecting pin 8 will give you a time delay of up to 128 × 100 seconds or more than 200 minutes!

The selected output of the XR-2240 is inverted by *Q1* and coupled through *C4* to the 555 monostable, a circuit essentially identical to the monostable in Figure 3. The timing period of the monostable is controlled by the time constant *R6 C5*.

The XR-2240/555 interval timer is far more versatile than the dual 555 or 556 version because intervals of several hours can easily be obtained. Calibrating the circuit, however, can pose problems if you attempt to perform the operation when output pin 8 is selected. Calibration is much easier if you select output pin 1. If, for example, you want a timing interval of one hour (3600 seconds), adjust *R1* until the interval at pin 1 is 28.13 seconds. Pin 8 will then output a pulse at 128 × 28.13 seconds or every 3600 seconds.

Incidentally, it's possible to select various combinations of XR-2240 outputs to achieve any time interval of from *T* to *255T* when the chip is operated in its triggered, monostable mode. However, this procedure does not give the desired results when the astable mode is used.

It might be possible to obtain the full versatility of the XR-2240 by operating the chip in its one-shot mode and triggering it externally. The XR-2240 would continue to trigger the 555 one-shot to provide the brief "on" time after each interval. The time delay would be selected by shorting combinations of outputs to a common bus. The delay would be the sum of the delays of the selected outputs. Thus, outputs *4T*, *8T* and *128T* will give a total delay of 4 + 8 + 128 or *140T*.

I'll leave the details to those readers who like challenges. See the XR-2240 data sheet for design tips. ■

3. Universal Tri-State Tone Generators

AMONG the most popular applications of tone generators are those of annunciators and alarms. The tri-state tone generator shown in Fig. 1 is more versatile in this use than most because it has three principle operating modes: steady, pulsating, and two-tone. It can be easily modified to produce a warbling sound and can generate a wide range of audible frequencies.

Each half of a 556 dual timer functions as an astable multivibrator. The first astable, whose timing components are *R1*, *R2*, and *C1*, oscillates at a frequency of slightly more than two hertz. The second astable, which drives a small 8-ohm dynamic speaker, is programmed by *R4*, *R5*, and *C4* to oscillate at a frequency of 2.5 kHz. Resistor *R6* governs the volume of the sound from the speaker.

The three principle operating modes are selected by *S1*, an spdt toggle switch with a neutral (off) center position. Position 1 connects the output of the first astable to the second astable through *R3*. The result is an attention-getting, two-tone signal whose frequency fluctuates between 2200 and 2500 Hz at a rate determined by the rate of oscillation of the first astable.

Position 2 disconnects the first astable from the second astable, allowing the second to operate independently. Consequently, the speaker emits a steady 2500-Hz tone. Position 3 connects the output of the first astable directly to the reset input of the second astable. This causes the 2500-Hz tone applied to the speaker to be interrupted at a rate determined by the first astable. The result is a series of tone bursts.

You can experiment with the timing components of both astable multivibrators to achieve a wider range of tone modes and frequencies than those described above. Increasing the capacitance of *C1* to 10 microfarads or more, for example will reduce the frequency of oscillation of the first astable to approximately 0.7 hertz. On the other hand, reducing *C1* to 0.45 microfarads will increase the first astable's oscillation frequency to about 15 Hz, causing a distinct warble to be heard when *S1* is in position 1 or a rapid series of tone bursts when *S1* is in position 3.

The frequency of the second astable can be made adjustable by replacing *R4* and *R5* with a 15,000-ohm potentiometer. To do so, connect the wiper of the potentiometer to pin 13 of the 556 and the stationary terminals to pins 8 and 14. You can make one or both astable responsive to changes in the level of ambient light by substituting a cadmium-sulfide photocell for one or more of the timing resistors.

Digital Tone Mode Selection. It's possible to select the tone mode electronically with the help of a 4051 CMOS analog multiplexer/demultiplexer. Figure 2 shows how the 4051 is introduced into the circuit

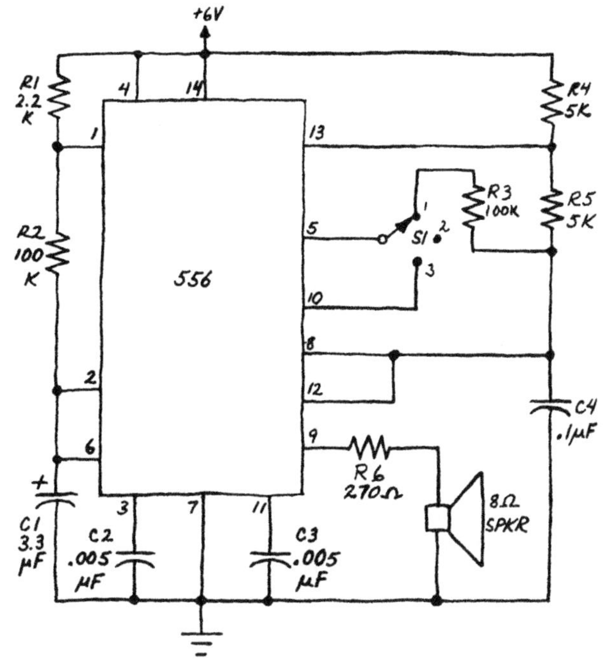

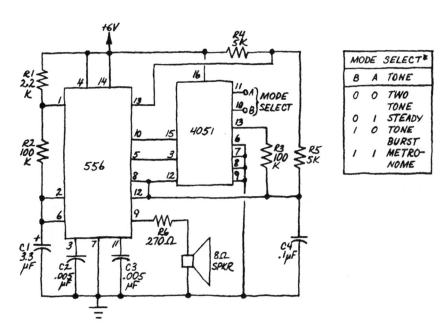

Fig. 1. Universal tri-state tone generator.

in place of *S1*.

A two-bit word selects the desired tone mode according to the truth table included in Fig. 2. Note the addition of a fourth tone mode, a ticking sound similar to that of a metronome. This sound represents the output of the first astable and its frequency can be altered by changing *R1*, *R2*, *C1*, or any combination of these components. When the metronome mode is selected, the second astable is disabled by the 4051.

You can modify the truth table in Fig. 2 by connecting any three of the eight analog switches in the 4051 to the rest of the circuit. Refer to the 4051 data sheet for details of the operation of this versatile chip.

Going Further. With a little ingenuity, you can select the desired tone mode—or turn the circuit on or off—with components other than the 4051. Try optoisolators, SCRs, transistors, or relays. You might even be able to add a tri-state LED to the circuit to provide simultaneous audible and visual outputs. ■

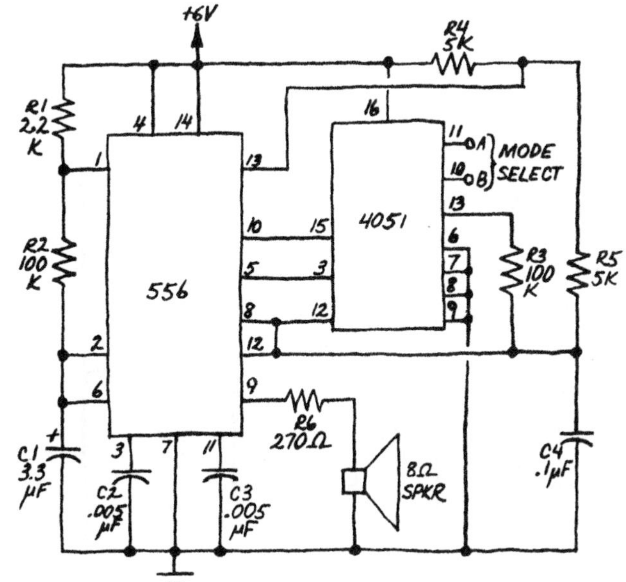

MODE SELECT*		
B	A	TONE
0	0	TWO TONE
0	1	STEADY
1	0	TONE BURST
1	1	METRONOME

Fig. 2. Programmable four-state tone generator

4. Pseudorandom Number Generator

MANY GAMES and statistical calculations require the generation of random numbers. Spinners and dice are often employed as mechanical random-number generators in games. Software routines are commonly used to generate random numbers for computer games and statistical computations.

A simple way to generate random numbers electronically is to manually apply a brief burst of high-speed clock pulses to a counter as shown in Fig. 1. Although this method utilizes electronic components, the "random" number selection is in large part dependent upon the interval of time that the switch allowing clock pulses to reach the counter is pressed. Ideally, the clock pulses will occur much too rapidly for the person closing the switch to anticipate the output when the switch is opened.

Figure 2 is a working version of the block diagram shown in Fig. 1. The counter is a 4017 CMOS chip with a built-in decoder that activates one of ten LEDs numbered 0 through 9. The clock could be a 555-timer or simple, two-inverter astable multivibrator. I decided to use an LM331 voltage-

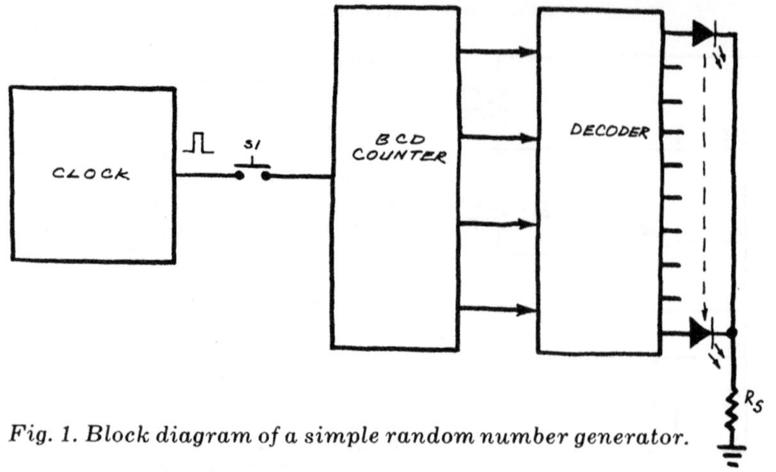

Fig. 1. Block diagram of a simple random number generator.

to-frequency converter to permit the addition of a gradual slowdown feature that reinforces the impression of randomness in the typical observer.

With a conventional clock circuit, the pulse train to the counter will be interrupted immediately upon the opening of *S1*, and the random number will be displayed before the operator's finger is lifted from the switch. In the circuit in Fig. 2, however, depressing *S1* for a second or two charges *C1* through *R1* to a voltage less than or equal to the supply voltage. The voltage across *C1* controls the output frequency of the LM331. Once *S1* has been released, *R2* begins to discharge *C1*, and the decreasing voltage across *C1* decreases the oscillation frequency of the LM331.

When the frequency of the LM331 is high, the LEDs connected to the counter switch on and off so rapidly that to the human eye they all appear to be glowing. As the clock slows down, however, the LEDs begin to flicker. Only one LED glows at any instant when the clock rate slows to a few pulses per second. Eventually, *C1* is completely discharged, the clock stops and a single LED remains glowing. If the LEDs are arranged in a circle, the overall visual effect is reminiscent of a wheel of fortune.

The critical components in this circuit are *C1*, *R1* and *R2*. Larger values of *C1* and *R1* will increase the time required to charge *C1* as well as the likelihood that *C1* will have charged to a random voltage after *S1* has been closed for an arbitrary time. Increasing the value of *R2* will increase the time required for the flickering LEDs to gradually settle down, thus enhancing the visual impression of apparent randomness. If *R2* is too large, however, *C1* may take a long time to fully discharge.

Is the output of this circuit genuinely random? The average of 100 trials should be 4.5 if the resulting numbers are perfectly random. I obtained an average of 4.38, a difference of 2% on the low side. The standard deviation of a perfectly random sample would be 3.03. Mine was 2.95.

Actually, a more careful analysis will reveal that the results are not nearly as random as might be desired. If the results were perfectly random, each of the ten LEDs would be selected an equal number of times or ten times each for a sample of 100 trials. Here are my results:

LED Number:	0	1	2	3	4	5	6	7	8	9
Observed*:	8	15	8	14	10	8	10	3	12	12
Expected*:	10	10	10	10	10	10	10	10	10	10

**Number of times observed or expected in 100 trials.*

As you can see, my operation of the circuit favored 1, 3, 8 and 9 and discriminated against 0, 2, 5 and especially 7. While Chi-square and other statistical tests can be used to determine how random the selected numbers are, clearly the results are not nearly as random as the simple averaging test initially indicated. Thus, the circuit is called a *pseudorandom* number generator.

Perhaps you can improve the randomness of the circuit's output by increasing the number of trials and experimenting with the values of *C1*, *R1* and *R2*. You might also want to add a digital readout to the circuit. This can be done by substituting a BCD counter, 7-segment decoder, and 7-segment LED display for the 4017 counter/decoder and string of LEDs. ■

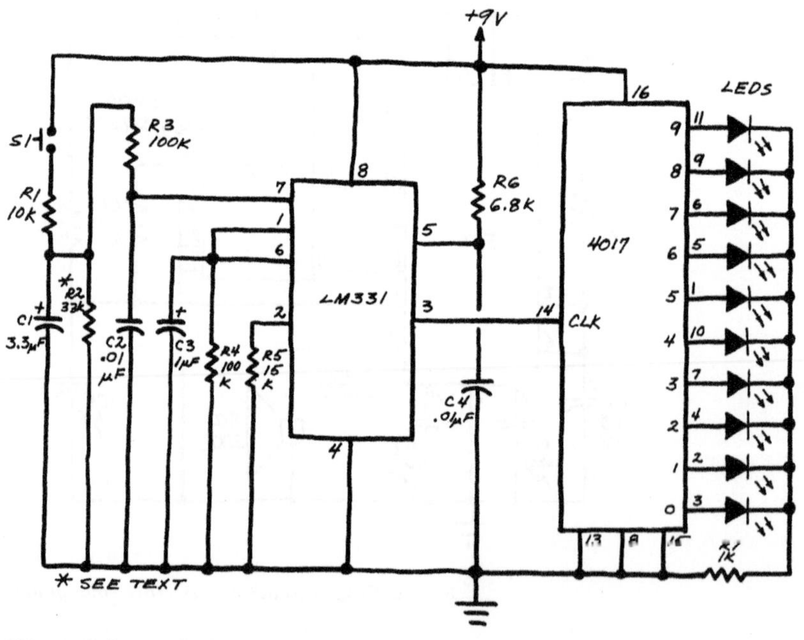

Fig. 2. Schematic diagram for a random generator circuit.

5. Pocket Color Organ

THIS LITTLE color organ has only a thousandth of the light power of its conventional counterparts, but it's ideal for solo viewing. I've also found it to be an effective attention-getting device at small gatherings.

As shown in Fig. 1, the circuit consists of three active filters which separate the audio input signal into low, medium and high frequencies. Each filter drives three series-connected LEDs. I chose red for the low frequencies, yellow for the middle range and green for the high frequencies.

The red and yellow LEDs are driven by bandpass filters. With the component values I chose, the red filter peaks at 20 Hz and has a total passband of 1 to 70 Hz. The yellow filter peaks at 80 Hz and has a

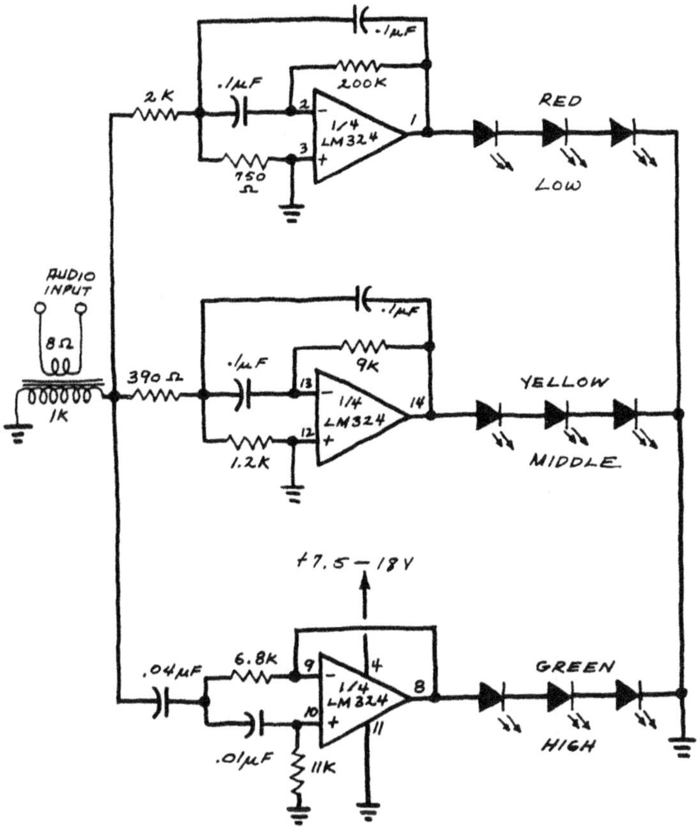

Fig. 1. Schematic diagram for a miniature color organ.

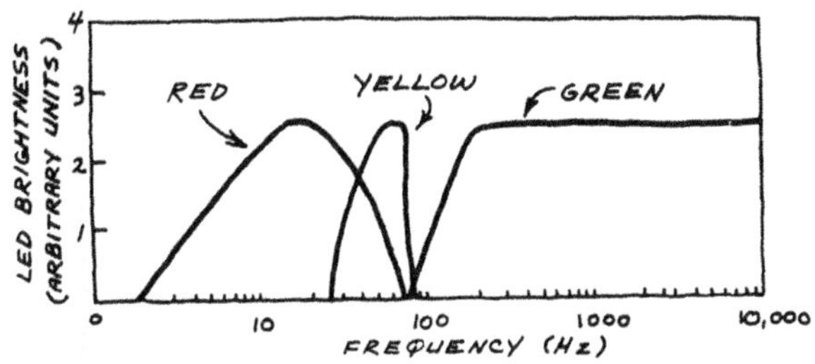

Fig. 2. Frequency response of LED color organ.

total passband of 25 to 85 Hz. The green LED is driven by a high-pass filter with a response which extends from about 75 Hz to beyond the limits of audibility.

Figure 2 shows the frequency response of the three filters superimposed on the same graph. The overlap of filter responses can be eliminated by altering the frequency responses of the filters. I intentionally included the overlap, however, to prevent the possibility of all the LEDs going dark simultaneously.

As clearly indicated in Fig. 2, the circuit favors the low end of the audio spectrum. While I've found this gives an excellent visual representation of most music, you might want to alter the frequency response of one or more filters for other effects.

For best results, assemble the circuit in a light-tight enclosure. Make a window for the LEDs from a thin sheet of translucent plastic or ground glass, materials sold at many camera stores.

Figure 3 shows the enclosure I selected, a French bicycle light sold under the trade name "Wonder." It or a similar bike light of domestic manufacture can be purchased from most bicycle stores.

I removed the light's reflector and switch assembly and installed a circuit board cut with a nibbling tool to fit the interior of the case. The LEDs were grouped in three red-yellow-green triangles as shown in Fig. 4. You can, of course, select your own color and pattern arrangements. The visual impact of the LEDs is much greater if they are viewed through a translucent screen. You can transform the plastic flashlight lens from transparent to translucent by lightly buffing it with fine emery paper.

I used point-to-point wiring to interconnect the various components. The flashlight case didn't have sufficient space for batteries, so I taped two 9-volt batteries to the back of the case and connected them to the circuit via a short cable and a couple of battery clips. The circuit will work when powered by a 9-volt supply, but 18 volts gives more light and provides better response at low volume settings.

A two-conductor cable with two miniature phone plugs soldered in parallel at one end and connected to the input transformer's 8-ohm primary is used to route audio signals to the color organ. Insert one plug in the phone jack of a transistor radio.

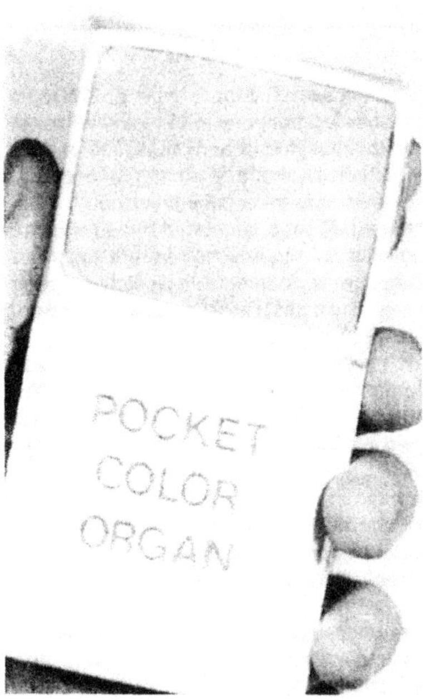

Fig. 3. Prototype assembled in a bicycle light.

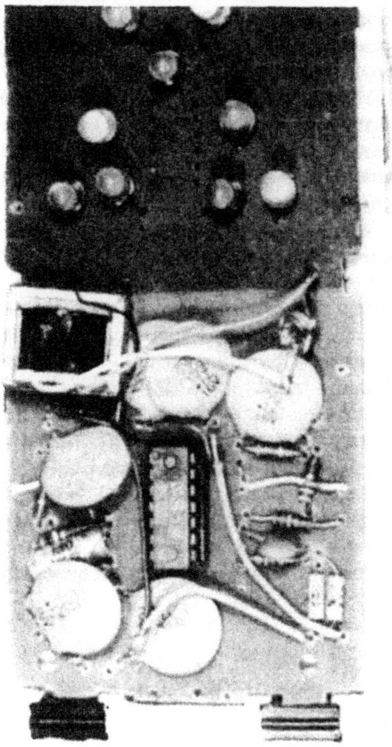

Fig. 4. Prototype internal assembly. Transformer is cemented to circuit board.

The other plug should go to a jack connected to a small monitor speaker box so you can hear the music while you're viewing it. Alternatively, you can defeat the radio's speaker cutoff by rewiring its earphone jack.

Next, dim the room lights, tune in some good music and enjoy the show. You can adjust the radio's volume and tone controls to alter the visual effects. And don't worry about power consumption. The circuit draws only 3 to 5 mA from a single 9-volt battery or 5 to 12 mA from two series-connected 9-volt batteries. ◇

6. More on Pseudorandom Number Generators

Coin Tosser. Figure 1 is a CMOS coin tosser which lights, with virtually the same probability, either of two LEDs. The oscillator, which is made from the two gates, operates at a frequency of about 1 MHz. Pressing *S1* for a second or two allows the oscillator to run, which in turn toggles the flip-flop at a rate far too fast to second guess which LED will glow when switch *S1* is released.

If you use red LEDs, label one heads and the other tails (or label one left and the other right). You can avoid labels by using a red LED and a green LED. Furthermore, you can add an additional element of chance by inserting a cadmium sulfide photoresistor between *R1* and point "x". Subtle variations in the light striking the cell will alter the oscillator frequency.

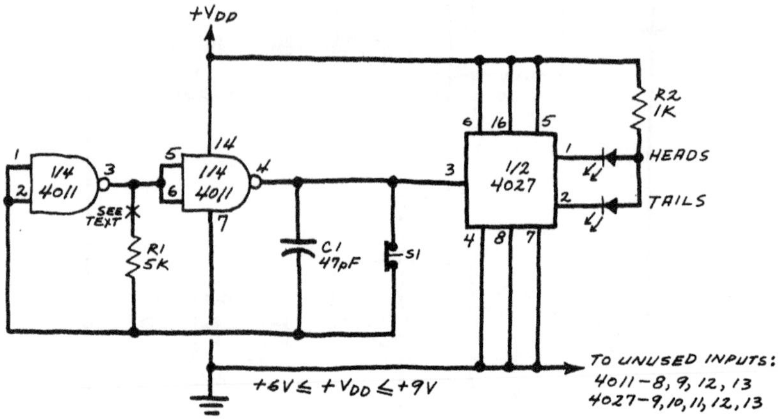

Fig. 1. Schematic diagram for a coin tosser using CMOS circuitry.

Pseudorandom Number Generator. If you can't find an LM331 for use in the December 1979 Project of the Month, try the circuit shown in Fig. 2. Although this circuit won't exhibit wheel-of-fortune operation (that is, press the button and all LEDs glow; release the button and the LED cycle gradually slows until only one remains on), it produces results that are just as random.

The oscillator portion of Fig. 2 is identical to the oscillator shown in Fig. 1. The remainder of the circuit is identical to the decade counter/decoder section of the December circuit.

Label the LEDs 0 through 9 (or 1 through 10). To operate the pseudorandom number generator, press *S1* for a second or two to activate the oscillator. All the LEDs will glow as they are scanned sequentially by the decoder. When *S1* is released, one of the LEDs remains on.

You can modify this circuit for digital readout by substituting a BCD counter/decoder and a seven-segment LED or liquid crystal display for the 4017 and its string of LEDs. Current consumption is low, so the circuit will operate for a long time when powered by a 9-volt alkaline cell. ■

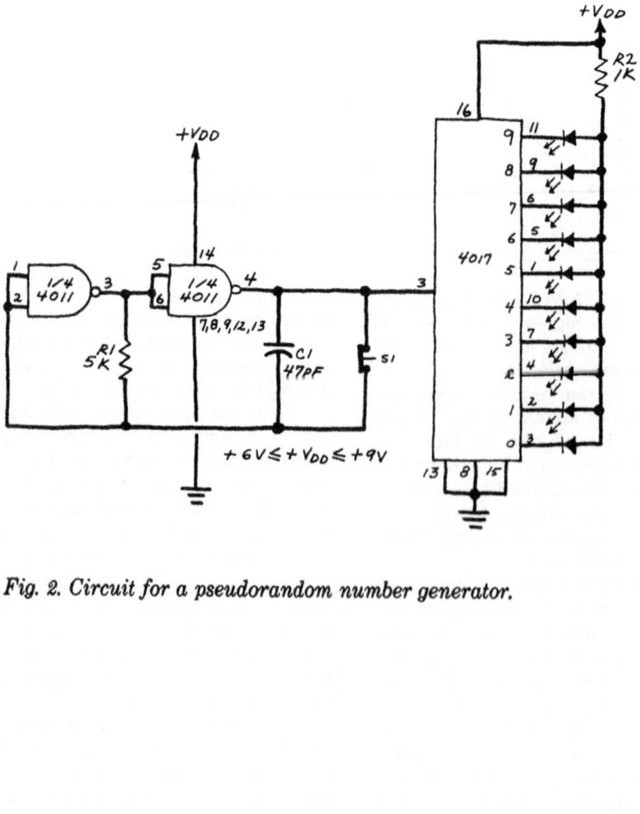

Fig. 2. Circuit for a pseudorandom number generator.

7. General-Purpose Utility Amplifier

A GENERAL-PURPOSE utility amplifier can be as useful to the experimenter as a VOM or an oscilloscope. Typical applications include signal tracing, listening to weak audio-frequency signals detected by an appropriate transducer, and monitoring subtle amplitude and frequency changes in an audio-frequency circuit undergoing test or adjustment.

Figure 1 is the schematic diagram of such an amplifier that you can assemble from readily available parts. In operation, the 741C serves as a high-gain preamplifier. Potentiometer *R1* controls the gain of the preamplifier and *C1* is the input coupling capacitor. The LM386 is a power amplifier which drives a small speaker or an earphone. Potentiometer *R2* serves as a level control. Capacitor *C2* sets the gain of the LM386 at 200. It can be reduced to 20 by omitting *C2*.

Signal Tracing. To use the amplifier as a signal tracer, connect a clip lead to the grounded side of the input and a probe or small alligator clip to *C1*. Use shielded cable if the probe lead is more than a few inches long. You can then follow a signal through an audio amplifier or the audio portion of a radio receiver by clipping the ground clip to the chassis or ground of the circuit under test and using the probe to follow a signal through the circuit. Be sure to reduce the gain of the amplifier when tracing a signal which has been subjected to a few stages of amplification. *Caution! Never* use this circuit to trace a line-powered ac/dc radio, phonograph or amplifier! Also, remember that touching the "hot" side of the ac power line or the terminals of a charged power-supply filter capacitor can be *fatal.*

Sound Detection. Connect a crystal, dynamic or electret microphone to the amplifier's input terminals by means of a shielded cable. For long-range, directional sound detection, mount the microphone at the focus of a plastic or metal parabolic reflector. Suitable reflectors include "saucer sleds" and certain plastic food containers, and hubcaps. For short-range, directional use, install the mike at one end of a hollow pipe or tube.

Induction Receiver. Connect a telephone pickup coil to the amplifier's input and you can detect electromagnetic signals from motors, switches, power lines and some electronic watches. This principle is used in some museums to broadcast taped messages to visitors equipped with a receiver comprising an audio amplifier, transducer, and pickup coil. The signal is received when the visitor walks near a large coil antenna driven by an endless loop tape player.

Omni-Frequency Radio Receiver. The addition of a simple tuned circuit and detector will enable the utility amplifier to receive AM radio signals. A suitable germanium diode connected to a simple loop antenna (Fig. 2) allows the reception of signals up to several gigahertz. Transmitters broadcasting in this region include direction finders, radars, earth-space telemetry, radiosondes, and various broadcast, amateur and other telecommunications.

The antenna can be a commercial uhf television loop (the type that attaches directly to the antenna terminals of a TV set) or a homemade version. For best results, experiment with different diodes and antenna configurations. You will find that the orientation of the antenna and its location with respect to large metal objects and electrical equipment greatly affect the receiver's performance.

Going Further. You can spend many entertaining hours experimenting with various other transducers connected to the input of the amplifier. A silicon solar cell will let you "hear" the modulated light emitted by a flickering candle, an automobile headlight (when the engine is running or the vehicle is on a bumpy road), fluorescent lights or a calculator's LED display. An old phonograph cartridge will convert the surface roughness of various objects into sound. ■

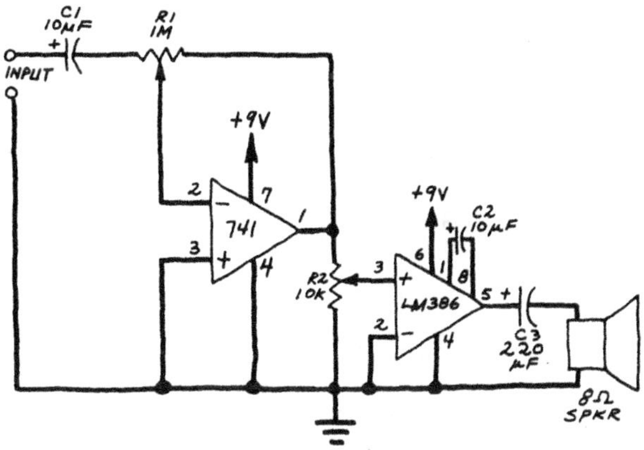

Fig. 1. A simple general-purpose utility amplifier.

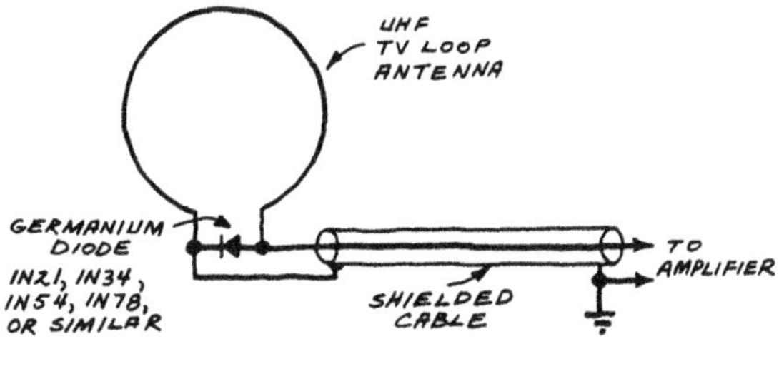

Fig. 2 A broadband pickup.

THE Hall effect, discovered in 1879, is the production of a voltage drop across a conductor or semiconductor through which a current is flowing under the influence of a magnetic field at right angles to the direction of current flow. Several types of semiconductor components that employ the Hall effect have been designed, one of which is the Hall-effect digital switch.

Figure 1 is the pinout and block diagram of a UGN-3020T Hall-effect switch. The chip, which is manufactured by the Sprague Electric Company, includes a self-contained amplifier that boosts the voltage generated by the Hall sensor and presents it to a Schmitt trigger. When the output of the amplifier exceeds a certain threshold, the Schmitt trigger turns on the output transistor.

The hysteresis of the Schmitt trigger prevents the circuit from oscillating when the amplifier output is near the turn-on threshold. In other words, the Schmitt trigger turns off only when the intensity of the magnetic field falls well below the level required initially to turn the Schmitt trigger on. Figure 2 summarizes the circuit's operation.

As you can see by referring back to Fig. 1, the UGN-3020T includes its own voltage regulator. This permits the chip to be powered by a supply furnishing from 4.5 volts to as much as 20 volts. Typical current consumption is 12 mA when the supply voltage is 12 volts.

Figure 3 shows a simple circuit that you can use to experiment with the UGN-3020T or similar Hall-effect switch. The UGN-3020T is available from Sprague distributors. Alternatively, you can use Radio Shack's No. 276-1646 Hall effect switch because its specifications are identical to the UGN-3020T.

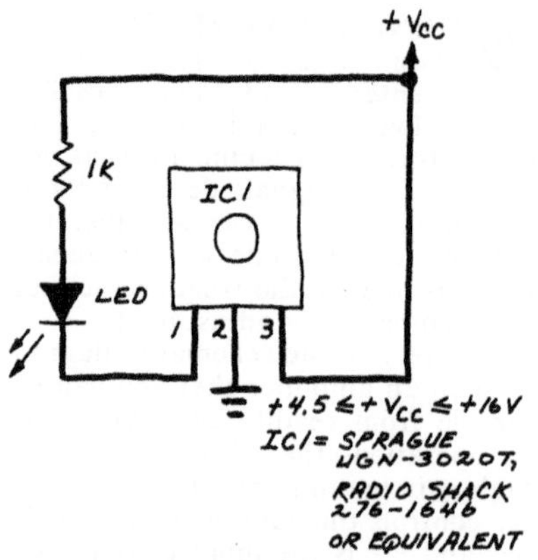

Fig. 3. A simple circuit showing operation of a Hall-effect switch.

Figure 4 shows the optimum orientation of the trigger magnet with respect to the Hall-effect switch. To actuate the circuit shown in Fig. 3, place a magnet close to the circle on the package of the Hall-effect switch and adjust its position with respect to the switch package until the LED begins to glow. When this occurs, the south pole of the magnet will be closest to the switch.

To turn the LED off, move the magnet away from the chip. Note that the hysteresis of the switch allows the magnet to be moved several times more distant from the chip before the LED goes dark than the spacing at which the LED began to glow.

You can use many different kinds of magnets to activate the switch. The flexible magnets used in refrigerator door gaskets work, but not very well. The samples I tried (which are more than ten years old) had to be placed within a millimeter of the chip package's surface before the LED would begin to glow.

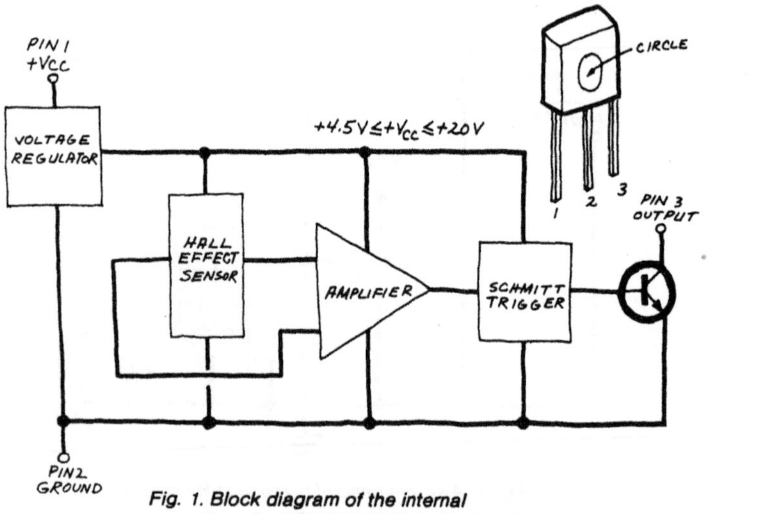

Fig. 1. Block diagram of the internal operation of a UGN-3020T Hall-effect switch.

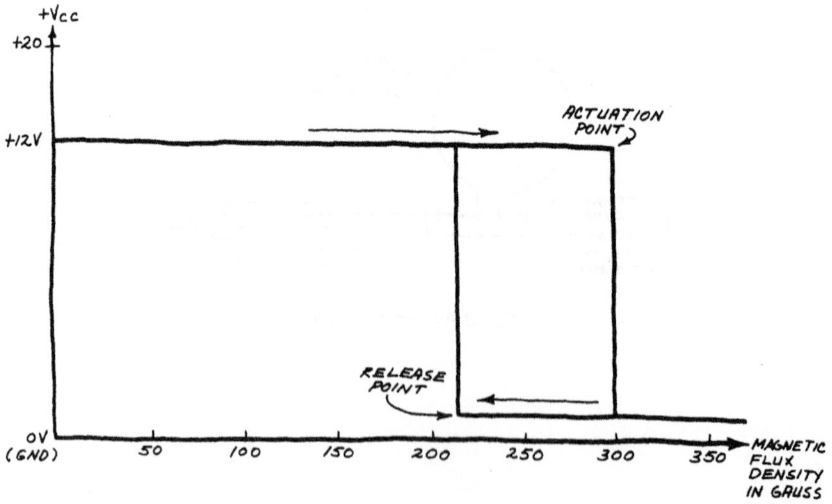

Fig. 2. Switching action, showing hysteresis, of a UGN-3020T.

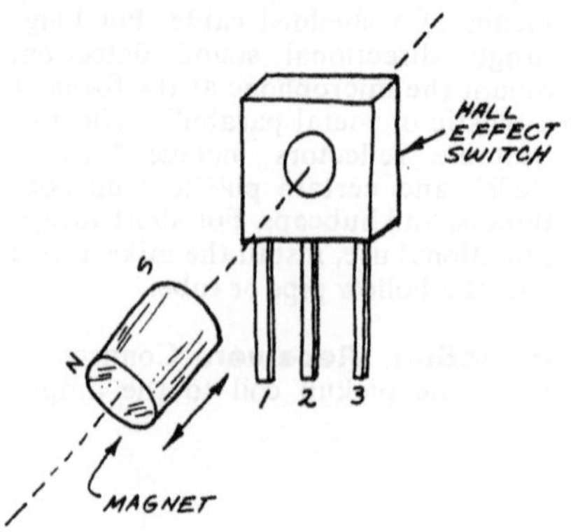

Fig. 4. How to orient a magnet with respect to a Hall-effect switch

Metal magnets work much better. Some that I have tried will cause the LED to glow when the magnet is several millimeters from the chip. The LED will continue to glow until the magnet has been removed as much as a few centimeters from the chip. Incidentally, I tried without success to activate the circuit with lodestones (naturally occurring pieces of magnetite).

Good sources for powerful magnets include small defective motors and discarded speakers. Try radio and television repair shops and automobile junk yards. You might be able to obtain at a refrigerator repair shop some flexible magnets from now-useless door gaskets. The cost of these items should range from nothing to minimal. Edmund Scientific Company (101 E. Gloucester Pike, Barrington, NJ 08007) stocks dozens of magnet types, some of which are rated at 8,000 Gauss!

You might want to activate the Hall-effect switch with an electromagnet. I tried several small electromagnets that I had purchased from Edmund Scientific several years ago (they're no longer in stock) and that worked moderately well. However, the core of a 6-volt relay failed to activate the switch even when it was operated at 12 volts and was placed in direct contact with the Hall-effect switch. You can make your own electromagnet by wrapping several hundred turns of small-diameter enamelled copper wire around a large nail or a bundle of several small ones.

Applications. Now that you know how to use a Hall-effect switch, you've probably begun to think about possible applications. Hall-effect switches provide bounce-free operation and are very fast—typical rise and fall times of 15 and 100 nanoseconds, respectively. Accordingly, they're ideal for use in keyboards and in mechanical switches connected to digital-logic circuits. Such switches and keys employ self-contained magnets that are moved toward Hall-effect switches by plungers or cams.

The company that pioneered many Hall effect switch applications, Micro Switch, supplies Hall-effect switches that replace the traditional breaker-points in the Plymouth Horizon automobile's ignition system. Hall-effect switches made by Micro Switch and other companies are also used in brushless motors, interlocks, telephone line-current sensors, tire-pressure monitors, sewing machines, flow meters and even miniature signal pickups for electric guitars.

How reliable are they? Micro Switch has been testing Hall-effect switches since 1968 and reports that such devices have logged nearly 20 *billion* successful operations! Hall-effect switches employed in experimental mechanical hearts that have been implanted in calves have performed more than 42,000,000 operations without failure. In short, the Hall effect switch is an exceptionally reliable component, especially when compared to conventional mechanical switches! ■

9. Programmable Countdown Timer

THIS month's project is a programmable digital timing circuit that has countless applications around the shop and home. You can learn how the circuit works by referring to the schematic diagram in Fig. 1. Astable multivibrator *IC1* provides clock pulses for *IC3*, a 74192/74LS192 decimal counter. The clock frequency can be altered by changing the value of either *R1* or *C1*.

The circuit can be programmed to generate any one of ten timing intervals. Any desired starting point from 0000 (0) to 1001 (9) is selected by means of switches *S1* through *S4*. When pushbutton *S5* is closed momentarily, the BCD number present at the parallel data inputs of the counter is loaded into it, and the chip's BORROW output (pin 13) goes to logic level 1.

Both the BORROW output of the counter and the output of the clock are combined in AND fashion by NAND gates *IC2A* and *IC2B*. The resulting logic signal is applied to the count-down input of the 74192. When the BORROW output is at logic 1, clock pulses pass through the gates to the counter's input. The clock pulses cannot reach the counter when the BORROW output is at logic 1.

As *IC3* begins its downward count, the current BCD number present at its outputs is decoded by *IC4* and is displayed on a common-anode, seven-segment LED readout. Once the count has reached 0000, the BORROW output goes to logic 0 upon receipt of the next clock pulse. Clock pulses are thereupon blocked and the counter outputs idle at 0000. The counting cycle can be initiated again by momentarily depressing *S5*. Of course, a new starting point can first be loaded into the counter by means of switches *S1* through *S5*.

The circuit shown in Fig. 1 includes two optional indicator LEDs. The diode designated *LED2* flashes each time a clock pulse is applied to the input of the counter. The other diode, *LED1*, is dark during the counting sequence, but begins to glow one clock pulse after the count has reached 0000. You can omit these LEDs if you choose, or you can connect other circuits in their place which will then be actuated by the counter.

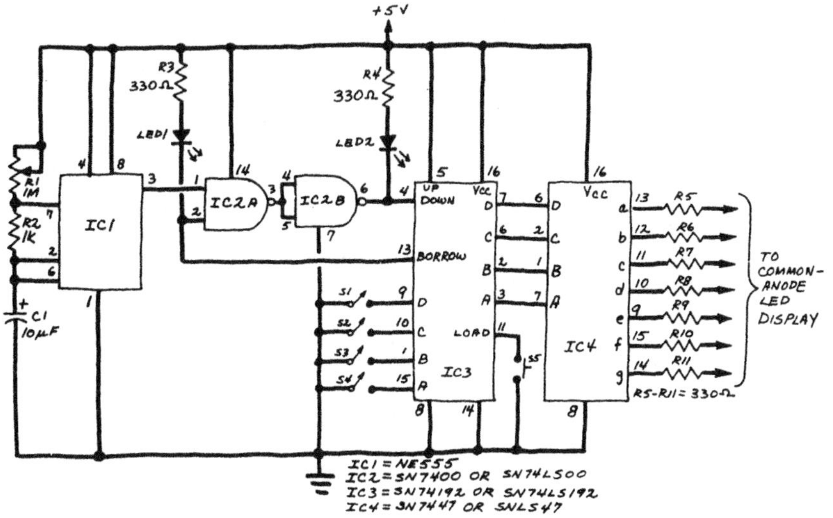

Fig. 1. Schematic diagram for a programmable countdown timer.

Model-Rocket Countdown Launcher. Having long been a model-rocket enthusiast, my favorite application for countdown timers is in automatic rocket-launching systems. Such systems are ideal for rocketeers who like to photograph their birds lifting off their launch pads. An audible beeper circuit triggered by the clock pulses applied to the counter input would be a helpful addition. It would allow the photographer to keep track of the countdown sequence while he aims his camera at the rocket as it sits on its launcher.

Figure 2 schematically shows how to add a rocket-motor ignition circuit to the basic countdown timer. A relay is used as the power-switching component because the resistance of its closed contacts is much less than the "on" resistance of an SCR or other solid-state switching device. This ensures that the highest possible current will flow through the nichrome-wire rocket-motor igniter. The diode connected across the relay coil absorbs the high-voltage spikes which can be generated by the collapsing magnetic field when the relay coil is deprived of current.

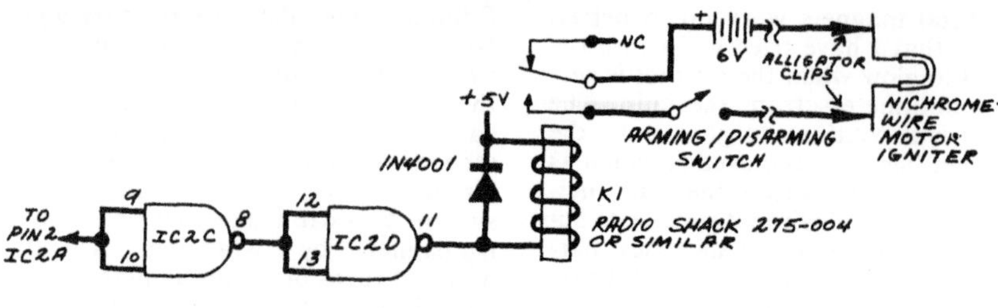

Fig. 2. How to add a relay to the timer for igniting model rockets.

If you use this circuit to launch model rockets, be sure to follow all of the standard safety precautions. Avoid the temptation to place the ignition-control system adjacent to the model-rocket launcher. Instead, connect the countdown circuit to the launcher with a 20′ length of lamp cord. Be sure to include a disarming switch so that you can safely and quickly deactivate the launch sequence should an unforeseen problem occur. You can use the ignition circuit's 6-volt battery to power the count-down circuit as well, if you so desire. This will simplify the overall launch system by eliminating the need for two separate power supplies. To do this, connect the +5-volt bus of Fig. 1 to the cathode of a 1N4001 diode. Connect the anode of the diode to the positive terminal of the battery, and the ground bus of Fig. 1 to the negative battery terminal, using a suitable length of lamp cord.

Other Applications. Of course, lauching model rockets is only one application for this versatile circuit. Many others are possible. For example, you can alter the timer's clock circuit to produce pulses at the rate of, say, 5 Hz. It can then be used as a darkroom timer. Slowing down the clock rate to one pulse per minute makes it possible to time a boiled egg, a phone call, or any other event lasting 10 minutes or less.

You can even extend the timing capacity of this circuit by cascading one or more additional counting stages. And you can reverse its operating mode by applying the gated clock pulse to the count-up input (pin 5) and connecting pin 2 of *IC2A* to the CARRY output of the counter (pin 12) instead of to pin 13, the BORROW output. Because the programmable countdown timer circuit is reasonably foolproof, it's an excellent first project for the novice experimenter. ■

10. Ultra-Simple Power Flasher

Many high-brilliance light-flasher circuits have been published over the years in this and other electronics magazines. Almost every one of these flashers has employed power transistors or SCRs, components which require heat sinking and careful circuit design.

Figure 1 shows an ultra-simple flasher circuit that I recently built around an FRL-4403 flashing LED, with only a few additional components. It will flash at a typical rate of 3 Hz any lamp whose current and voltage requirements fall within the ratings of the relay's contacts. The particular relay specified in Fig. 1 has contacts that can handle up to 1 ampere at 125 volts. The lamp shown is rated at 150 mA at 6.3 volts.

Incidentally, *FRL* stands for *Flashing Red LED.* The FRL-4403 is a Litronix product (19000 Homestead Road, Vallco Park, Cupertino, CA 95014). This novel LED, which incorporates a flasher integrated circuit, is also available from Radio Shack (stock No. 276-036).

The FRL-4403 LED in the circuit shown in Fig. 1 does *not* produce a visible flash. You can modify the circuit as shown in Fig. 2 if you want the red LED in the FRL-4403 to flash each time the relay coil is energized. Potentiometer *R2* must be adjusted until the relay starts to oscillate. Although the LED will flash, it will not be as bright as if it were powered directly from a 5-volt supply. The circuit shown in Fig. 2 might occasionally cease to oscillate. When this occurs, it is necessary to readjust potentiometer *R2.* For this reason, use the circuit shown in Fig. 1 for such applications as emergency beacons in which high reliability is essential.

Both versions of the circuit might operate erratically or even fail to operate if both the oscillator circuit and the flashing lamp are powered by the same battery. These difficulties are due to the large current demand placed on the battery when the lamp is switched on. In some cases, the circuit will oscillate at much higher than its normal rate.

The oscillator portion of each circuit consumes only 20 to 35 miliamperes. Therefore, it's feasible to power it with a small 6-volt battery if a silicon diode is connected in series with the positive terminal of the battery to drop the voltage to approximately 5 V. A much larger 6-volt lantern battery can be used to power the flashing lamp.

Of course, *both* the oscillator and flashing lamp can be powered from a common supply if it can source a sufficiently large current. You can experiment with different supplies and lamps to determine if you will need more than one supply.

Finally, the circuits presented in Figs. 1 and 2 can both be used to apply power pulses to devices other than lamps. Typical applications include gating power to warning horns, alarm sirens, etc. ■

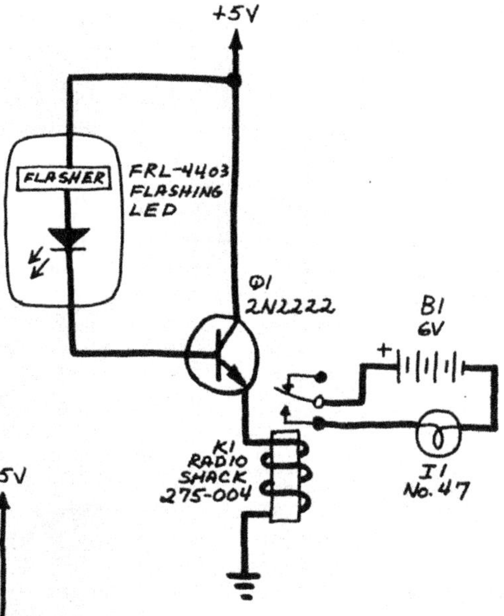

Fig. 1. Ultra-simple high-brightness lamp flasher.

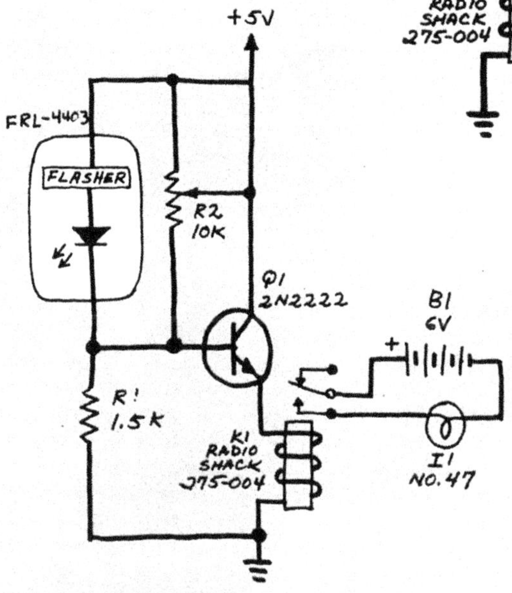

Fig. 2. Relay driver circuit with a flashing LED.

11. A Simple Wind-Speed Indicator

You can make a very simple but effective, portable anemometer (wind-speed indicator) using only a small dc motor, a propeller and a voltmeter. Figure 1 shows a typical arrangement that works quite well. When wind causes the propeller to rotate, the motor behaves as a dc generator. The voltage produced by the generator is relatively linear with respect to the wind speed.

Not all motors will work in this application. For best results, use a low-friction motor such as one intended for use in photovoltaic solar-energy demonstrators. Some motors of this type are supplied with plastic propellers attached to their shafts. Suitable motors are also available from hobby shops that cater to enthusiasts of radio-controlled model aircraft. Generally, if a motor produces a few tenths of a volt across its power terminals when the shaft is rolled between your thumb and forefinger, it will work well in this application. I've found Mabuchi motors to be particularly effective wind-powered generators.

Hobby shops also sell propellers. Alternatively, you can make your own prop using Fig. 1 as a guide. Cut four blades from light-gauge aluminum or from one end of a metal food or beverage can.

Blade dimensions are not specified in Fig. 1 because motors having a wide range of sizes can be used in this project. For best results, the blades should extend somewhat beyond the housing of the motor as shown. Use care when cutting the blades, and round off their sharp ends to prevent cuts. Twist the base of each blade 45-60° and push the tab at the narrow end of each blade into the side of a balsa model-rocket nose cone (another hobby-shop item) that has been sanded smooth. For best results, use a nose-cone whose base has a diameter close to that of the motor housing. Spread some white glue around the base of each blade and, when it dries, paint the entire propeller assembly.

Next, attach the propeller to the motor's shaft. The easiest way is to force the shaft into the center of the base of the balsa cone. Secure the

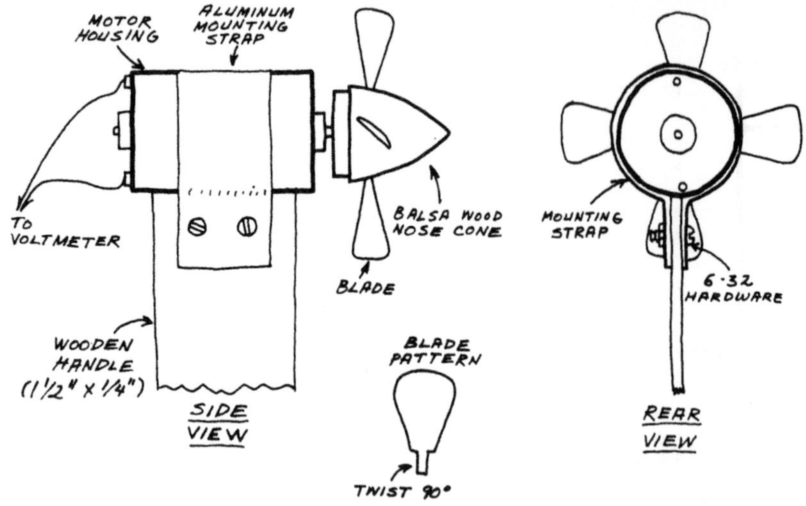

Fig. 1. Design for a portable anemometer.

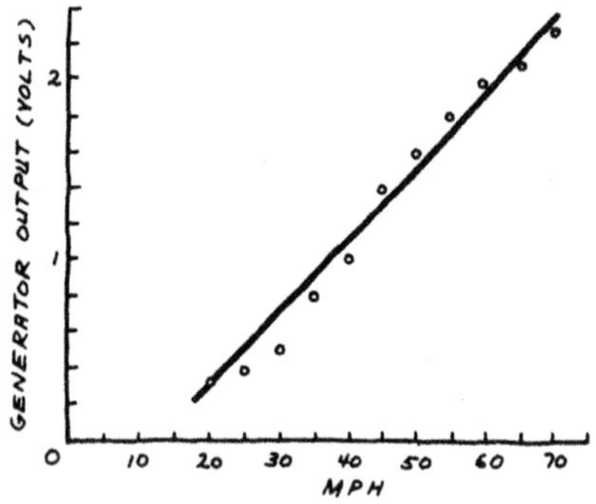

DATA

MPH	VOLTS
<10	0
20	.25
25	.40
30	.55
35	.80
40	1.00
45	1.40
50	1.60
55	1.80
60	2.00
65	2.10
70	2.30

Fig. 2. Typical anemometer calibration curve.

cone in place with some white glue. Mount the motor on a handle as shown in Fig. 1. Then connect its power leads to the voltage-input terminals of a multimeter that has been set to read dc volts.

You can now calibrate your anemometer. One way is to extend the generator from the window of a moving car. Record the voltage from the generator at increments of 5 or 10 mph and plot the results on a graph. Be sure to perform the calibration on a windless day and to have a friend drive while you hold the generator and record the measurements!

Figure 2 is the calibration curve for an anemometer that I built for use in a homemade, model-rocket wind tunnel. Note the reasonable linearity of the curve.

Going Further. You can easily modify this project. To measure low wind velocities, for example, you can replace the propeller with a four-cup rotor made from two ping-pong balls sliced in half and the multimeter with a solid-state bargraph readout. Use an LM3914 LED driver for best results. ■

12. Transistorized Light Flasher

A TRANSISTORIZED light-flasher circuit for model rockets played a key role in the development of the Altair 8800, the POPULAR ELECTRONICS hobby-computer breakthrough project of January 1975.

In 1969, H. Edward Roberts, Stanley Cagle, Robert Zaller, and I formed a company to manufacture and sell a commercial version of my model-rocket light flasher. I wanted our company to have a name using the initials M., I., and T. because, at the time, the Massachusetts Institute of Technology was an important center of model-rocket research activity. Stan came up with Micro Instrumentation and Telemetry Systems, or M.I.T.S.

Eventually, M.I.T.S. was simplified to MITS, and we made telemetry transmitters, light flashers, and a lightwave communicator. The company also manufactured calculators and test equipment—which led to the conception of the Altair 8800 by Ed Roberts. I left MITS in 1970 to become a full-time freelance writer and my first article was about the flasher that was to become MITS's first product.

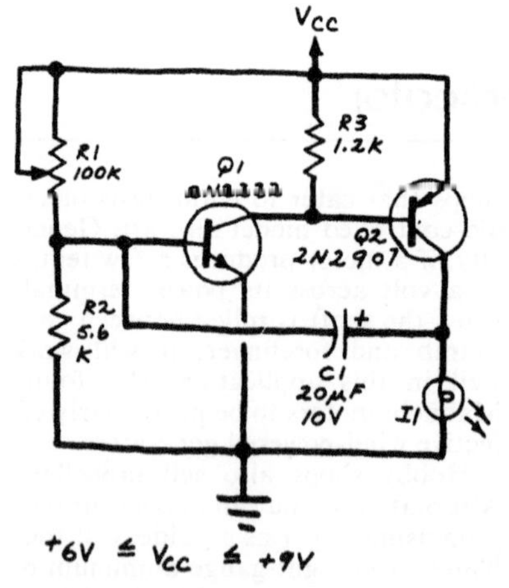

Fig. 1. Schematic diagram of miniature transistorized light flasher.

Figure 1 is the schematic diagram of the circuit, commonly referred to as a regenerative amplifier, that was responsible for both MITS and my present profession. When lamp *I1* is off, *Q2* must be cut off. Since *Q1* controls the behavior of *Q2*, *Q1* must also be cut off. The voltage divider *R1R2*, however, supplies enough base drive to turn *Q1* on, which in turn causes *Q2* to conduct. When *Q2* conducts the lamp is connected directly across the power supply.

Now, *C1* begins to charge through *Q2* and *R2*. At some point, enough charge has accumulated in *C1* to turn *Q1* off. This, in turn, cuts off *Q2* and the lamp darkens. When the lamp is dark, *C1* keeps *Q1* cut off while the capacitor discharges through *R2* and the lamp. When *C1* has discharged completely, *Q1* is biased into conduction by *R1R2*. This turns on *Q2* and the lamp, and the cycle begins again.

The circuit is easily assembled on a

miniature circuit board measuring only about one-half inch square. Potentiometer *R1*, which controls the flash rate, can be any standard trimmer. The lamp should be a low-voltage bulb such as the No. 122, No. 222, or a similar, miniature incandescent pilot lamp.

Peak currents of two or more amperes flow through the lamp's filament, so the filament will glow much more brightly than normal. If operated continuously at such current levels, the filament would be quickly destroyed. This does not occur, however, because the flasher circuit supplies current pulses lasting only tens of milliseconds in duration.

Actually, the circuit has broad applications. The flash is bright enough to be an effective personal warning light for cyclists and pedestrians at night. For such uses, you can build the entire circuit inside a plastic automatic lamp housing.

For model-rocket applications, install the flasher in a transparent payload capsule, preferably with the battery mounted *below* the flasher circuit. This keeps the rocket's center of gravity from moving too far forward. It also prevents the comparatively heavy battery from pressing against the flasher circuit during the acceleration phase of the flight.

I hope you enjoy experimenting with this flasher circuit. It may not impact on your career the way it did on mine, but it will certainly brighten up your life! ■

13. A Liquid-Level Indicator for the Blind

HAVE you ever wondered how a blind person knows when to stop pouring liquid into a glass or cup? The blind people I have known usually insert a finger into the cup to the desired depth. Then they pour until they feel the liquid with their fingertip. Unfortunately, this procedure can cause discomfort if the liquid is hot, and can be socially awkward if a blind host is serving sighted guests.

The simple circuit shown in Fig. 1 will help your blind friend or relative pour liquids with relative ease and without having to touch the liquid. The circuit is inexpensive to build and uses an LM3090 LED-flasher chip as an audio oscillator. The circuit emits a tone whose frequency in part depends on the value of *C1*. When the probe leads, which parallel *C1*, are bridged by a low-to-moderate resistance, the frequency of oscillation suddenly increases. Such a probe, shown in Fig. 2, can be connected to the audio generator by a suitable length of flexible two-conductor wire.

With the value of *C1* as shown, the frequency of oscillation is about 1.6 kHz. When the probe clips were attached to the rim of a plastic cup and tap water was poured into the cup to the level of the probes, the frequency of the tone increased to approximately 3 kHz. The frequency change that your circuit exhibits might differ from these values. Different beverages have various values of conductivity and can produce different frequency changes. In any event, you will hear a distinct tone change when the liquid bridges the probe tips.

If you build this circuit for a blind friend, house it in a small plastic box. Make sure the battery holder is readily accessible, as the cells must be easily replaced by one who cannot see.

Be sure to place a raised marker at the positive end of the battery holder to indicate the correct battery orientation. A drop of paint or glue will do. Your blind friend should have little difficulty replacing the cells if *you* can do it blindfolded.

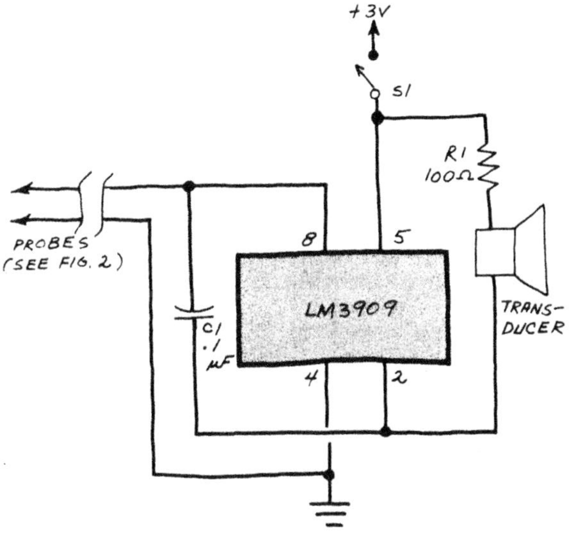

Fig. 1. Schematic diagram for a simple liquid level indicator circuit for the blind.

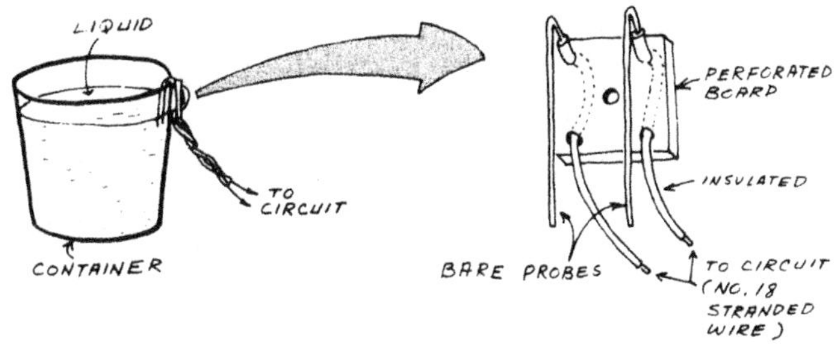

Fig. 2. How to assemble a liquid level indicator probe to fit over the edge of a container.

Incidentally, the liquid-level indicator can be designed to emit a tone *only* when liquid bridges the probes. I prefer the two-tone approach because the blind user always knows when the unit has been switched on and therefore will remember to turn it off when he has finished using it.

Going Further. You can add a volume control or alter the frequencies that the circuit generates or both. The former is easily accomplished by reducing the value of *R1* to 47 ohms and inserting a 1-kΩ potentiometer in series with *R1*. To alter the tone frequencies, try different values for *C1*. Increasing the value of *C1* will lower the frequency and reducing the capacitance will raise it.

The entire circuit should be self-contained and complete with battery and miniature acoustic transducer. The LM3909 can even be powered by a single 1.5-volt silver-oxide cell of the type used to power digital watches. This will allow you to assemble a miniature unit. The audio output could be provided by a miniature earphone salvaged from a discarded hearing aid. Alternatively, you can use a midget transistor-radio earphone. ■

14. Steam Engine and Whistle Sound Synthesizer

Originated at Texas Instruments, the circuit is designed around the SN76477 sound-effects chip. In operation, the output of the chip's noise generator is switched on and off by its super-low-frequency (SLF) oscillator. Potentiometer *R2* controls the switching rate, hence the speed of the engine sound.

When *R2*'s resistance is high, the sound resembles that of a stopped train whose engine is idling. As the potentiometer's effective resistance is reduced, the sound speeds up and resembles that produced by an accelerating train.

The sound of the train's whistle is derived from the output of the voltage controlled oscillator (vco) in the SN76477. The values of *C2* and *R3* control the whistle's pitch. Pressing *S1* activates the whistle.

The output of the SN76477 is amplified by *Q1*, which in turn drives a small 8-ohm speaker. Resistor *R11* controls the amplitude of the sound from the speaker. If you prefer, you can drive an external audio power amplifier with the signal voltage appearing between pin 13 of the IC and ground.

For a little more money, you can buy the SN76488. This chip has everything that the SN76477 has, as well as a built-in amplifier, but it has a different pinout. If you use this chip, omit *Q1* from the circuit in Fig. 1 and connect pin 13 directly to one terminal of the speaker. Connect the second speaker terminal to ground through *C4*. Resistors *R10* and *R11* should be omitted.

A drawback of the circuit in Fig. 1 is that the steam-engine sound generator is disabled when the whistle is activated. This problem can be remedied by adding a simple whistle-multiplexer circuit (Fig. 2) and by removing *S1* from the circuit of Fig. 1.

When activated, the whistle multiplexer, which was also suggested by Texas Instruments, switches the whistle on and off at a rate of 26 kHz. Even though the steam-engine sound is turned off when the whistle is on, the switching rate is far too fast for the ear to detect. Consequently, the whistle seems to be superimposed on

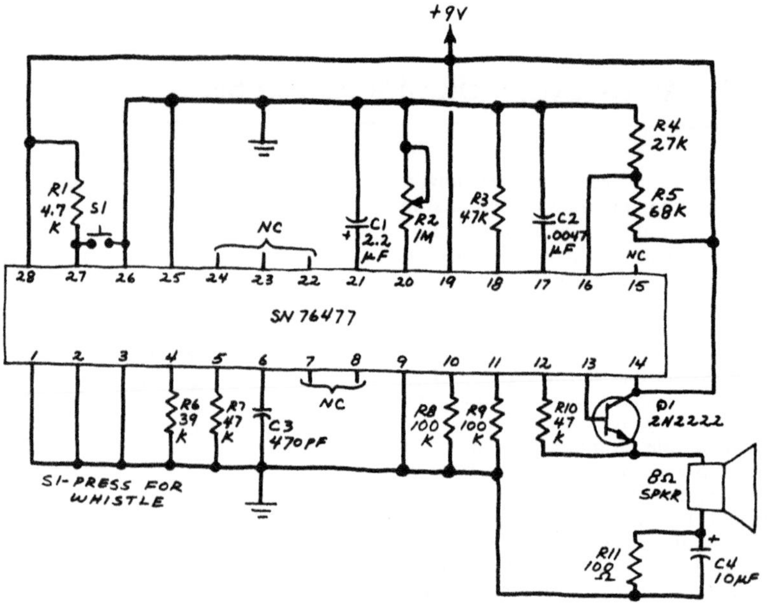

Fig. 1. Schematic using the SN76477 sound-effects chip to generate sounds of a steam locomotive.

the sound of the engine. The only audible effect of the whistle multiplexer on the steam-engine sound is a slight reduction in volume when the whistle

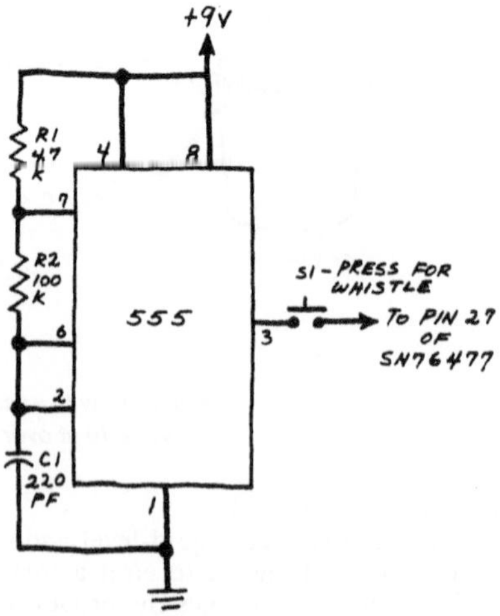

Fig. 2. Whistle multiplexer for steam-engine simulator.

is activated.

Model railroaders might want to modify this circuit so that the engine sound speeds up *automatically* when a model train is accelerating. This can be done with the help of a homemade optoisolator made from a small lamp and a cadmium-sulfide photocell. Use black electrical tape or heat-shrinkable tubing to mount the lamp adjacent to the photocell and to block ambient light.

Connect the lamp in the optoisolator to the train's transformer. Remove *R2* from the circuit of Fig. 1 and connect the photocell in its place. As the train's speed is increased, the lamp will glow more brightly. This will reduce the resistance of the photocell and increase the rate at which the sound-effects generator is switched on and off by the SLF oscillator.

It might be necessary to add a series resistor between the photocell and the circuit to match the sound of the engine with the speed of the train. You can achieve the same result by blacking out part of the photocell's window. ■

15. A Simple, Low-Cost Timer

FOR simple timer applications, transistor circuits are as good as more complicated and expensive IC designs. Figure 1, for example, shows a simple timer made from only three components—a field-effect transistor (FET), a timing capacitor and a discharge resistor. A LED, protected by a current-limiting resistor, indicates when a timing cycle is complete.

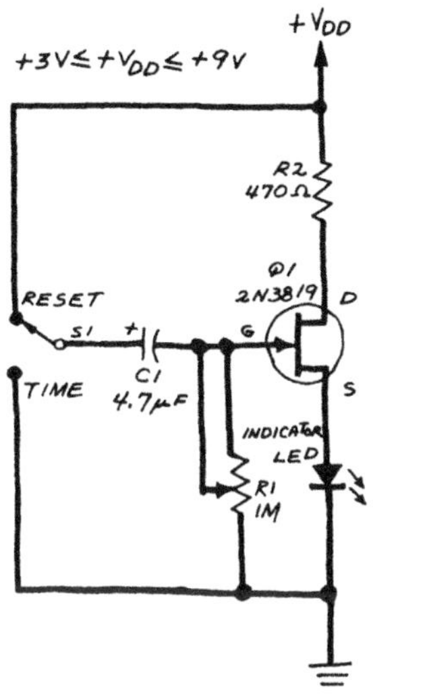

Fig. 1. Ultra-simple FET timer circuit.

The FET is responsible for the simplicity of this circuit. Its very high input impedance places a negligible load on *C1* during a timing cycle. By contrast, a bipolar transistor would quickly discharge the capacitor and prematurely end the timing cycle. In effect, the FET serves as a high-impedance buffer between timing capacitor *C1* and the output LED, which provides the timing indication.

When the circuit is connected to a single-ended positive supply, *C1* is charged to $+V_{DD}$ by momentarily placing *S1* in its RESET position. A timing cycle is initiated by placing the switch in its TIME position. This causes *Q1* to turn off and extinguish the LED. At the same time, *C1* begins to discharge through *R1*.

When the voltage across *C1* decreases to approximately 0.6 volt, the FET conducts, and the LED glows to indicate completion of the timing cycle. A new timing cycle can be initiated by momentarily toggling *S1* to its RESET position. Figure 2 is a plot of the voltage across *C1* during a typical timing cycle.

When I used a 4.7-µF miniature aluminum electrolytic capacitor as *C1*, I could obtain a maximum time delay of approximately 10 seconds. Shorter delays can be achieved by adjusting *R1* so that its effective resistance is decreased. Longer delays are available by using a component with more capacitance and less leakage for *C1*. Tantalum capacitors are well suited to such applications. Substituting a higher resistance potentiometer for *R1* can also give longer delays.

Adding a Relay. Figure 3 shows a more practical, expanded version of the basic timer of Fig. 1. Here each of three (or more) separate capacitors can be switched into the circuit to provide different time delays without any adjustment of *R1*. Of course, *R1* can also be adjusted if desired.

The most important addition to the circuit shown in Fig. 3 is *K1*, the output relay. Normally, *R2* keeps *Q2* conducting, which in turn energizes the relay coil. When a timing cycle is complete, however, *Q1* grounds the base of *Q2*, cutting off the bipolar transistor. This causes the relay to drop out. Diode *D1* absorbs any high-voltage inductive kick which might be generated during the keying of the relay coil.

With this circuit, delays of ten minutes or more are possible if quality capacitors are selected. Low-leakage capacitors are a *must* for time delays of this magnitude. ■

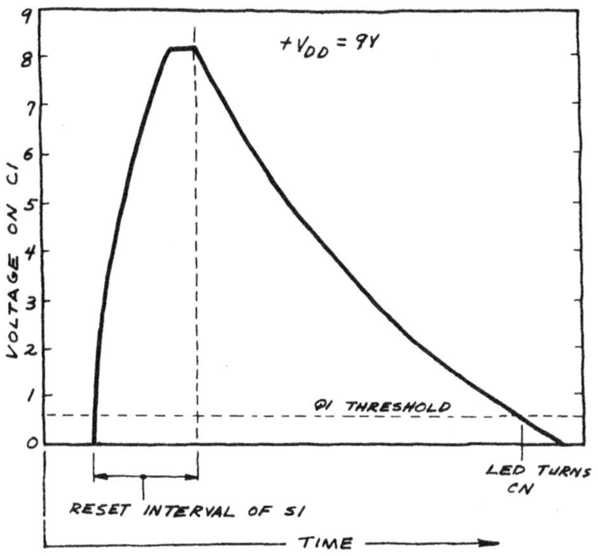

Fig. 2. Voltage on C1 during a timing cycle.

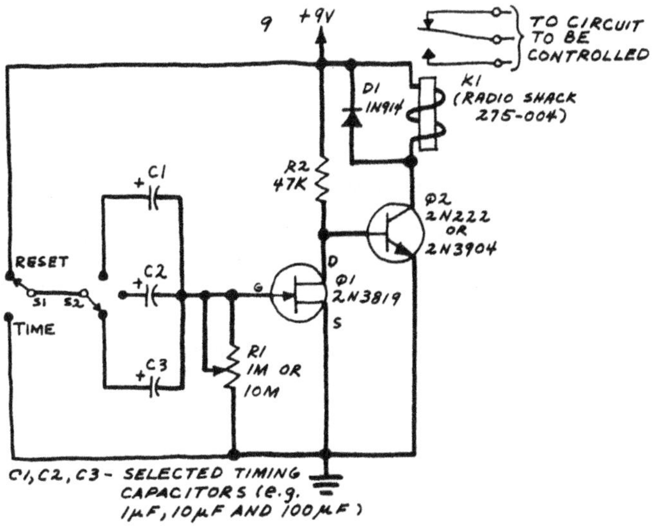

Fig. 3. A variable-delay FET timer.

16. Experimenting with an Air Pressure Switch

RECENTLY, I learned that an ultrasensitive air pressure switch is available from Edmund Scientific (101 East Gloucester Pike, Barrington, NJ 08007). I immediately ordered one and have been impressed with its capabilities.

The switch, a Honeywell Model PSF 100A, is actuated (closed) by an air pressure of only 0.02 pounds per square inch (psi). This is equivalent to the pressure of about 0.5 inch of water or a gentle puff of air from a distance of a few inches.

You might be able to purchase the PSF 100A directly from Honeywell. Otherwise, you can buy one from Edmund (Cat. No. 41,623) for $7.00, plus $1.30 for postage and handling.

The PSF 100A has two differential control ports—one for low—and the other for high-pressure operation. If one port is at atmospheric pressure (*i.e.*, open), the other will trigger the switch on pressure (high port) or vacuum (low port). If both ports are connected to external gas sources, the switch will close when the pressure difference between the two sources exceeds 0.02 psi.

Fairchild assigns a life of 1,000,000 on-off operations to the PSF 100A. Contact resistance of closed switch is 0.5 ohm.

The major drawback of the switch is its current rating of only 10 milliamperes dc. This means that, in many applications, external buffering is required. We will look at several buffering methods, as well as some practical applications for the PSF 100A shortly. First, let's review some of the applications listed in the Edmund data sheet:

1. Replacement of vane-type flow switches.
2. High-wind detector.
3. Proximity sensor.
4. Counting sensor.
5. Clean-air system pressure-drop detector.
6. Edge sensor.
7. Fan or cooling system failure sensor.
8. Fixed-point temperature detector (in a closed system dependent upon the contraction and expansion of a fixed volume of gas).
9. Respiration rate sensor.
10. Venturi tube sensor.
11. Pressurization sensor for inflatable structures.

These applications in turn suggest others. For example, the high-wind detector idea could be used as a fixed-point air-speed indicator for a model rocket, aircraft, bicycle, or automobile. In each case, the input ports of the sensor require constriction to permit the switch to operate at higher air pressures. Or a higher threshold sensor switch can be used. Honeywell's PSF 100A-3, for example, has a switching threshold of 0.1 psi.

Buffer Circuits for a Pressure Sensor. As long as the current to be switched is less than 10 mA, the PSF 100A needs no buffering. This means the switch can directly actuate LEDs and some solid-state warning devices and alerters. For many applications, however, the rated current capacity of the PSF 100A is insufficient.

Figure 1 shows how to connect a low-current, inexpensive relay to the PSF 100A to increase its switching capability from 10 mA to a full ampere (at 125 volts). Since the relay coil current can safely exceed the 10-mA maximum rating of the PSF 100A's contacts, it is necessary to limit the current flow with an external resistor (R_S). Figure 1 gives the values of R_S for power supplies of both 6 and 9 volts which will allow the relay to pull in without exceeding the 10-mA rating.

I arrived at these values by actual measurements, and you may wish to verify my results. Though the relay coil is specified to have a resistance of 500 ohms, the unit I used actually measured 480 ohms. At 6 volts, this relay pulled in at 5.5 mA and dropped out when the current fell below 4.5 mA. Therefore, the currents given in Fig. 1 provide ample margin for proper operation of the relay.

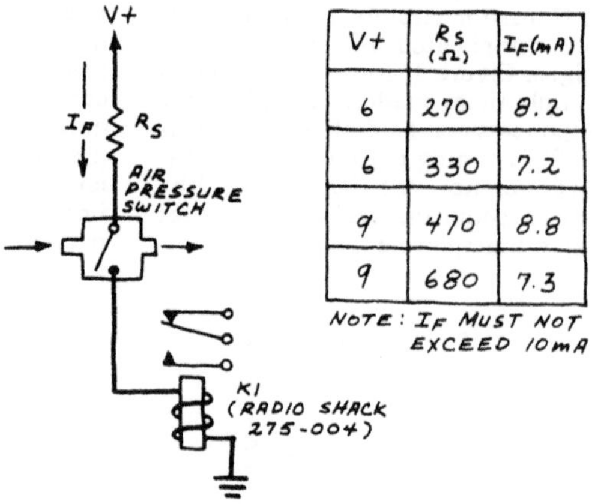

Fig. 1. Using a relay to increase current capacity.

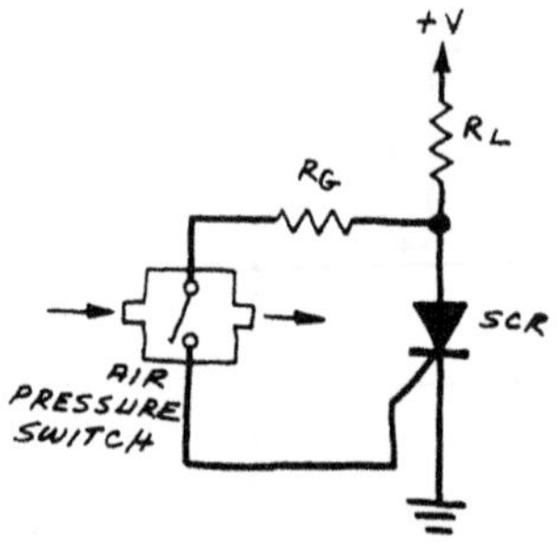

Fig. 2. An air pressure switch can be used to trigger an SCR as shown here.

Figure 2 shows how the PSF 100A can be used to trigger an SCR. The pressure switch is simply inserted in the SCR's gate circuit. Resistor R_G should provide ample SCR gate current while limiting the current through the switch.

Incidentally, remember that a triggered, dc-powered SCR stays on even after the gate signal is removed. Only when the forward current falls below what is termed the minimum required *holding current* does the SCR turn off. This occurs, of course, when the load is temporarily disconnected. It also occurs on the negative transition of an ac voltage.

Optoisolating the PSF 100A will electrically isolate the sensor from the circuit being controlled. Figure 3 shows how the PSF 100A is connected to the LED portion of a LED-phototransistor optoisolator (also called an optocoupler).

Current-limiting resistor R_S must be selected to limit the current through the LED, and therefore the PSF 100A, to less than 10 mA. The appropriate series resistance can be found with the simple formula: $R_S = (V_F - V_{LED})/I_F$, where V_F is the forward voltage, V_{LED} is the LED forward voltage, and I_F is the desired current in amperes.

GaAs LEDs having a forward voltage from 1.2 to about 1.8 volts are used in most optoisolators. Inserting a typical V_{LED} of 1.5 volts and a desired I_F of 5 mA into our formula gives the following values of R_S for a range of forward voltages:

V_F	R_S
3	300
4	500
5	700
6	900
7	1,100
8	1,300
9	1,500
10	1,700
11	1,900
12	2,100

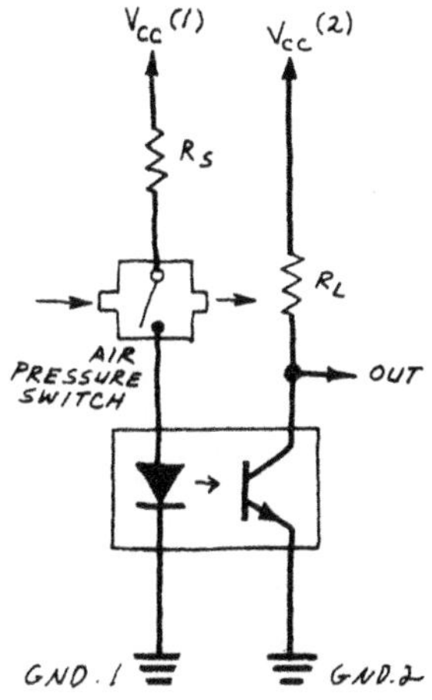

Fig. 3. The air pressure switch can be isolated from the controlled circuit by an opotoisolator.

Application Circuits. Having explored the operation of the PSF 100A and seen how its contacts can be buffered, we can now use the switch in practical applications. I've designed three circuits with biomedical applications in mind. Remember that these circuits are merely representativé of the ways the PSF 100A can be applied. You can use the same techniques for applications of your own.

Puff/Sip Multi-Channel Controller. Several years ago I read about an electric wheelchair that could be controlled by puffing or sipping on one or more tubes connected to air pressure switches. The same method was used to turn on lights and appliances.

Figure 4 shows one way to implement a "puff/sip" controller. The circuit provides up to five channels of on-off control. More channels can be added by expanding the basic circuit.

The CMOS decade counter (*IC2*) is a 4017 with self-contained 1-of-10 output decoding. In operation, a clock formed by two NAND gates in *IC1* repeatedly cycles *IC2* through each of its ten outputs. The five control channels, only one of which is shown in Fig. 4, are provided by adjacent pairs of decoded outputs from *IC2*.

Channel 1 is controlled by pins 3 and 2 of *IC2*. At the beginning of a count cycle, pin 3—the lowest order decoded output from *IC2*—goes high while all other outputs remain low. This turns on Channel 1's ON LED, notifying the operator that the device or appliance controlled by Channel 1 can be turned on by puffing or sipping on the plastic tube connected to the circuit's single PSF 100A air switch. Depending upon the value of *C1*, the operator has up to a second to operate the air switch. If more time is required, the value of *C1* can be increased at the expense of slowing down the control cycle.

Whether or not the air switch is closed when Channel 1's ON LED is glowing, the clock eventually advances *IC2* to decoded output two (pin 2). This turns on Channel 1's OFF LED and notifies the operator that the device or appliance controlled by Channel 1 can be turned off by puffing or sipping on the air switch's tube. Again, whether or not the switch is closed, *IC2* continues to advance through the decoded outputs as the clock supplies pulses. If the switch is *not* closed, the controlled device or appliance remains either on or off.

The four transmission gates in a single 4066 analog switch (*IC3*) provide the necessary control logic for a single channel. If the air switch is closed when Channel 1's ON LED is glowing, *IC3A* closes, firing the SCR and pulling in the relay.

If the air switch is closed when Channel 1's OFF LED is glowing, *IC3B* closes. This, in turn, closes *IC3C*. Switch *IC3D* is normally in the closed state due to the voltage drop across *R6*, but when *IC3C* closes, the control pin (6) of *IC3D* goes to ground. This opens the current path through the SCR, turning off the SCR and allowing the relay to drop out.

When *IC2* advances to the next decoded output, *IC3B* and *IC3C* open and *IC3D* is again closed by the drop across *R6*. The SCR can then be triggered by a puff or sip the next time Channel 1's ON LED is glowing.

I used a low-current relay (Radio Shack 275-004) in the prototype of the circuit. The SCR can be any low-voltage, economy-grade unit.

Follow the circuit used for Channel 1 to add additional control channels. The PSF 100A in Fig. 4 should be connected to pins 5 and 13 of each additional channel's 4066. This permits one switch to control all channels. Connect the mouth tube to the switch's HIGH port for puff operation or the LOW port for sip operation.

Caution: Do not exceed the relay's contact ratings. Avoid shock hazards by powering the circuit with a 9-volt battery and carefully insulating connections to the relay's contacts.

Fig. 4. A "puff-sip" single-switch, multi-channel controller circuit.

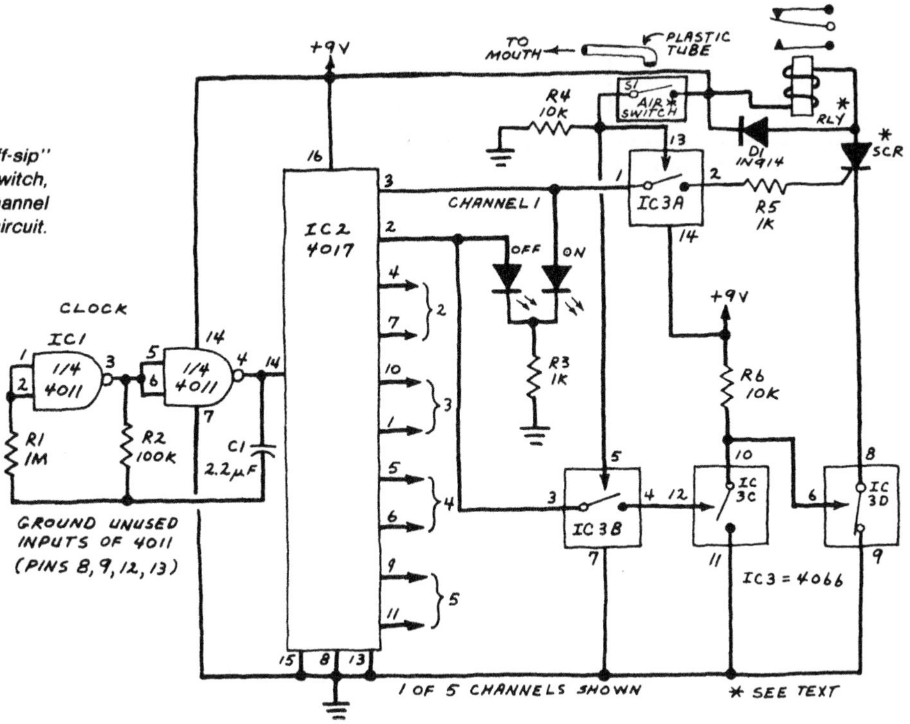

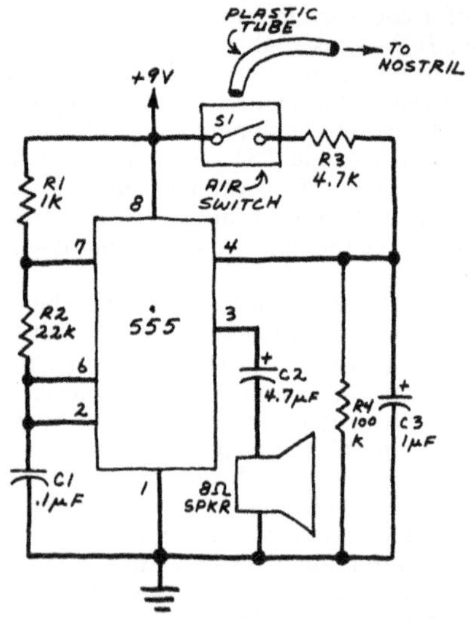

Fig. 5. Respiration indicator provides audible signal.

Respiration Indicator. The circuit in Fig. 5 provides a brief tone burst each time a person or animal being monitored inhales or exhales. The circuit is a straightforward 555 astable oscillator whose frequency is controlled by *C1*.

When the PSF 100A air pressure switch, *S1*, is open, the 555's reset input (pin 4) is held low by *R4* and the oscillator is disabled. When *S1* is closed, pin 4 of the 555 is made high via *R3* and the oscillator is enabled. Simultaneously, *C3* is charged through *R3* to the battery voltage. When *S1* is opened, pin 4 is held high by the charge on *C3* until it discharges through *R4*. The oscillator is then disabled.

The tone frequency of this circuit can be increased (or decreased) by reducing (or increasing) the value of *C1*. The length of the tone burst can be extended by increasing the value of *C3*, or the extended tone burst can be eliminated entirely by removing *C3*. The circuit will continue to provide a tone for each respiration cycle.

I tested the circuit by taping a length of flexible aquarium tubing under one nostril. When the remaining end of the tubing was attached to the PSF 100A's LOW port, the circuit beeped each time I *inhaled*. When the tube was connected to the HIGH port, the circuit beeped when I *exhaled*. Try both operating modes if you build the circuit.

This circuit provides a simple way for recording the breathing rates of animals for study and evaluation. It can also be used with human subjects such as athletes. It should be used with seriously ill patients only under medical supervision. In any case, power the circuit with a 9-volt battery or an isolated line-operated supply to avoid electrical shock.

Respiration Failure Alarm. When used under proper medical supervision, the circuit in Fig. 6 can save a life. It continuously monitors respiration and emits a warning tone when the time between breaths exceeds a predetermined interval.

The portion of the circuit involving *IC1* is a missing pulse detector. If you prefer, you can use a 7555, the CMOS version of the 555.

In operation, the missing pulse detector is reset each time *S1*, a PSF 100A air switch, is closed. Resistor *R2* controls the maximum time allowed between reset pulses. If the circuit is not reset before the allowed time expires, pin 3 of the 555 goes low. This actuates an astable oscillator made from *IC2B* and *IC2C*. If the 555 is subsequently reset, the oscillator will be disabled. Otherwise the oscillator will provide a continuous warning tone. The frequency of the warning tone can be changed by changing the value of *C2*.

You can test this circuit by using a length of aquarium tubing as described in the previous section. Be sure to power the circuit with a 9-volt battery or isolated line supply to avoid the possibility of electrical shock. Like any biomedical electronic device, the respiration failure alarm should be used with seriously ill patients only under medical supervision. ■

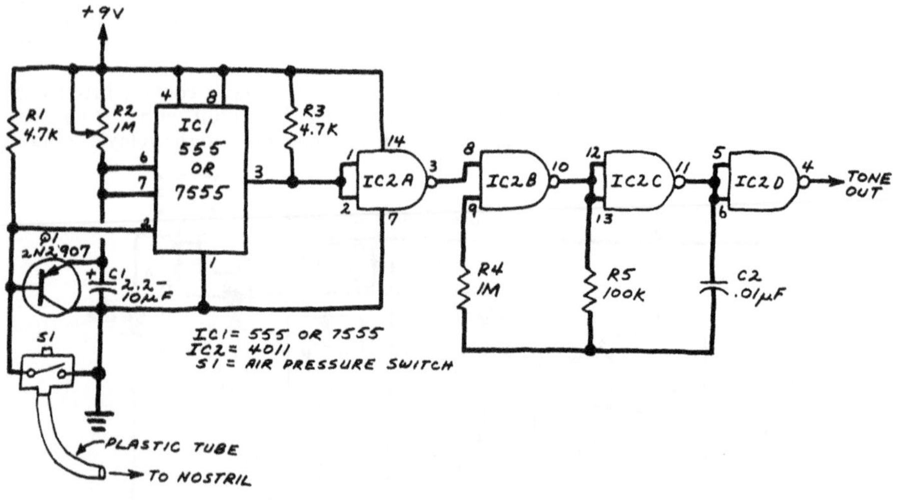

Fig. 6. This circuit monitors respiration and emits a warning tone when time between breaths exceeds a predetermined interval.

Printed and bound by CPI Group (UK) Ltd, Croydon, CR0 4YY
03/10/2024
01040342-0020